THE MAPPING OF NEW SPAIN

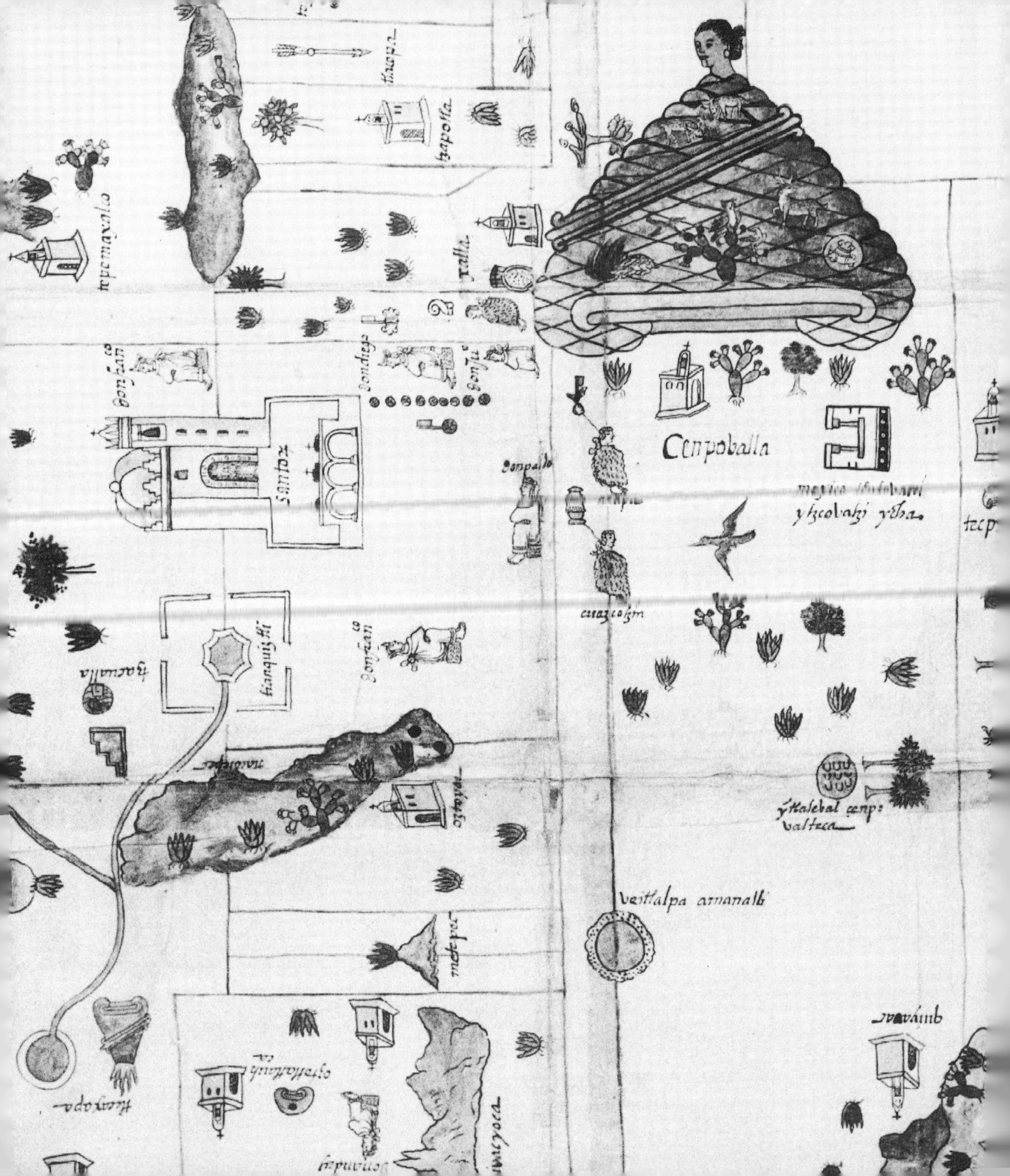
Cenpoballa
veitlalpa amanalli
ytlalebal cenpo:
valteca

THE MAPPING OF NEW SPAIN

Indigenous Cartography and the Maps of the Relaciones Geográficas

Barbara E. Mundy

The University of Chicago Press ❦ *Chicago and London*

The University of Chicago Press, Chicago 60637
The University of Chicago Press, Ltd., London

Paperback edition 2000
Printed in the United States of America

09 08 07 06 05 04 03 02 01 00   2 3 4 5 6
ISBN 0-226-55096-6 (cloth)
ISBN 0-226-55097-4 (paperback)

Library of Congress Cataloging-in-Publication Data

Mundy, Barbara E.
The mapping of New Spain : indigenous cartography and the maps of the relaciones geográficas / Barbara E. Mundy
p. cm.
Includes bibliographical references (p. –) and index.
ISBN 0-226-55096-6 (cloth : alk. paper)
1. Cartography—Mexico—History. 2. Cartography—New Spain—History. 3. Indian cartography—New Spain—History. 4. Aztec cartography—History. I. Title
GA481.M86 1996
912.72—dc2096-15824   96-15824
CIP

This work is supported in part by a grant from the Program for Cultural Cooperation between Spain's Ministry of Culture and United States Universitites.

This book is printed on acid-free paper.

Contents

List of Illustrations *vii*
Preface *xi*
Acknowledgments *xxi*
Author's Note *xxxi*

CHAPTER ONE
Spain and the Imperial Ideology of Mapping 1

CHAPTER TWO
Mapping and Describing the New World 11

CHAPTER THREE
Colonial Spanish Officials and the Response to the Relación Geográfica Questionnaire 29

CHAPTER FOUR
The Native Painters in the Colonial World 61

CHAPTER FIVE
The Native Mapping Tradition in the Colonial Period 91

CHAPTER SIX
Language and Naming in the Relaciones Geográficas Maps 135

CHAPTER SEVEN
The Relaciones Geográficas and Other Viceregal Maps in New Spain 181

CHAPTER EIGHT
Conclusion 213

APPENDIX A
Catalogue of Maps Studied 217

APPENDIX B
The Questionnaire of the Relaciones Geográficas 227

APPENDIX C
The Nahuatl Inscriptions of the Macuilsuchil Map 231

APPENDIX D
A Typical Viceregal *Acordado* 233

Notes 235
Bibliography 247
Index 269

Illustrations

COLOR PLATES

(following page 144)

1. Relación Geográfica map of Teozacoalco, 1580
2. Relación Geográfica map of Guaxtepec, 1580
3. Relación Geográfica map of Cholula, 1581
4. Relación Geográfica map of Texupa, 1579
5. Relación Geográfica map of Cempoala, 1580
6. Relación Geográfica map of Amoltepec, 1580
7. Relación Geográfica map of Misquiahuala, c. 1579
8. Relación Geográfica map of Macuilsuchil, 1580

FIGURES

1. Cortés map of Tenochtitlan, 1524 xii
2. Codex Mendoza map of Tenochtitlan, fol. 2r, c. 1542 xv
3. Map of Mesoamerica xvii
4. Map of gobierno of New Spain xviii
5. View of Barcelona from the south, by Anton van den Wyngaerde, 1563 1
6. "Key" map from Escorial atlas, attributed to Juan López de Velasco, c. 1585 6
7. Relación Geográfica questionnaire, 1577 21
8. Relación Geográfica map of Coatzocoalco or Villa de Espiritu Santo by Francisco Stroza Gali, 1580 24
9. Relación Geográfica map of Los Peñoles, attributed to Diosdado Treviño, 1579 25
10. Relación Geográfica map of Teozacoalco, 1580 26
11. Schematic rendering of Relación Geográfica map of Los Peñoles 36
12. Relación Geográfica map of Tecuicuilco, 1580 37
13. Relación Geográfica map of Atlatlauca-Malinaltepec, 1580 38

14. Relación Geográfica map of Meztitlan by Gabriel de Chávez, 1579 39
15. Native calendar from Relación Geográfica text of Meztitlan by Gabriel de Chávez, 1579 40
16. Map of Florence, 1550 41
17. Region of Meztitlan 42
18. Detail of Relación Geográfica map of Meztitlan, 1579 43
19. Relación Geográfica map of Tescaltitlan, 1579–1580 45
20. Frontispiece of Alonso de Molina's *Vocabulario*, 1571 46
21. Relación Geográfica map of Santa María Ixcatlan A, 1579 47
22. Relación Geográfica map of Santa María Ixcatlan B, 1579 48
23. Escutcheon of Archbishop Alonso de Montúfar, 1566 49
24. Relación Geográfica map of Tlacotalpa by Francisco Stroza Gali, 1580 51
25. Relación Geográfica map of Tehuantepec B, attributed to Francisco Stroza Gali, 1580 52
26. Details of three Relaciones Geográficas maps, 1580 53
27. Relación Geográfica map of Tehuantepec A, 1580 54
28. Relación Geográfica map of Ixtapalapa by Martín Cano, 1580 63
29. Relación Geográfica map of Culhuacan by Pedro de San Agustín, 1580 64
30. Relación Geográfica map of Guaxtepec, 1580 68
31. Relación Geográfica map of Acapistla, 1580 69
32. Relación Geográfica map of Cuzcatlan A, 1580 70
33. Relación Geográfica map of Cuzcatlan B, 1580 71
34. Relación Geográfica map of Cholula, 1581 72
35. Relación Geográfica map of Chimalhuacan Atengo, 1579 73
36. Relación Geográfica map of Acambaro, 1580 73
37. Facade of Monastery of Oaxtepec, Morelos 79
38. Relación Geográfica map of Texupa, 1579 80
39. Codex Kingsborough, fol. 204r, c. 1555 81
40. Relación Geográfica map of Mizantla, 1579 83
41. Map of New Spain 93
42. Relación Geográfica map of Cempoala, 1580 95
43. Relación Geográfica map of Tetlistaca, 1581 96
44. Relación Geográfica map of Minas de Zumpango, 1582 98
45. Drawing of indigenous place-name of Xilotepec (redrawn from Codex Mendoza, fol. 31r) 101
46. Codex Zouche-Nuttall, page 36 102
47. Drawing of Codex Zouche-Nuttall, page 36, with annotations 103
48. Schematic map of Apoala Valley 104
49. Lienzo of Zacatepec 1, c. 1540–1560 109
50. Historia Tolteca-Chichimeca, fols. 32v–33r, c. 1547–1560 110
51. Relación Geográfica map of Amoltepec, 1580 113
52. Diagram of genealogies of Teozacoalco Relación Geográfica map 115
53. Comparison of Teozacoalco region as shown on Relación Geográfica map with modern topographical map 117
54. Historia Tolteca-Chichimeca, fols. 26v–27r, c. 1547–1560 120

55. Diagram of fols. 26v–27r of Historia Tolteca-Chichimeca 121
56. Map of Chichimec History 122
57a. Drawing after Map of Chichimec History, with transcription of inscriptions 123
57b. Drawing after Map of Chichimec History, identifying sociopolitical units 124
58. Diagram of Chalco's sixteenth-century social structure 125
59. Diagram of social structures in the Cempoala region 129
60. Diagram of Cempoala Relación Geográfica map 130
61. Diagram of Cempoala Relación Geográfica map, showing rulers portrayed on map 131
62. Relación Geográfica map of Misquiahuala, c. 1579 136
63. Drawing after Relación Geográfica map of Misquiahuala, with translations of Spanish inscriptions on map 137
64. Mixtec warlord 8 Deer Jaguar Claw. Drawing after Codex Zouche-Nuttall, page 43 139
65. Codex Mendoza, fol. 43r, c. 1542 141
66. Relación Geográfica map of Muchitlan, 1582 146
67. Toponyms on Relación Geográfica map of Muchitlan 147
68. Relación Geográfica map of Gueytlalpa, 1581 150
69. Relación Geográfica map of Jujupango, 1581 151
70. Relación Geográfica map of Matlatlan and Chila, 1581 152
71. Relación Geográfica map of Papantla, 1581 153
72. Relación Geográfica map of Tecolutla, 1581 154
73. Relación Geográfica map of Tenanpulco and Matlactonatico, 1581 155
74. Relación Geográfica map of Zacatlan, 1581 156
75. Relación Geográfica map of [illegible], 1579 157
76. Logographic place-name of Teozacoalco. Drawing after Relación Geográfica map of Teozacoalco 159
77. Logographs of Amoltepec 160
78. Logograph from Lienzo of Yolotepec 161
79. Relación Geográfica map of Macuilsuchil, 1580 162
80. Relación Geográfica map of Suchitepec, 1579 163
81. Relación Geográfica map of Teutenango, 1582 168
82. Relación Geográfica map of Gueguetlan, 1579 172
83. Relación Geográfica map of Tlaxcala, c. 1584 174
84. Relación Geográfica map of Ameca, 1579 177
85. Relación Geográfica map of Xalapa de la Vera Cruz, 1580 183
86. Land grant map, 1585 185
87. Drawing after 1585 land grant map from Tarímbaro, Michoacán 186
88. Details of maps made by Xalapa artist 190
89. Land grant map from Actopan, 1578 191
90. Land grant map from Atezca, 1587 192
91. Map of Xalapa region showing areas covered by Relación Geográfica map and two land grant maps of 1578 and 1587 193
92. Details of logographs made by Tehuantepec artist 196
93. Map of Tehuantepec region 197

94. Details of maps made by Tehuantepec artist 198
95. Land grant map by artist of Relación Geográfica map of Tehuantepec A, 1573 199
96. Land grant map by artist of Relación Geográfica map of Tehuantepec A, 1580 200
97. Map of Tehuantepec region showing areas covered by Relación Geográfica map and six land grant maps made by Tehuantepec artist 201
98. Land grant map from Ixtapalapa by Martín Cano, 1589 204
99. Map of Ixtapalapa region 205
100. Topography and inscriptions of Martín Cano's land grant map of Ixtapalapa 206
101. Humboldt Fragment II, detail of lower right corner, after 1565 208

Preface

In February of 1519, a motley group of Spanish conquistadores in eleven caravels set off from Cuba, leaving the Punta de San Antón behind them as they headed west. Their brief journey, which led them across the narrow Canal de Yucatán, along the shores of the southern Gulf, and into the port of San Juan de Ulúa on April 21, would have momentous consequences: the imposition of their language, their social hierarchies, and their political systems on the millions of peoples who inhabited the mainland of the New World. With this voyage, the Spanish, and later the Portuguese, began the widescale colonization the continental New World.

At the same time that these Spaniards were entering and "discovering" the American continents, their fellow countrymen back in Spain were discovering themselves, creating images of their countries, their provinces, their cities, making their surrounding world visible through maps. Not only Iberian, but also European cartography burgeoned in the sixteenth century, this "age of exploration" reaching an ever-widening audience through engravings and woodcuts. As a result, Europeans came to know what the rest of the world looked like. With European world maps reaching new levels of geometrical precision, Europeans were given a vehicle for envisioning the New World, in particular. As the century progressed, the outline of their possessions in the New World, once tentative and imprecise, gained strength and certainty.

The colonizers were not the only ones to map their surrounding world, making it visible through maps; the colonized also did this. Amerindians in the New World were also mapmakers. They too sought ways to represent the landscape surrounding them, to push beyond the tiny slice of the world that their eyes could see. While their maps, like their

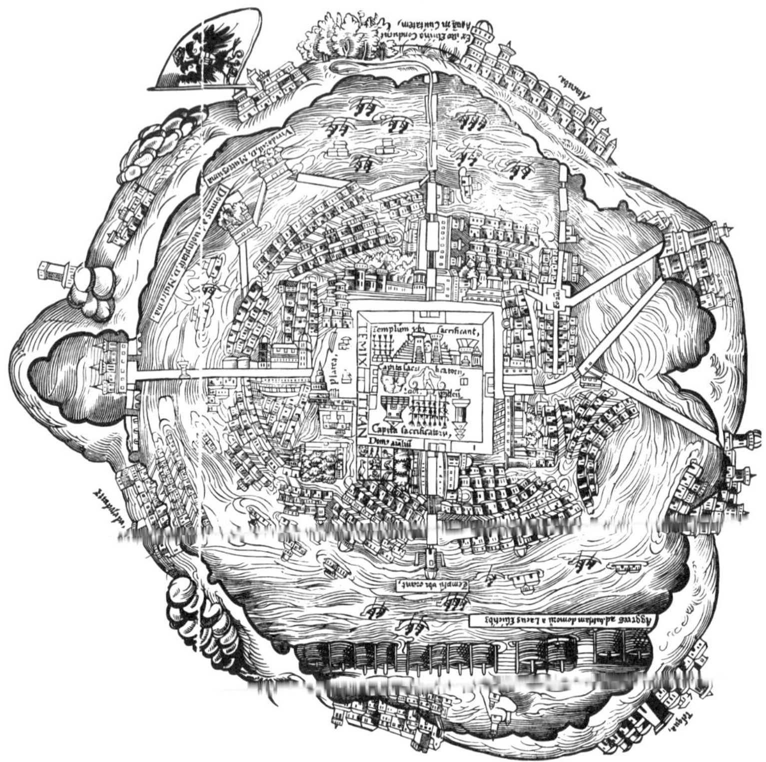

Figure 1. The Cortés map of Tenochtitlan, 1524. This woodcut map, made by an anonymous European engraver, purported to show the Aztec capital city as it looked when Spanish conquistadores entered in 1519. While many details may have been drawn from the accounts of Cortés and his native informants, the architecture and overall design follow the conventions of coeval European city plans. This map was published in Nuremberg in 1524 to illustrate Cortés's second letter to Charles V, which appeared under the title *Praeclara Ferdinandi de Nova Maris Oceani Hispania Narratio. . . .* Size of the original: approx. 48 x 48 cm. Photograph courtesy of the Rare Books and Manuscripts Division, The New York Public Library, Astor, Lenox and Tilden Foundations.

European counterparts, were rooted in a spatial understanding of the human world,[1] they shared few of the same formal and conceptual constructs. No wonder: European and Amerindian maps had wholly independent points of origin, being born and developing on continents thousands of miles apart. Their differences also point to differences in the mind's inner landscape, which Jill Ker Conway has described as the "unspoken, unanalyzed relationship to the order of creation" that people of different continents, and different cultures, maintain (Conway 1990: 218). The map, by definition, arises out of a particular culture's understanding of space, which in turn is presaged on a culture's own construction of reality;[2] when cultures both understand and encode space differently, their maps will vary as well.[3] Nowhere is the difference between two maps and two realities better seen than in two coeval maps, one from Europe, one from the indigenous New World, both showing the Aztec capital city of Tenochtitlan, both painted in the first half of the sixteenth century.

In 1524, a European map of Tenochtitlan by an unknown draftsman was published to accompany the second letter of Hernán Cortés, a letter in which the Spanish conquistador described his entry into this Amerindian metropolis in 1519 (fig. 1). The Cortés map (sometimes called the Nuremberg map, after its city of publication) shows the Aztec capital in a fashion common enough in European city plans at the time. The viewer is allowed to see the whole expanse of the Valley of Mexico, with the city of Tenochtitlan at center. A lake surrounds this city, and adjacent towns nestle on its shores. The viewer can also glimpse Tenochtitlan's famed system of canals, and its blocks of "floating" gardens and houses, that led Cortés himself to liken Tenochtitlan to another floating city, Venice (Cortés 1986: 68). At center, the European viewer would see the city's dark heart, for here lay the main temple precinct of Tenochtitlan, with its temples and skull-racks, where human sacrifice was once celebrated.

In creating the map of Tenochtitlan, this unknown artist was attempting to represent the city with a model, since the city, like the larger world, was visible to the human eye only in parts, never in its entirety. The artist employed two systems of projection, one that we will call Euclidean, the other Albertian, to represent space, patching them together, like an expert tailor, into a seamless whole.[4] The mapmaker's method seemed so natural as to have escaped mention then and now, but if we dismantle the seamed and overlapping projections, we can see the artifice. The base structure of the map is Euclidean, wherein the distances between points in space are set out on a grid or similar graticule and then their dimensions are reduced geometrically. In the case of the Cortés map, the artist did not use the right-angled grid that we are most accustomed to on modern city maps but, rather, ended up with a graticule that looked more like a system of concentric circles that grew closer together as they rippled out from a central point. At the time the map was made, European knowledge of the city was guesswork more often than not, so the artist can hardly be faulted for not adhering to a standard projection; inconsistencies aside, it is through this level of projection that viewers grasped the idea of the city surrounded by a lake, surrounded in turn by more cities. An Albertian projection, and its close relative, the panoramic projection, overlie and modify this basic Euclidean projection. The Albertian represents space as if seen from a single viewpoint, and the panoramic shows it as if that single viewpoint captured a 360° arc, like the view surrounding the curious public standing within a nineteenth-century panopticon. The lakeshore is shown with the panoramic projection, as is much of the city, with houses and temples shown as if seen by a viewer passing along a Tenochtitlan street or canal.

This parsing of the map, analyzing the structures through which it represents space, allows us entry into how this artist, and the culture out of which he emerged, understood the surrounding world and how that reality should be represented on the paper. Both the Euclidean and the Albertian projections grew out of an understanding of space as primarily a series of connected points, reducible by geometric means. On this map the Albertian system is used to single out architecture and other man-made constructions, thereby assigning these as the defining and constituting features of space. This map is a fitting image for sixteenth-century Europe, whose high culture espoused a scientific rationalism, holding man as the measure of all things.

Compare this European map of Aztec Tenochtitlan with one from the Codex Mendoza.[5] This second view of the capital was created by an indigenous artist, probably a resident of Tenochtitlan, around 1542, or a little less than twenty years after the Cortés map (fig. 2). It is a historical map: the Mendoza mapmaker pictured the island city of Tenochtitlan as he or she believed it to be at the time of its official founding in A.D. 1325. This landscape is hardly identifiable with that of the Cortés map: the encircling lake is reduced to a rectangular band forming the inner border of the page. Two of the cities that stand at the edge of the lake, Culhuacan and Tenayuca, are here represented beneath the lower edge of the lake band by conventional depictions of temples and with their hieroglyphic names. Of the many canals within the city of Tenochtitlan itself, only four main ones are shown, their intersection forming a blue "X" that carves the city up into four roughly equivalent triangles. Clearly, there is no geometric projection ruling this map of the city, whether it be Euclidean or panoramic. This Aztec artist's view of reality did not distill into a planimetric map.[6]

But the Aztec map is not just a random concoction; it likely reflects a long tradition of city maps, stretching back well into the pre-Hispanic past. Its forebears are the maps made by community governments in Central Mexico, discussed in chapter 5. What, if not geometry, structures this Aztec map of Tenochtitlan? If we look at other details of this map, we see that the space of Tenochtitlan is occupied by ten men: three of the four quadrants hold

Figure 2 *(opposite)*. The Codex Mendoza map of Tenochtitlan, fol. 2r, c. 1542. This hand-painted map shows Tenochtitlan, the capital city of the Culhua-Mexica, who led the Aztec empire, as it looked at the time of its founding in A.D. 1325. The design is schematic: the large X represents the city's crossing canals; the interior rectangle is the lake surrounding this island city. The ten founding fathers of the new city are seated on the island. Below it, Culhua-Mexica warriors defeat two nearby towns, Culhuacan and Tenayuca, in the first of the many conquests that would bring the Aztec empire into being. Size of the original: 32.7 x 22.9 cm. Photograph courtesy of the Bodleian Library, Oxford (MS. Arch. Selden. A. 1, fol. 2r).

A. Theuet cosmographe du Roy
numero de años . L1 .
tenuchtitlan
colhuacan. pueblo.
tenayucan. pueblo

two men, while four men occupy the left quadrant. The indigenous residents of Tenochtitlan, had they seen the map in 1542, would have quickly recognized these ten as the initial founders of the city, the most important of whom, Tenoch, sits on a woven mat to the left of center. In addition, they would have immediately connected the four areas wherein the founders sat to the four neighborhoods or parts that composed their contemporary city—Atzaqualco, Teopan, Moyotlan, Cuepopan. Indigenous viewers in 1542 would also have understood that within each of these quarters lived the *calpolli,* or neighborhood groups, each one claiming its own founding father from among the group shown. Thus, the map pictures the founder Xomimitl at center right (his name means "Foot Arrow" in Nahuatl, the Aztec language, and his name is depicted, behind his head, by a foot pierced by an arrow). Xomimitl sits in an almost empty quadrant of the newly founded city of 1325; two hundred years later, his affiliated calpolli—Yopico—populated that same space.[7]

The Codex Mendoza map of Tenochtitlan is not based on a geometric projection like the Euclidean one we saw guiding the Cortés map. Perhaps it is better thought of as being a humanistic or social projection—that is, the physical space of the city has not been filtered through and reduced by an overlying graticule; rather, its structuring device is the human or social layout of the city—four constituent parts, populated by the calpolli of ten founders.[8] Through this map, we can begin to glimpse how residents of Tenochtitlan understood their surrounding space: their spatial reality was one defined and structured by social relationships, which were strengthened by their endurance through time. This use of social organization to structure both the understanding and the representation of space was as common in ancient Mexico as the use of Euclidean geometries to define space was in Europe.

What we witness in the two city maps, one European, one indigenous, both of the same place, is a difference in how space is represented, reflecting, in part, how space was understood. The Cortés map encodes the city's space with rational, geometrically based projections, whose norms were understood throughout Europe, where the map was widely legible—residents of Paris or Venice, say, could read and interpret the Cortés map of Tenochtitlan, and might see their own cities mapped in the same fashion (Snyder 1993). Likewise, the humanistic projection of the Codex Mendoza map, although absolutely particular to that community, also had wide legibility in contemporary Mesoamerica, as the region with a cohesive indigenous culture, now falling within Mexico and Central America, is often called (fig. 3; Kirchhoff 1943). Urban elites from Cholula or Culhuacan would have understood that Tenochtitlan's space was being filtered through the structure of Tenochtitlan's society. Members of both these cultures, European and Mesoamerican, given the choice, would have certainly chosen *their* map of Tenochtitlan to be the true representation. Europeans, in fact, held the Cortés woodcut map to be such a fitting picture of

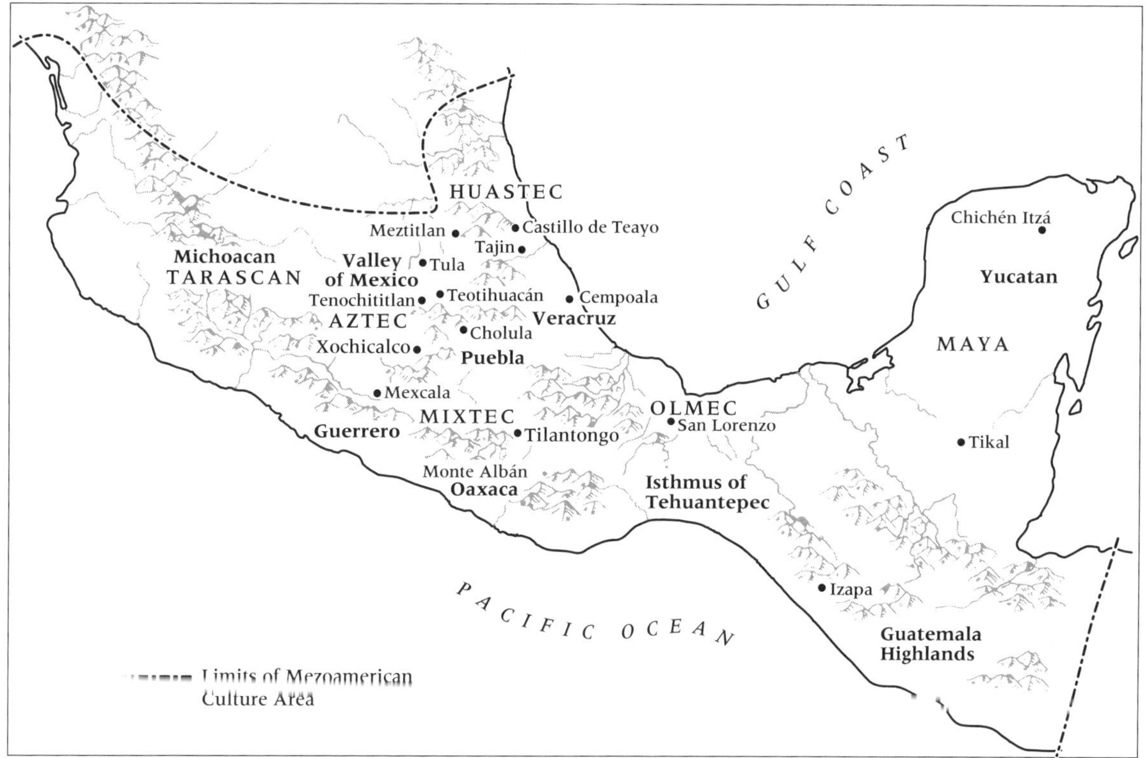

Figure 3. Map of Mesoamerica. This region of pre-Hispanic America held a population of linguistically diverse Amerindians. They nonetheless shared many cultural traits, such as the use of a 260-day calendar and a diet of maize and beans.

the Aztec city that they made it one of the most widely circulated pictures of the New World in the sixteenth century. While the Cortés map gained in authority as its reach grew wider, the Codex Mendoza map was authoritative at its genesis. It was created by one of a cadre of elite artists to represent the city for the benefit of, we believe, its first Spanish viceroy, Antonio de Mendoza.

In these two different, yet both complete and accurate, maps of Aztec Tenochtitlan, we see that both are the "right" way to map, following Stephen Toulmin's suggestion that "the fundamental map [is] complete only if it [shows] all the things which in that region [are] the cartographer's ambition to record" (Toulmin 1960: 116–7). In their acuity, the maps show us how two very different cultures held that space should be understood and then represented. Should space be primarily represented as a set of points in relation to one another? Or should it be first and foremost a stage created out of and for human action? And what happens, then, when these two very different ways of understanding and mapping space, one rational, the other humanistic, come together?

Perhaps nowhere can we find better answers to this last question than through an examination of the maps of the Relaciones Geográficas from the *gobierno* of New Spain.[9] This group of sixty-nine maps was painted around 1580—some sixty years after Spanish conquistadores had brought down the indigenous empire of the Aztecs—in response to a questionnaire dispatched by the Spanish crown.[10] Each maps a different city, village, or small province of the gobierno of New Spain (fig. 4), a region largely coincident with modern-day Mexico.

Each was drawn by a local talent. Many sectors of colonial society—professional sailors, local Spanish officials, indigenous artists—took up pen to draw a Relación map. The corpus gives us a spectrum of the rational and humanistic ways—of the Old World and the New—of structuring space on local maps, as well as the multifold improvisations that denizens of the colony drew up in response, no doubt, to the particulars of the colonial moment. In its variety of authors and approaches, the maps of the Relación corpus

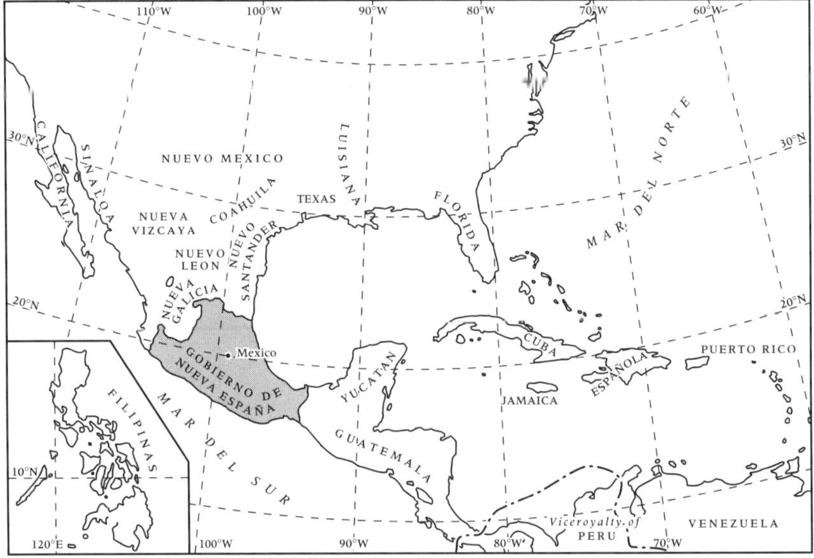

Figure 4. Map of the gobierno of New Spain. After the Spanish conquest of the Aztec empire in 1521, Spain carved up its new territories in America into various "gobiernos." The gobierno of New Spain coincided with much of the culture region of Mesoamerica and was the nucleus of the modern country of Mexico.

best capture the viewpoints on a local level as one world collided with another and New World inhabitants grappled with change.

While many maps were being made in New Spain during the sixteenth century, this corpus stands out among them. It is accompanied by texts, because written "geographic reports" were compiled along with the maps, so we know the dates, the provenance, and the actors involved in making the maps. No such wealth of circumstantial knowledge accompanies any other large group of maps from this time and place. But most important are the maps themselves.

What the maps best show us is the indigenous world—how its inhabitants, still reeling, no doubt, from the blows of conquest, reshaped their once insular maps to keep pace with the rapid changes in their understanding of the surrounding world. In using visual images to reconstruct the realities of indigenes in colonial Mexico, I am returning to the methods of the first historians of the New World, be they Spanish, mestizo, or native, who drew on, as their primary sources, the "picture-books" and painted manuscripts of the pre-Hispanic world. For unlike texts, wherein the indigenous voice is muffled and distorted in its translation into Spanish or its conversion into Spanish genres of alphabetic writing, maps, along with other images, give us the viewpoint of sixteenth-century Amerindians who were using a familiar, indigenous idiom. This being the case, the map image has a value without compare.

The main focus of my inquiry has been how change, both on the surface of indigenous maps and on the underlying strata of spatial understanding, is effected during a colonial moment. Why did indigenous people of New Spain make maps differently in 1518, before that flotilla set out from Cuba, than they did in 1580? Traditional art history has posited that indigenous art gradually took on the coloring of European style, before being completely subsumed by it (Kubler 1961; Robertson 1972a). But this explanation neglects the special quality of maps, which is that they are dependent upon how a culture envisions space. The introduction of, say, European perspectival drawing hardly gave indigenous people who saw it a profound reason to abandon a deeply rooted way of understand their surrounding world (Harvey 1980; Jervis 1937). Historians of science might argue that Europe introduced technological advancements into the New World that allowed for more "accurate" maps. But, as I have argued above, accuracy is in the eye of the beholder. Moreover, few if any technological advancements that might have had a bearing on geometric projections were introduced into New Spain in the sixteenth century; if anything, scientific knowledge in the New World seems to have declined in the sixteenth century.

Rather, we will see with the Relaciones Geográficas maps new reasons for the cultural shift occurring as Amerindians recast their spatial understandings. Change was gradual: their humanistic ways of mapping had great durability as long as the community that fostered them still existed, still promoted and consecrated traditional ways of understanding the surrounding landscape. Indigenous maps began to change when the understanding of space held by their makers did, most visibly when Spanish programs of land use and urbanization forced them into different relationships with their environment. In addition, both within the indigenous community and outside of it, new types of writing and literacy undercut the authority that native maps once had.

This study of the Relación Geográfica corpus is also meant to address the history of early Mexican cartography, to fill a lacuna in our knowledge of the history of maps. Most writers on this subject have portrayed Mexican cartography—indeed, all New World cartography—as attempts by the European imperial powers to represent their New World domain through maps (Cline 1962; Harvey 1980). These narratives suppress those of indigenous communities who made maps well before European conquests; in Mexico, these peoples were shorn of the power to represent themselves. In covering both Spain's projects in the New World and the effects upon indigenous cartography, I try to show how the dazzling story of possession is trailed by the dark shadow of dispossession, and to dispel the myth that the path of cartography in the New World was an ascent towards an apex of an ever more perfect and dispassionate rendering of space (Harley 1988). In Mexico, the evolving perfection ascribed to European maps of the colony may have been little more than an effective suppression, indeed, a leveling, of all other points of view.

We begin in Europe, as this study traces the Relación Geográfica project back to its genesis in Spain, where it was one of Spain's many and ongoing attempts to map the New World. By following the project back to its beginnings, we can meter the atmosphere of expectations that surrounded the Relación Geográfica project and our needle will single out the kinds of spatial encodings that Spanish officials concerned with the Indies—as they called the New World—desired and expected. One legacy of European success in implanting itself in the New World is that Europe, its history and cartography, remains terra cognita to readers today. Most of us will find firm footing in the courts of imperial Europe, and in the lives of kings, that are discussed in the opening chapters. We then move, as did the questionnaire that gave rise to the Relaciones Geográficas, across the Atlantic and into the New World, first examining the maps that the Spanish and Creole (American-born Spaniard) colonists made in response to the questionnaire, and then stepping into the indigenous colonial world. Here we will see what pre-Hispanic maps would have looked like, and how they, and the process of spatial envisioning that lay behind them, registered the impact of the colonial moment. This latter project is the central preoccupation of this book; in it, I try to lead the reader to see the contextual truth of indigenous colonial mapping and use the maps as a window onto the reality that colonial indigenes were in the process of creating for themselves.

Acknowledgments

This work is supported in part by a grant from the Program for Cultural Cooperation between Spain's Ministry of Culture and United States' Universities. I also received University Fellowships from Yale University, Josef Albers grants for Pre-Columbian research, an Evelyn A. Jaffe Hall Charitable Trust Fund grant, a John F. Enders Fellowship for dissertation research, and a Philip L. Goodwin Fellowship. A Junior Fellowship in Pre-Columbian Studies at Dumbarton Oaks and a Whiting Fellowship in the Humanities allowed me ideal environments for writing. I am indebted to Yale University, Dumbarton Oaks, and the Mrs. Giles Whiting Foundation for such support.

In my research, I was aided by the patient and knowledgeable staffs of the Yale University Libraries, the Columbia University Libraries, the New York Public Library, the Real Academia de la Historia in Madrid, the Archivo General de Indias in Seville, the Escorial Library, the Archivo General de la Nación in Mexico, and the Nettie Lee Benson Latin American Collection at the University of Texas at Austin. Rosario Parra, the former director of the Archivo General de Indias, allowed me to see the originals of the Relaciones Geográficas at the Archivo. Laura Gutiérrez-Witt, the director of the Benson Collection, was remarkably generous with her ideas. She also allowed me to examine and to photograph all the Relaciones Geográficas maps in the Texas collection; her staff, especially Anne H. Jordan, was particularly helpful and obliging.

Louise Burkhart first opened up the world of Nahuatl to me, and Frances Karttunen, R. Joe Campbell, and José Alberto Zepeda were my Nahuatl teachers in a National Endowment for the Humanities summer seminar; many thanks to these *nahuatlacah*. James Boyden guided me across the Atlantic into the world of sixteenth-century Spain. Mary Elizabeth Smith was magnanimous in discussing ideas and sharing her research; she and Dana Leibsohn read the entire manuscript, and it is a better work for their suggestions. Anthony Aveni was particularly helpful with chapters 2 and 3, and Richard Kagan enriched my understanding of local Spanish maps. Michael Coe and George Kubler nurtured the project as a dissertation, along with Mary Ellen Miller, my advisor, who fostered my intellectual growth and my pursuits. Elizabeth Boone was likewise unstinting in her support. During my year at Dumbarton Oaks, her door was always open, and her mind was always welcoming to new ideas. She has been both mentor and friend, and is unparalleled in both roles.

My husband, Gerald Marzorati, has lived with the project almost as long as with me. His support of my work has never flagged, even when tried by late nights and long separations. Other members of my family, especially my mother, Elaine Mundy, and parents-in-law, Eugene and Ursula Marzorati, have been quick to respond when I needed their help. My father, Norman Mundy, was my gateway to Spanish-speaking America and helped with translations. Without members of my family, this book would not have been; it is with them that my greatest debt lies.

Author's Note

Throughout the book, I use the sixteenth-century spellings of Mexican town names that are found in the Relaciones Geográficas. Although at times these spellings are different from modern ones, the reader can correlate them easily to Howard F. Cline's census of the Relaciones Geográficas (1972a) and Donald Robertson's census of the Relaciones Geográficas maps (1972a), which use the same sixteenth-century spellings.

To assist the reader, the first use of a sixteenth-century place-name is followed by, in parentheses, the modern name, if different, and the name of the modern Mexican state in which the place now lies. In addition, in the index modern names are cross-referenced to their sixteenth-century versions.

In the spelling of Nahuatl words, I have used the versions found in Frances Karttunen's *An Analytical Dictionary of Nahuatl* (1983) but have omitted diacritic marks; thus I use *Moteuczoma* rather than *Montezuma.* Nahuatl plural forms are noted in the text. Some Nahuatl words, such as *calpolli* and *altepetl,* that appear frequently in the text take no plural in the sense in which they are used and thus retain their singular forms.

I often refer to codices in the text. Full bibliographic information on them can be found in the first section of the bibliography under the heading "Native Pictorial Sources."

BARCELONA

CHAPTER ONE

Spain and the Imperial Ideology of Mapping

THE KING AND THE MAP

If one could trace the genesis of the sixty-nine Relaciones Geográficas maps, following the narrative threads of the story along the globe of the earth, the filaments coming out of towns and villages in New Spain would twine together, stretching across the Atlantic. From the shores of Spain, this thick cord would reach into the Alcázar, the royal palace in Madrid. For ultimately it was Philip II's (r. 1556–1598) interests in making his territories visible through maps that resulted in these local maps being made of New Spain. Exploring the dimensions of his interest, which he inherited from his father, Charles V (r. 1517–1556), will explain how the Relaciones Geográficas maps came to be made and, more important, show how maps, and the visions of territory that they promoted, were shaped and animated by politics and ideology (Alpers 1983; Kagan 1986). Royally sponsored maps from sixteenth-century Spain in particular allow us valuable insight into Philip's attempts to shape the nascent nation-state, to harness the competitive forces of nationalism and regionalism.

Philip shared an interest in mapping and cosmography with many other Renaissance princes, but as one of the most powerful rulers of his time, he had the means to sponsor mapping projects and patronize cartographers on a level nearly unparalleled in the rest of Europe (Akerman and Buisseret 1985; Kagan 1986; Kagan 1989; Parker 1992). No doubt, as ruler of the Spanish Netherlands (or Low Countries, as they were called), Philip fell

Figure 5. View of Barcelona from the south (detail), by Anton van den Wyngaerde, 1563. Van den Wyngaerde drew this picture of Barcelona as part of a series of Spanish cityscapes he created at the behest of King Philip II. While the city seems to be viewed from one vantage point beyond Montjuich, van den Wyngaerd actually blended together a number of viewpoints in this drawing. Size of the original: 39.5 x 159 cm. Photograph courtesy of the Österreichische NationalBibliothek (Cod. Min. 41, fol. 12).

under the sway of the technical and artistic brilliance of cartography in these northern realms. In 1558, two years after he ascended the throne, Philip sponsored Jacob van Deventer (1500–1575) to produce a series of drawings of his towns in the Low Countries. At that same time Philip's lieutenant in the Netherlands, the third duke of Alba, commissioned Christopher 'Sgrooten to make topographical maps of the same territory (Akerman and Buisseret 1985; Parker 1992: 143). Another one of Philip's many mapping projects was a fifty-item questionnaire that was circulated widely both within Spain and in the New World (López Piñero 1979: 217–19; Viñas y Mey y Paz 1949–1971). The questionnaire replies from the New World were dubbed the "Relaciones Geográficas," or "geographic reports," because they dealt with human history and geography. The maps that concern us are those created in New Spain, where indigenous respondents drew on traditions of painting and logographic writing unparalleled elsewhere in the New World. However, we will save these maps for later chapters. For now, the parameters of Philip's interest, and the kind of maps he desired to be made, are nowhere better seen than in his two most important commissions to map Spain, both of which he launched in the early decades of his reign, in which we find contrasting expressions of the nature of the Spanish state, one regional, the other national.

With his Spanish commissions, Philip seemed to have been trying to do for Spain what he was successfully doing for the Low Countries—bring it into the embrace of the most skilled and advanced cartography of its day. Thus he called upon both the Flemish artist Anton van den Wyngaerde (c. 1512–1571) and a group of Spanish cartographers led by Pedro de Esquivel (Becker 1917: 113; Kagan 1986; Kagan 1989; Reparaz Ruiz 1950: 75). Van den Wyngaerde was asked to draw a series of topographical views of Spanish cities, which he did from the 1560s until his death in 1571; these city views were to be much like the views van Deventer had drawn of their Dutch counterparts (Kagan 1986, Kagan 1989). Esquivel's group, which began its countrywide survey in about 1570, was to draw up a set of small-scale maps of all of Spain, a project parallel to 'Sgrooten's in the Netherlands.

To this end, both van den Wyngaerde and Esquivel traveled around Philip's realm, taking stock of the land that they were to portray. The divergence of their projects is revealed in their working methods. Van den Wyngaerde worked in the Albertian mode: his preparatory sketches reveal that he picked various vantage points outside of and around one of his assigned cities and then sketched the views in front of his eyes. To create his finished maps, he seems to have combined sketches into a composite cityscape that, although it was more than ever could be seen from a single vantage point, still upheld the artistic fiction that it was within an eye's view (Kagan 1986: 126). One of his completed works, of the port city of Barcelona, can be seen in figure 5. We know from existing sketches that van den Wyngaerde sketched many views before creating this scene of Barcelona as seen from beyond Montjuich, identified on the foreground of the map as "Mont giovi."

Esquivel's method was wholly different and reveals the way Euclidean space was constituted: he toured the country taking direct measurements derived from measuring chains, wooden goniometers (for measuring angles), and compasses, although he failed to compensate for the vagaries of magnetic north (Aveni 1980: 49, n. 1; Multhauf 1958; Price 1955; Singer et al. 1957–1958; Taylor 1968). He also depended on triangulation and distance estimation (Cortesão and Teixeira da Mota 1960–1962, vol. 2: 84; Goodman 1988: 66; Reparaz Ruiz 1950: 77). A contemporary viewer, Felipe de Guevara, commented that Esquivel's map "was without exaggeration the most careful and accurate description ever to be undertaken for any province since the creation of the world. . . there is not an inch of ground in all of Spain that has not been seen, walked over or tramped upon by [Esquivel], checking the accuracy of everything (insofar as mathematical instruments make it possible) with his own hands and eyes" (Guevara 1788: 219–21, cited in Parker 1992: 130).

It seems that Philip wanted both van den Wyngaerde's artistic cityscapes and Esquivel's area maps to be brought together to create two atlases of maps: these two would make visible the two natures of his Spanish realm, one regional, the other national. The former would show the various cities of Spain; the latter, the geographic expanse of his Iberian kingdom (Haverkamp-Begemann 1969). Only this second atlas was eventually compiled, not by Esquivel himself, but probably by a later royal geographer named Juan López de Velasco, for it bears marginal notes in his hand (fig. 6; Reparaz Ruiz 1950; Parker 1992: 132). Called the Escorial atlas, this work contains not only the area maps, but also an overall map, or "key map," that was created some time after 1585 by combining Esquivel's survey maps with an earlier survey map of Portugal, thus reflecting the pan-Iberian kingdom that Philip created after his annexation of Portugal in 1581 (Reparaz Ruiz 1950). Philip probably never intended the Escorial atlas to become public—he secreted it in his library at the Escorial, where it remains today—but he sent the van den Wyngaerde drawings to the Plantin press to be engraved. For reasons not fully understood, the press never carried out this project (Haverkamp-Begemann 1969; Kagan 1986: 118).

PTOLEMY

The most influential factor in determining the scope of Philip's projects was undoubtedly the work of Claudius Ptolemy, whose widely disseminated *Geography,* first published in 1475, distinguished between chorographic and geographic maps. The former, which includes cityscapes, deals with partial and particular views of a whole; the latter deals "only with regions and their general features" and includes area maps (Stevenson 1932: 26–7). The two modes of mapping were necessary complements—one needed the geographic map to understand the larger whole, but would be unable to understand the specific and remarkable without maps made in the chorographic mode. Ptolemaic concepts have colored how we think of spatial representations up to the present; for instance, the chorographic and geographic dovetail neatly with the Euclidean and Albertian projections that we have discussed.

As patron of complementary mapping projects, the cosmopolitan Philip was swayed not only by Ptolemaic prescription but also by other royal mapping projects that were being carried out by princes and nobles across Europe. The Esquivel survey found a forerunner in the survey maps Philipp Apian made of Bavaria between 1554 and 1561 at the behest of Duke Albrecht V, and had as its contemporaries the 1579 maps of England that Christopher Saxton produced under Elizabeth I and the area maps of Italy made by Egnazio Dante that Pope Gregory XIII proudly displayed in the Vatican (Bönisch 1967; Brown 1979; Cortesaõ and Teixera de Mota 1960–1962, vol. 2; Helgerson 1986; Schulz 1987). Van den Wyngaerde's work found its closest parallel in van Deventer's Dutch cityscapes. The number and scope of these mapping projects tell us how European princes, in holding up a mirror to their realms, found in maps their true reflection.

While Esquivel's regional maps and van den Wyngaerde's cityscapes appear to stare at each other across the divide of the geographic and chorographic, they are more alike than different; both are European cultural products colored by contemporary European ideas of scientific rationalism (Helgerson 1986). This rational approach to the surrounding world manifests itself in the uniform geometric projections—Euclidean and Albertian—that we find guiding these maps—Esquivel imagined a (nearly) rectilinear grid of latitudinal and longitudinal lines, cast like a net over Spain. After he measured the contours and distances in the landscape, he reduced them exponentially to be able to plot them on his Euclidean grid. Van den Wyngaerde's works followed the rules of Albertian perspective, and were perhaps additionally influenced by the mathematical rules guiding ichnographic city plans that had been expressed by Leonardo da Vinci (Pinto 1976). In van den Wyngaerde's maps, objects in space are aligned, as Alberti dictated, along a series of axes that converge on a central vanishing point. In both kinds of maps, man defines his relation to the world through his ability to measure it.

In addition, both series of maps, in seeking a way to define and describe space, center on combinations of topography and the built environment—the same combination the artist of the 1524 Cortés map of Tenochtitlan used as he blended a panoramic projection of landscape with elevations of architecture. The van den Wyngaerde maps show the buildings of a city set within the embracing contours of the landscape; in Esquivel's maps, topographic features such as river networks are brought to the fore, but cities and towns are also marked on the maps with circles and names. These maps, then, are aligned in that both their makers and their audience agreed on two basic premises: the map is ruled by a mathematically consistent projection, and the space to be represented on a map is that of topography and human settlements.

Since European viewers of these maps (as well as contemporary ones) accepted these premises, they are hidden from view by their obviousness. What would have struck the coeval European viewer were the differences between chorographic and geographic. Ptolemy's distinctions, as illustrated by these maps, would only have been made more evident by the difference in framing between the van den Wyngaerde and Esquivel maps. Van den Wyngaerde centers the frame of each of his maps on the built environment of one city—in his finished compositions, he stopped his picture where buildings tapered off and landscape began. Esquivel, on the other hand, used the squares of the large grid he had envisioned over Spain as frames for each of the area maps in the Escorial atlas.

These opposing frames, and the corresponding Ptolemaic distinctions, should not be passed over lightly. The chorographic/geographic distinction was perhaps the most important classifying scheme for maps in sixteenth-century Europe; generations of educated Europeans learned of it through one of the many editions of Ptolemy published in the sixteenth century and also in the many collections of published maps, such as Georg Braun and Frans Hogenberg's *Civitates Orbis Terrarum* (1965 [1572]) and Abraham Ortelius's *Theatrum Orbis Terrarum* (1991 [1579]). The influence of the Ptolemaic schema ran two ways. First, it allowed Europeans to classify existing maps. Second, as in the case of Philip's two commissions, it created a standard dual model of how space should be mapped, and in doing so it largely dictated the scope of many sixteenth-century mapping projects. Why did these distinctions in ways of understanding and mapping the world resonate so widely with a European audience? After all, there are other ways of classifying maps, but no other seemed to have had the same hold on the elite imagination as did the Ptolemaic.

The geographic/chorographic model of the world, I believe, was so influential because this dual model neatly corresponded to sixteenth-century models of the nature of the state—indeed, its contours followed the fault lines between regionalism and nationalism. Richard Kagan has pointed out how chorographic cityscapes flowered in northern Europe at the same time that powerful cities in Flanders and Germany asserted their independence by resisting attempts by Holy Roman Emperors to curb their autonomy: the individual cityscape became the representative image of the autonomous city (Kagan 1986: 123). Such cityscapes were engraved and widely disseminated; their popularity on the local front was due in no small part to the correlation between the image that the cityscape offered and the citizens' vision of their independent polis.

Figure 6. "Key" map from the Escorial atlas, attributed to Juan López de Velasco, c. 1585. The Escorial atlas is a group of small-scale survey maps of Spain created by the Spanish cartographer Pedro de Esquivel under the patronage of King Philip II. Some time after Esquivel had finished his mapping project, his maps were bound together and this composite "key" map was created as the atlas frontispiece. This key map seems to have been the work of the royal cosmographer Juan López de Velasco. Size of the original: 30.5 x 45 cm. Photo courtesy of the Newberry Library, Chicago. Photography granted and authorized by the Patrimonio Nacional, Spain (Monasterio de El Escorial, Real Biblioteca, Atlas del S. XVI, sign. K. I. 1, fol. 1).

But a chorographic map overwhelms us with the specifics of an individual city; in van den Wyngaerde's paintings of Spanish cities we risk losing sight of Spain itself (Helgerson 1986: 75). Its antidote is the geographic map that shows the territory of the whole country. In the key map of Spain found in the Escorial Atlas (fig. 6), the space of the nation was not pictured as autonomous and competitive cities, but as a continuous and politically undifferentiated geographic expanse. This map speaks of centralized power, since the production of this image could only be possible through the patronage of the king, who controlled it all. His interests were not provincial, for his power resided not locally, but uniformly across the nation. And it was *this* image, the Escorial key map, rather than van den Wyngaerde's collection of chorographic city views, that called upon its Spanish viewers, like the tolling bell summoning the faithful, to observe their national loyalties. For it revealed the nation in its entirety, a peninsula bounded by sea and ocean and tightly webbed by rivers. It suppressed (by making nearly invisible) the parochial entities of Spain, its kingdoms, provinces, and fiefs, and, by association, their claims on the viewer's loyalty. This body of the land suggests a corollary, the body of the king, since the dimensions of the land shown in the key map were Philip's creation: he and he alone controlled its scope through the reach of his conquests and extent of his inheritances.

In commissioning both geographic and chorographic views of Spain, Philip was registering these two ideologies about the nature of his country. One, rooted in its feudal heritage, saw Spain as a group of realms, or as a federation of "comunidades" or city states, an ideology that reached its sixteenth-century apex in the Comuneros movement (1517–1522). The other ideology posited Spain as a nation-state that owed its loyalty above all to its king in Madrid and that was governed by a centralized bureaucracy. Philip, of course, favored the latter. He tried throughout his reign to suppress the rancorous competition between cities and the corrosive nationalism of Spain's composite kingdoms by building national institutions. In this he followed his father's lead. Charles crushed the Comuneros soon after his ascent to the Spanish throne; Philip, likewise, acted to define Spain as a nation-state by choosing Madrid, in the very center of Spain, as the seat of his centralized government, soon after his own ascent.

Despite his fealty to the nation-state, Philip perpetuated the debate between regionalism and nationalism by promulgating two wholly different views of his country in his commission of the two atlases. The Spain glimpsed in van den Wyngaerde's atlas was one of autonomous cities held together tenuously by their own recognizance and in their traditional obeisance to a king in Madrid. Such an atlas was a fitting product of the Flemish van den Wyngaerde, "whose political loyalties were steadfastly local" (Kagan 1986: 128). In the Escorial atlas, Spain is set out by Spaniards whose loyalty was first and foremost to their patron, the king. Their particular breed of rational perspective allows no region or town to be superior to another, making each part of the kingdom subservient to the grid that only the king—who had the resources to survey and the access to cover the whole—could have sponsored. Spain's geographic unity across the Iberian peninsula, as seen in this map, seemed to bear witness to its national unity, one body whose heart was its monarch in Madrid.

Since Philip stood to benefit from the latter view, why would he have sponsored the van den Wyngaerde atlas as well as the Escorial one? Like others of his time, he appreciated the artistic achievements made in these cityscapes and no doubt wanted to join the ranks of his fellow princes as a patron. His patronage of the Flemish van den Wyngaerde, paradoxically, would have had the added value of placating local interests in Spain. When Philip wanted to have maps of Spain decorate the walls of the entrance hall of the Alcázar, he chose, it is believed, copies of the van den Wyngaerde cityscapes. No doubt Philip understood that as Spain's proud grandees and privileged nobles filed into the palace, the image of Spain that greeted them would be one that underscored the importance of their local fiefdoms. Yet in the Escorial atlas, which showed a united Iberian peninsula, the body of the nation was divided into sections only by the grid imposed by the king's mapmaker, not by the political boundaries of kingdoms and the fratricidal urges of its peoples. In this Philip would see the image that he held most dear; this was *his* Spain made visible.

THE NEW WORLD

The map projects that Philip sent out of the Alcázar—and the stream of map images that returned as part of the royal diet—allowed this monarch to reach a realm that was no longer physically accessible to him. Having his realm within his reach was crucial to this conscientious king. He had inherited, after all, the kingdom and the standards of his great-grandparents, Ferdinand (r. 1474–1516) and Isabella (r. 1474–1504). This pair traveled constantly through their fractured realm, understanding that it was only their physical presence before courtiers and commoners that gave meaning to the idea of a united Spain. Their incessant travels, from city to city, across province after province, knit together a state out of its component parts. But both Ferdinand and Isabella had been born into small Iberian kingdoms, where traveling through their realms by horse or coach, however arduous, was still possible and practical. On these journeys, their gazes, perhaps distracted or bored by the interminable trip, had swept over much of the territory that was their royal domain: they saw and they knew.

In a sliver of time, a few decades, the perimeters of this kingdom expanded outward.

Spain in 1474 was a mere thumbprint on the globe, but fifty years later its reach spanned hemispheres. By the time Philip acceded the throne, his empire stretched across the Atlantic to lay claim to much of the New World and its peoples; it embraced the Low Countries, on the other side of a hostile France. The voyage to the Spanish Netherlands along the Spanish Road took about seven weeks (Parker 1987: 96–7); it took three months to get from Seville to Veracruz (McAlister 1984: 206). Philip's realm, then, could no longer be traversed by the journeys of the monarch—his people could no longer count on a glimpse of their king to inspire their loyalty to the Spanish crown. Most important, Philip ruled over vast areas, even continents, that were out of his reach and far from his gaze. Of course, he was able to read about them through the hundreds, perhaps thousands, of written accounts that reached his desk, but for him, as for other educated men of his time, the written found its ideal complement in the pictorial. Knowing was predicated on seeing.

Philip, a monarch who insisted on reading and personally signing every significant paper dealing with his realm, knew that such a distance, both physical and psychic, between himself and his subjects hindered to his ability to rule. But travel was no longer the solution—Philip wanted to stay home, at least in his later years, when he was to be found in Madrid or his new palace at the Escorial. One almoner accused him of "sitting forever over your papers, from your desire, as they say, to seclude yourself from the world" (Kamen 1983: 146). By the end of Philip's life, he had concluded that "traveling about one's kingdom is neither useful or decent" (Kamen 1983: 147).

While his numerous mapping projects were no doubt spurred on by those of his noble contemporaries, it seems plausible that for Philip the maps filled a void, and created a bridge between this king and his far-flung and absent subjects. In the same way that Philip gave as gifts copies of his portrait so that his subjects and relatives would at least have an image of this cloistered king (Kubler and Soria 1959: 205-6), Philip himself commissioned maps so as to have an image of his unreachable realms. His peripatetic great-grandparents had seen and had been seen; the sedentary Philip received and sent out images instead: just as the portrait stood in for the king's own presence, so the map substituted for his own vision.

But what to do with the New World? Lying thousand of miles away, the New World had been portrayed by European writers as being full of magical possibilities—Columbus wrote of islands peopled by the grotesque and fantastic, while Francisco Vázquez de Coronado set out in search of the fabled Seven Cities of Cíbola—yet it was a world that Philip would never see. Maps, once again, were Philip's conduit. Just as he aimed to, and did, bring Spain into the embrace of the most sophisticated cartography that Europe had known, he, like a generous and equitable father, wanted the same for his New World territories. They, too, would be made visible for him.

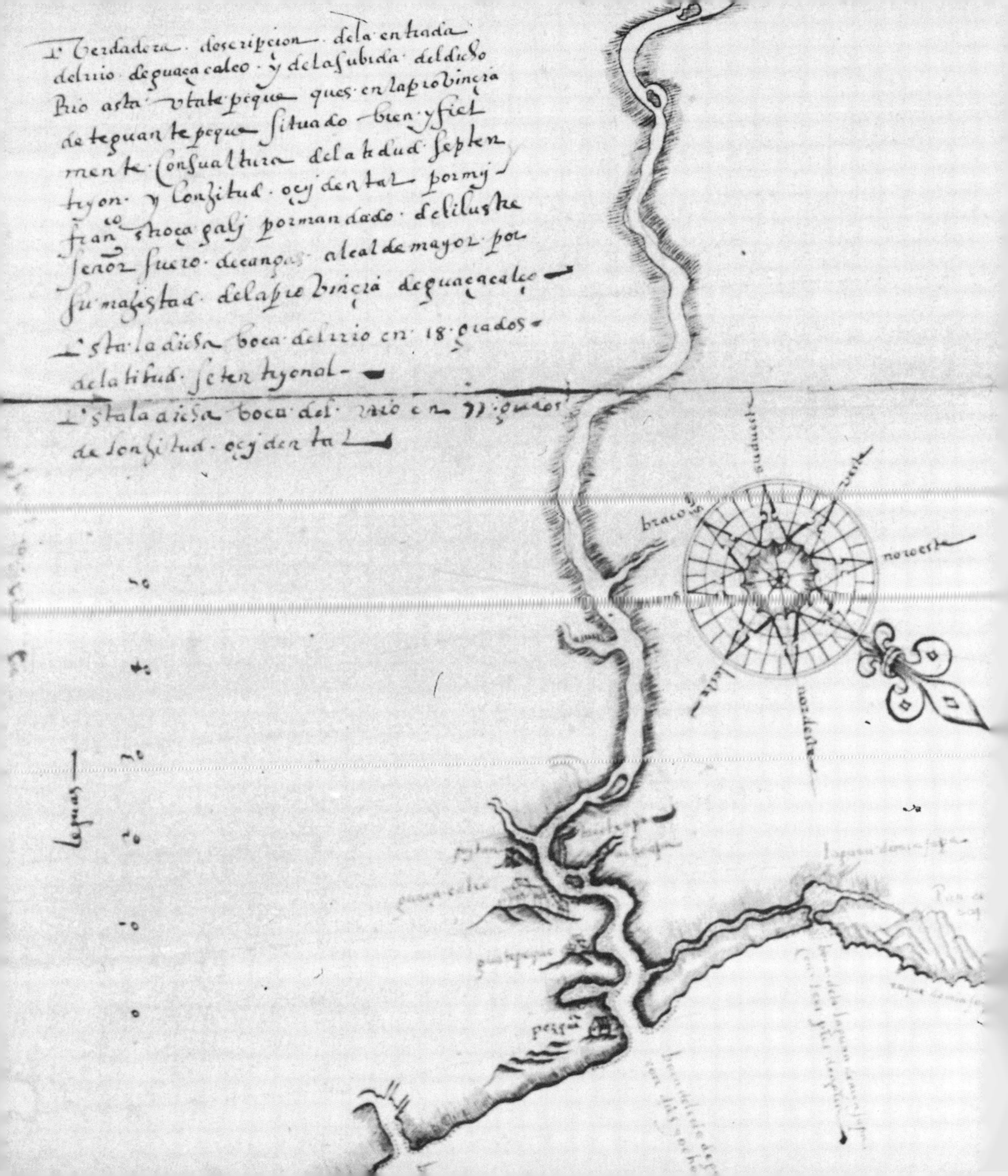

Verdadera descripcion dela entrada
delrrio de guaçacalco y dela subida del dicho
Rio asta utatepeque ques en la provinçia
de tequantepeque situado bien y fiel
mente con su altura dela latitud septen
tryon y longitud ocydental por my
fran[co] stroçagali por mandado del ilustre
señor suero decangas alcal de mayor por
su magestad dela provinçia de guaçacalco

Esta la dicha boca del rrio en 18 grados
dela titud setentryonal

Esta la dicha boca del rrio en 77 grados
de longitud ocydental

leguas

braço

noroeste

nordeste

CHAPTER TWO

Mapping and Describing the New World

Since Philip II desired to have a map of the New World made, the task fell to two of the realm's most prominent cosmographers, Alonso de Santa Cruz (c. 1505–1567) and Juan López de Velasco (d. 1598). The challenge that faced them was parallel to Philip's: just as he had to rule without seeing, they had to map without seeing. For unlike Pedro de Esquivel, a cartographer who tramped over every inch of ground that he was to map, Santa Cruz and López de Velasco were to map the New World while remaining an ocean away in Spain. As cosmographers—a word that derives from the Greek *kosmos,* meaning "order" or "universe"—they were attempting to organize knowledge about the whole world. To them and to other sixteenth-century Europeans, the New World was the missing part of their cosmos. Santa Cruz and López de Velasco's attempts to make the New World knowable through maps rank among the high cosmographic achievements of the sixteenth century. Among the various phases of their grand project to map the New World was the Relaciones Geográficas.

One after the other, these royal cosmographers planned questionnaires to be sent to local officials across Spain's domain in the New World. They anticipated that their questionnaires would elicit hundreds of responses in the form of texts and maps (Cline 1972d; Jiménez de la Espada 1965). Both Santa Cruz and López de Velasco, like alchemists, planned to take the written responses and distill them into a descriptive chronicle of the New World. In this respect, their aims were conservative, for their chronicle would be like others of the time. Their mapping projects in contrast, were unique, and aimed to make the New World visible through state-of-the art (and art-of-the-state) cartography.

In theory, their New World maps were possible since Santa Cruz and López de Velasco posited rational, repeatable principles underlying cartography, ones that could be taken and imposed on any landscape, anywhere in the world.[1] As their first phase, Santa Cruz and then López de Velasco planned to establish a geometric projection of the New World, based on a grid of latitude and longitude that had long been theorized but never realized. A second phase was initiated by Philip, who commissioned a surveying project for New Spain that would supply him with geographic maps of the colony that were similar to those that Esquivel produced of Spain. Having both a structuring grid and an enveloping survey would enable the cosmographers to draw up geographic maps of the New World that were of the kind envisioned by Ptolemy. As a third and final phase, these cosmographers planned to cull local maps from respondents in the New World so that they could create a complementary chorographic album of New World cities and towns, like the one van den Wyngaerde produced of Spain. Fired by such ambitions, the cosmographers aimed to make the New World visible to the eyes of its king—to put it, and themselves, on the map.

SANTA CRUZ AND MAPPING THE NEW WORLD

As one of Spain's leading cosmographers under both Charles V and Philip, Santa Cruz spent much of his career mapping, or trying to map, the New World (Cuesta Domingo 1983). He was a royal cosmographer from 1553 to 1567 (Schuller 1912: 415–6), and during these years he seems to have been planning an ambitious atlas of the New World, as he wanted to bring the New World into the realm of the visible, into the light of the known. Among Santa Cruz's papers at the time of his death were the rudiments of such an atlas, described in a coeval inventory as "a large bundle covered with parchment in which there are 169 pieces of good quality paper . . . upon which are drawn in colors many provinces, islands, mainlands, and ports of the Indies as well as of other places" (Jiménez de la Espada 1885–1897, vol. 1: xxxiii). Santa Cruz's papers are now scattered, and three maps of the Gulf coast that recently have been attributed to him may be the only remaining fragments of his New World atlas (RGS 5: 303–8).

For Santa Cruz, this project of "making visible" was indissolubly linked with Spain's exploration and conquest of the New World. That is, the maps he knew followed the creeping tide of Spain's acquisition of New World land and peoples. Each year that Spain explored and laid claim to more ground, more appeared on maps. In the 1530s, Santa Cruz served as cosmographer to the Casa de Contratación in Seville, the government body that oversaw trade with the Indies (Pulido Rubio 1923; Schäfer 1935–1947; Schuller 1912: 415–6; Stevenson 1927). The Casa de Contratación was a clearinghouse for New World maps. It assigned maps to ships' pilots at the beginning of voyages to the Indies; during the voyage the pilots were expected to note mistakes and corrections on the maps in their charge (Pulido Rubio 1923). Upon the ships' return to Seville, the pilots' corrections would be included in the *padrón real*—a master map mandated by the king in 1508—to keep it up to date with each new European discovery (Jiménez de la Espada 1965: 48; Lamb 1974; Latorre y Setien 1913: 32; Stevenson 1909). In the 1530s, the crown summoned Santa Cruz, along with other cosmographers, among them Sebastian Cabot, to update the padrón (Pulido Rubio 1923: 45–6; Jiménez de la Espada 1965: 20). Thus as each new flotilla from the New World was welcomed into Seville's harbors, it brought with it maps with coastlines more solid and confirmed as Spain expanded and consolidated its control. In fashioning a master map from all these, Santa Cruz was creating for both his proud king and any jealous competitors a map that acknowledged new explorations and proved possession.

While Santa Cruz could make the New World—and the extent of Spain's domain—visible with his maps, he still could not position it in relation to Europe. This positioning was both figurative and literal. For the New World contained, as one sixteenth-century writer despaired, "things of the natural world which depart from the philosophy received from and discussed in ancient times" (Acosta 1940: 7). Its entry into the order of things, as Europeans understood them at the time, was unsure. As was emblematic of its stand outside the known order of things, the New World had yet to be reigned in by cosmographers with lines of latitude and longitude, lines that would make rationally visible its global position in relation to Europe and the precise extent of its land mass. Santa Cruz, like other European cosmographers, lacked established latitude and longitude lines. The cosmographers, of course, were quite familiar with the theory of latitudes and longitudes from Ptolemy's *Geography,* which had described the system of latitude lines—imaginary circles rippling outward from the two Poles toward the equator—that would intersect with the longitudinal arcs that divided the globe into equivalent sections, like the segments of an orange (A. Aveni 1992: personal communication). Santa Cruz found this grid particularly compelling, as did other cosmographers of the time, because it posited a geometrical system for measuring and mapping the entire globe.

In drawing a world map in 1542, Santa Cruz must have chafed at the position of the New World continents, for in the end they had been imaginatively, rather than rationally, located (Dahlgren 1892). He was also like the child who had discovered his father's clay feet. Ptolemy, the patriarch of European cosmographers, had assigned latitudinal and longitudinal lines to different places in the world, but at the time he wrote, he thought that the narrow reaches of the classical world covered much of the globe. By the sixteenth century, after Ptolemy's work had been republished, it was clear that his "scientific" prescriptions of latitude and longitude were mainly guesswork, that his placement of the equator was inaccurate, and that his distances between meridians of longitude were too short (Bunbury and Beazley 1911: 623–4; Santa Cruz 1921: 18–9). To the European mind the world had gotten bigger. And while Santa Cruz could rail against the shortcomings of Ptolemy's own measurements of longitudes, when he came to draw a world map, he was forced to estimate rather than calculate, for he had yet to determine them himself.

As a royal cosmographer, Santa Cruz took on the burden of longitude, which was as much a political problem as a scientific one. It had begun to be an issue during the reign of Isabella and Ferdinand, when Spain and Portugal between them carved up the globe. Under the treaty of Tordesillas of 1494, the yet unknown line of longitude falling 370 leagues west of the Cape Verde islands off the coast of Africa would divide the world in two (Goodman 1988: 53–65). Those lands to the east of this line (which would eventually include Brazil) would fall to Portugal, and those to the west (which would include Mexico and Peru) to Spain. The problem of longitude became more pressing during Charles V's reign when, in 1529, he agreed that an additional line of longitude should separate Spain's and Portugal's domains in Asia. The establishment of this second line was even more crucial than the first, for while the former meant that some stretch of the untamed Amazon would fall either to Spain or to Portugal, the latter held the promise of such prizes as the Philippines or Indonesia. Thus Santa Cruz, along with other Spanish cartographers and cosmographers, jockeyed with the Portuguese to find an accurate way of measuring longitude, hoping for favorable results, that is, for the prizes of the eastern Pacific to reveal themselves to be on the Spanish side of the line.

Santa Cruz had certainly begun to grapple with the longitude problem by the time he drew up his 1542 world map. At this time, latitudes were easily found. Any mariner could locate the Pole Star, which lies almost exactly along the line of the earth's polar axis. With a simple instrument of two rods joined at one end, somewhat like straight-armed calipers, the mariner could measure the vertical distance between the Pole Star and the horizon, this angle being almost exactly the same as the latitude of the observer.[2] Mariners in sixteenth-century Spain had appropriate equipment at hand, and used cross staffs and astrolabes (the latter could be fitted with a quadrant or shadow square) to make such observations (Goodman 1988: 77; Schulz 1978: 431–41). And even without complicated measuring devices, latitude can be found within the Tropics by observing the day of the sun's zenith passage.[3]

While latitudes were easy to find, longitude still remained out of Santa Cruz's grasp; no one had the means to calculate the circumference of the earth, or the cumulative distance between each of the 360 imaginary arcs dividing the terrestrial globe from the North Pole to the South. What was the distance between Seville and Santo Domingo? Between Acapulco and Manila? Mariners who traveled these routes could provide cosmographers like Santa Cruz with only crude calculations that were based on estimated speed and time in transit. World maps of the time capture the confusion brought about by longitude measurement, or lack thereof. In his 1542 world map, Santa Cruz could show precise relationships along lines of latitude: both Madrid and Mexico City are shown the correct distance north of the equator. But longitudinal relationships are skewed. In his writings, Santa Cruz calculated the longitudinal distance between Mexico City and Genoa to be 217°30″ (Goodman 1988: 55); the actual figure is closer to 108°. Any inaccuracies had international import; a calculation of Atlantic distances that was too small allowed Spain to claim more of the eastern Pacific, while one that was too large favored Portugal.[4]

With political, ideological, and scientific interests driving him, Santa Cruz attacked the problem of longitude. One result was a manuscript he wrote on the subject in which he proposed collecting simultaneous readings of the moon's position from observers across the globe. From these he could mathematically ascertain where lines of longitude lay (Cuesta Domingo 1983: vol. 1; Santa Cruz 1921). His inventive method called for observations of a lunar eclipse, with directions to laymen on how to record its position in the sky (Santa Cruz 1921: 22–3). Santa Cruz could easily predict the eclipse because the moon's eclipses and its orbit around the earth were well known to sixteenth-century cosmographers. Eclipses were also easily observable, if not perfectly understood, by laymen. From a body of observations of a single eclipse, Santa Cruz could mathematically determine longitudes and construct the model of an enveloping corset for the earth. With lines of latitude and longitude established, Santa Cruz could outstrip the father of European cartographers, Ptolemy. He could finally fit the New World where it belonged on the globe and bring to life the first mathematically precise world maps known to mankind.

SANTA CRUZ AND THE NEW WORLD CHRONICLE

While Santa Cruz was at work on world maps, he clearly felt that a graphic image of space alone was not sufficient to describe the New World. The map, a description of space, found its ideal complement in the chronicle, which described the passage of time (Helgerson 1986: 72). Text and image, time and space: together these amounted to a full description of the New World. At the same time that Santa Cruz was at work drawing maps of the New World, he was also writing chronicles. Under Charles V, Santa Cruz was charged with writing "an account of 'all the islands of the world'" (Schuller 1912: 431). To this end, Santa Cruz collected and copied many maps and composed chronicles of places. His fulfillment of the king's request, the *Islario general de todas las islas del mundo,* was left among his papers at his death (the manuscript is now in the Biblioteca Nacional in Madrid). This *isolario*—island book—was one among many produced in the fifteenth and sixteenth centuries (Harvey 1987: 482–4). Santa Cruz's *Islario* intersperses maps of the world's islands with textual accounts that give a brief description and history of the discovery and settlement of these places, often drawn from the published first-hand accounts of Spanish conquistadores (Santa Cruz 1918; Wieser 1908).

Text and image, time and space: Santa Cruz's combinations were not new ideas. Texts were commonly paired with illustrative maps in encyclopedias dating from the late antique and early medieval periods (Schulz 1978: 446–7). In revitalizing the old encyclopedia form, Santa Cruz was fitting the new information gathered from the colonies onto the frames established by the ancients. He was not alone in this attempt, for other mid-century royal chroniclers were trying the same, and Santa Cruz acknowledges his fellowship with them in the prologue to his *Islario,* where he mentions Pietro Martire d'Anghiera, author of *De Novo Orbe Decades,* and Gonzalo Fernández de Oviedo y Valdés, who wrote *Historia general y natural de las Indias.*[5] Oviedo was forthright in his devotion to the ancients, declaring that "I [illegible] Pliny" (1944–1945 [1535, 1557], vol. 1: 37), referring to the encyclopedic *Naturalis historia* (of the first century A.D.), well known to chroniclers of the New World after its publication in 1492 in Rome (Garibay 1953–1954, vol. 2: 68–9; Sandys 1911).

But Santa Cruz knew first-hand from his work on the problem of longitude that the ancients could not be the final arbiters: he, like other chroniclers of his day, prized direct experience and observation over the rusty opinions of ancient authorities (Elliott 1970: 28–53; O'Gorman 1940: ix–lxxxv). Yet in the case of the New World, direct experience lay half a world away, beyond the grasp of the European chronicler. Thus chroniclers like Santa Cruz, Oviedo, and Pietro Martire refreshed Pliny and Ptolemy with information gathered directly from the accounts being sent back from the New World. In his *Islario,* as in Pietro Martire's *Decades,* Santa Cruz drew heavily on published accounts, such as those of Hernán Cortés, as well as unpublished ones. This reliance on individual accounts was not without shortcomings: for example, Columbus, in his letters, described New World islands as being filled with creatures born of his own imagination. To account for the vagaries of the eyewitness account, Santa Cruz proposed a solution.

Santa Cruz may have thought of his *Islario* as a forerunner of a wider inquiry, one that would allow him to make a larger, more comprehensive chronicle-atlas that would cover not just islands, but the entire New World (Latorre y Setien 1916: 11). However, the information he would use for his New World chronicle would not only be drawn from published accounts that he used in the *Islario,* but also would be gathered expressly for this purpose from colonial officials themselves. Some time around 1546 or 1547, Santa Cruz wrote to a high official, probably the marqués de Mondéjar, who was president of the Council of the Indies, advising that a questionnaire be sent out to "captains and officials of his majesty" as well as to viceroys and governors in the colonies (Cuesta Domingo 1983, vol. 1: 67–72). Notably, Santa Cruz asked for respondents to take paper, mark an orientation as in a navigational chart, and then make a *padrón de leguas*: a sketch map showing league distances (Cuesta Domingo 1983, vol 1: 70). His proposal of a questionnaire was not unusual: Spanish administrators had long used questionnaires to gather information from their huge overseas colonies (Gerhard 1968; Jiménez de la Espada 1965; Menéndez-Pidal 1944: 5–6). However, Santa Cruz proposed a questionnaire that asked for both texts and maps, with an eye towards creating a chronicle-atlas.

Thus Santa Cruz was part of a larger shift that was tipping the balance of authority away from classical models and toward the eyewitness accounts of humble local officials, whose numerous responses would have a collective authority that other individual accounts lacked. Using scores of responses culled from questionnaires as the basis for his new chronicle-atlas was Santa Cruz's novel way of reviving an old and authoritative form the encyclopedic combination of texts and maps. The questionnaire replies would fuel him with the fresh, sharp insights of contemporary observation and give him enough data to avoid the pitfalls of the individual account.

Santa Cruz's plans to collect a wide body of maps and accounts by questionnaire proved to be beyond his reach; this grand and novel scheme was unfulfilled at the time of his death in 1567. His parallel project, the determination of longitudes by questionnaire, was unrealized, and the many world maps in his collection were still marred by inaccuracies of scale and distance. The New World had yet to be captured in his global net or tamed by his prose.

LÓPEZ DE VELASCO AND MAPPING THE NEW WORLD

Four years after Santa Cruz's death, his projects were rekindled when, in 1571, Juan López de Velasco was named to the newly created double office of *cosmógrafo-cronista mayor,*[6] "main cosmographer-chronicler," of the Council of the Indies, the branch of the Spanish government that oversaw the New World colonies.[7] It was as if the crown, by establishing the cosmógrafo-cronista mayor office, was trying to bring Santa Cruz back to life, for its first occupant, López de Velasco, was officially charged with creating the definitive chronicle-atlas of the New World, the very work that Santa Cruz had never been able to complete (Jiménez de la Espada 1965: 42).[8] In this task he enjoyed the direction, as well as the patronage, of Juan de Ovando y Godoy, the president of the Council of the Indies. It was López de Velasco's good fortune to inherit Santa Cruz's papers, wherein he found numerous maps drawn on parchment, books Santa Cruz had written on geography, cosmography, and practical mechanics, as well as diagrams of geometry, genealogies, and commentaries on Ptolemy (Jiménez de la Espada 1885–1897, vol. 1: xxx–xxxvi).[9]

Upon inheriting these papers, López de Velasco seems also to have assumed Santa Cruz's ambitions as if they were unpaid debts. For soon after he was named to the post of cosmógrafo-cronista mayor, López de Velasco spearheaded a three-pronged project that aimed to bring Santa Cruz's dearest projects to completion. To begin, López de Velasco sent out a questionnaire to Spanish colonies to gather eclipse observations, thereby amassing the data he would need to establish longitudes. It is likely that López de Velasco also came across Santa Cruz's own plans for a definitive chronicle-atlas based on questionnaires, for shortly after issuing the eclipse directive, he sent out the Relación Geográfica questionnaire, which elicited both texts and maps and closely resembled the one that Santa Cruz had proposed to Philip years before. Such a questionnaire would supply him with the raw material for an ambitious chronicle-atlas of the New World. Concurrently, Philip commissioned a survey of New Spain, which was meant to result in a map like the one Esquivel made of Spain, and which López de Velasco certainly would have used.

Without the inspiration of Santa Cruz, López de Velasco would probably never have aimed so high. For one of the works López de Velasco did compose soon after he was named to his post, the *Geografía y descripción universal de las Indias* of 1574, reveals its author to be dutiful rather than inventive: it is mostly a compilation of dry items—population counts, geographic locations—gathered from the work of other men, a main source being Santa Cruz (González Muñoz 1971: xix; López de Velasco 1894; López de Velasco 1971).[10]

THE ECLIPSE QUESTIONNAIRE

The first part of the Santa Cruz program that López de Velasco carried out was the eclipse questionnaire, a project that no doubt owed its genesis to Santa Cruz's manuscript on longitudes, which had proposed using just such a questionnaire to gather eclipse observations (Santa Cruz 1921). In 1577, López de Velasco sent his printed *"instrucción,"* as it was titled, to the colonies, first to the viceregal seats, from which it was then disseminated across the land.[11] In it, López de Velasco attempted to turn local officials into momentary cosmographers (Edwards 1969). To do this, López de Velasco chose, as Santa Cruz had advised, the simplest of many complex ways of determining longitude—the observations of a lunar eclipse. He singled out the lunar eclipses of September 26th, 1577, and September 15th, 1578, events that happened simultaneously whether it was midnight in Mexico City or sunrise in Seville. The instrucción told local officials to observe one of these two prominent eclipses. On the appointed night, they were to record the direction of the moon and its elevation above the horizon both at the onset of the eclipse and at its close. To help these officials, the instrucción provided detailed instructions on how to make the simple devices needed to carry out these observations. Although the officials may not have known it, cosmographers were well aware that the eclipse would happen at the same time wherever it was seen, but the direction and elevation of the moon would vary according to where one stood on the earth; it might be overhead in Mexico City while on the western horizon in Seville. Once López de Velasco had numerous sightings of this eclipse from across the land, he could reconstruct the relative positions of the viewers on the surface of the earth and thereby figure out where observers in Mexico City stood in relation to those in Madrid. With this information, he would be the first to determine the long-awaited dream of Santa Cruz: longitude.

THE SURVEY

In addition to the eclipse questionnaire, Philip II had sponsored a countrywide survey some six years before López de Velasco sent out his Relación Geográfica questionnaire, dispatching a Portuguese cosmographer, Francisco Domínguez, to New Spain in 1571. Domínguez traveled in the party of Francisco Hernández, a royal physician charged with writing a botanical account of the colonies (Somolinos D'Ardois 1960: 253; Goodman 1988: 66). The cosmographer's commission was much like that of Esquivel, who began work in Spain at about the same time. Domínguez was to cover Spain's colonies, measuring latitudes and ground distances to create survey maps. If Domínguez had completed these survey maps, López de Velasco might have been able to use them to impose a regularity of scale on subsequent maps—including the responses to the Relación questionnaire—coming from New Spain. In New Spain itself, this survey might have provided local Spanish officials with a compelling model of a kind of map desired by the crown.

Domínguez's methods in mapping New Spain were probably comparable to those of his contemporaries in Europe, described in the previous chapter. He could have used either a sea-astrolabe or quadrant to measure the angle of elevation of the Pole Star, thus determining striations of latitude (Taylor 1964: 545). For ground measurement, he could have used goniometers to measure angles and may have coupled these angular calibrations with measurements by ropes and estimations of distances. All such measurements would have been reduced according to a fixed scale and plotted on paper.

Despite the royal commission, Domínguez produced little or nothing. Although he seems to have spent five years traveling through New Spain to make surveys, none of them are known today (Somolinos D'Ardois 1960: 252–8; Goodman 1988: 66–7).[12] A number of the Relaciones' writers mention the physician Hernández's visit to their towns, but none says anything about an accompanying cosmographer or geographer.[13]

THE RELACIÓN GEOGRÁFICA QUESTIONNAIRE

The year after López de Velasco sent out his eclipse instrucción to the New World, he sent out a similar missive, the Relación Geográfica questionnaire. Again, it owed much to Santa Cruz's lead, for it closely resembled the list of queries Santa Cruz had proposed sending out some thirty years earlier; it was also molded, in part, by Ovando (Cline 1972d; Jiménez de la Espada 1965). In his final questionnaire, also called an instrucción, sent to local officials in the New World, López de Velasco had carefully honed fifty questions that would be essential to the definitive chronicle-atlas that he was charged with making (appendix A). The questions clearly reveal what López de Velasco, like his contemporaries, felt that a chronicle, or modern history, should include. Many items dealt with local history, inquiring about preconquest rulers, warfare, and religion. Other items concerned natural history, asking about flora and fauna. Some items treated matters of direct economic interest to the crown: one asked about gold and silver mines, and another about quarries for valuable stones. Yet others dealt with trade and navigation, quizzing the respondents about trade routes, the depth and the dangers of the waters, and the quality of the ports.

In the Relación Geográfica questionnaire, López de Velasco's curiosity about the landscape is palpable. Items 6, 7, 8, 12, 16, 19, 34, 39, 41, 43, and 45 asked for written information about geography; items 10, 42, and 47 explicitly requested maps. Taken together with the eclipse instrucción, the Relación Geográfica questionnaire was certainly meant to

Figure 7 *(opposite)*. The Relación Geográfica questionnaire, 1577. A three-page printed questionnaire was disseminated widely in Spain's New World colonies; this first page instructed Spanish crown officials on how to proceed with the fifty items on the two following pages. Photograph: B. Mundy, reproduced courtesy of the Benson Latin American Collection, The General Libraries, The University of Texas at Austin.

Instructiõ, y memoria, de las relaciones que se han de hazer, para la descripcion de las Indias, que su Magestad manda hazer, para el buen gouierno y ennoblescimiento dellas.

PRimeramente, los Gouernadores, Corregidores, o Alcaldes mayores, a quien los Vireyes, o Audiẽcias, y otras psonas del gouierno, embiaren estas instructiones, y memorias impressas, ante todas cosas haran lista, y memoria de los pueblos de Españoles, y de Indios, que vuiere en su jurisdiction, en que solamente se pongã los nombres de ellos escriptos de letra legible, y clara, y luego la embiaran a las dichas personas del gouierno, para que juntamente, con las relaciones que en los dichos pueblos se hizieren, la embien a su Magestad, y al consejo de las Indias.

Y distribuyran las dichas instructiones, y memorias impressas por los pueblos de los Españoles, y de Indios, de su jurisdictiõ, donde vuiere Españoles, embiandolas a los concejos, y donde no, a los curas si los vuiere, y sino a los religiosos, a cuyo cargo fuere la doctrina, mandando a los concejos, y encargando de parte de su Magestad, a los curas y religiosos, que dentro de vn breue termino, las respondan, y satisfagan como en ellas se declara, y les embien las relaciones que hizieren, juntamẽte con estas memorias, para que ellos como fueren recibiendo las relaciones, vayan embiandolas a las personas de gouierno que se las vuieren embiado, y las instructiones y memorias las bueluan a distribuyr si fuerẽ menester por los otros pueblos a dõde no las vuierẽ embiado

Y en los pueblos, y ciudades, dõde los Gouernadores, o Corregidores, y personas de gouierno residieren, haran las relaciones dellos, o encargarlas han a personas intelligentes de las cosas de la tierra: que las hagan, segun el tenor de las dichas memorias,

Las personas a quien se diere cargo en los pueblos de hazer la relacion particular de cada vno dellos, responderan a los capitulos de la memoria, que se sigue por la orden, y forma siguiente.

Primeramente, en vn papel a parte, pondran por caueça de la relacion que hizieren, el dia, mes, y año de la fecha de ella: con el nombre de la persona, o personas, que se hallaren a hazerla, y el del Gouernador, v otra persona que les vuiere embiado la dicha instruction.

Y leyendo atentamente, cada Capitulo de la memoria, screuirá lo que huuiere q̃ dezir a el, en otro capitulo por si, respondiendo a cada vno por sus numeros, como van en la memoria, vno tras otro y en los que no huuiere que dezir, dexarlos hã sin hazer mẽcion de ellos, y passaran a los siguientes, hasta acauar los de leer todos, y responder los q̃ tuuieren que dezir: como queda dicho, breue y claramente, en todo: afirmando por cierto lo que lo fuere, y lo que no, poniendolo por dudoso: de manera que las relaciones vengan ciertas, conforme a lo contenido en los capitulos siguientes.

provide López de Velasco with enough information to make accurate maps of the New World, and an overall continental map. The Relación Geográfica questions also gave López de Velasco a way to cross-check some of the information he planned to reap with the eclipse questionnaire. For instance, item 6 of the questionnaire asked for either the latitude of the town or of its near equivalent: the elevation of the Pole Star, information also requested in the eclipse instrucción. Item 6 also requested that respondents state the days in which the sun cast no shadow at midday, in other words, to find the sun's zenith passage, a phenomenon that López de Velasco could easily correlate to his observers' latitudinal positions.

Just as some questions reveal what López de Velasco felt was an appropriate scope for the chronicle he planned, many others unwittingly show the parameters López de Velasco constructed for his complementary maps. The questions about latitude and zenith passage show the inescapable print of Ptolemy's geographical grid. Others were clearly meant to fulfill chorographic aims. Item 10, in particular, asked for a sketch map of the respondent town showing the layout of the streets and placement of plazas and monasteries. Here we see that López de Velasco envisioned space on the chorographic map to be largely defined and represented by means of the human-built environment—just like the Cortés map which shows Tenochtitlan to be a nexus of streets and important buildings. López de Velasco's questions concerning other kinds of nonurban spaces make it clear that he had the kind of images found in isolarii or on mariner's maps in mind. Items 42 and 47 were directed at coastal towns and asked for a sketch of the coast showing ports and unloading points and a map of nearby islands.

When López de Velasco first sent out his freshly minted questionnaires in 1577 and 1578 (fig. 7), no doubt his optimism ran high as he beckoned residents of the New World to join him at the avant garde of European mapmaking. If colonial officials rose to his call, they would supply López de Velasco with accurate topographical and coastal maps, as well as precise lunar observations. When the responses to these two questionnaires from New Spain were combined with the survey map that Philip had sent Francisco Domínguez to make, López de Velasco would have had ample information to make both geographic and chorographic maps of this crown jewel of Spain's colonial empire, thus making this part of the New World visible to its absent king. The geographic maps would be ordered by the newly established grid of longitude and latitude and could show topographical features that Europeans put on maps: lakes and waterways, mountain ranges, volcanos, and caves. This geography would be bounded by the thin line that marked the coast, and here the limitless sea would begin, occasionally pocked by reefs or coastal islands. Scattered among this topography would be symbols, small dots or churches, to call to mind the human settlements: the bustling port cities, the *pueblos de españoles*—the segregated towns in which only Europeans were supposed to live—and the *doctrinas*—centers for the evangelization of natives run by the three monastic orders. The chorographic maps would shift gears by bringing to the fore the spaces defined by an urban environment; they could detail layouts of the burgeoning Spanish towns of New Spain: Mexico City, Puebla, or Oaxaca.

With these maps of New Spain, López de Velasco would have been able to fashion an atlas that would have fulfilled the plans laid by Santa Cruz, even outstripping the earlier cosmographer by achieving his unyielding aspirations. López de Velasco's maps would have combined fresh and direct observations of modern man with the revamped schemes of the classical age to create a comprehensive visual description of New Spain for his king. As Philip thumbed through the atlas pages, each map, be it chorographic or geographic, would have animated for him the unseen world of New Spain. López de Velasco's feat of bringing New Spain into the realm of the visible would have rivaled the combined achievements of Esquivel and van den Wyngaerde in their creation of maps of Spain. Most important, if López de Velasco had been the one to determine longitude and measure the earth, his lasting fame would have been assured, and his hold on the royal graces irrevocable.

RESULTS

The moment López de Velasco put his Relaciones questionnaires and his eclipse instrucciones onto the ships that would carry them across the Atlantic to the New World, his project careened out of his control. The beauty of these questionnaires, of course, was that they were to draw on a wide network of respondents, but this also meant that their success ultimately rested, not with López de Velasco, but with faceless men an ocean away, petty officials of the empire, over whom López de Velasco had little immediate influence. While López de Velasco did instruct local colonial officials to draw on accounts of "people knowledgeable about the things of the land," did he ever envision that his Relación questionnaire would be discussed, and replied to, by dark-eyed "indios" speaking Nahuatl, Otomi, and Cuicatec? Could he ever have foreseen that, in response to item 10, native artists who had never heard of Ptolemy would send him maps that reflected their vision of their communities? Some of the responses to the Relación questionnaire brought López de Velasco face to face with a culture that was not his own and with perceptions that were alien. For López de Velasco, many of the Relaciones maps must have been absolutely discordant with his aspirations.

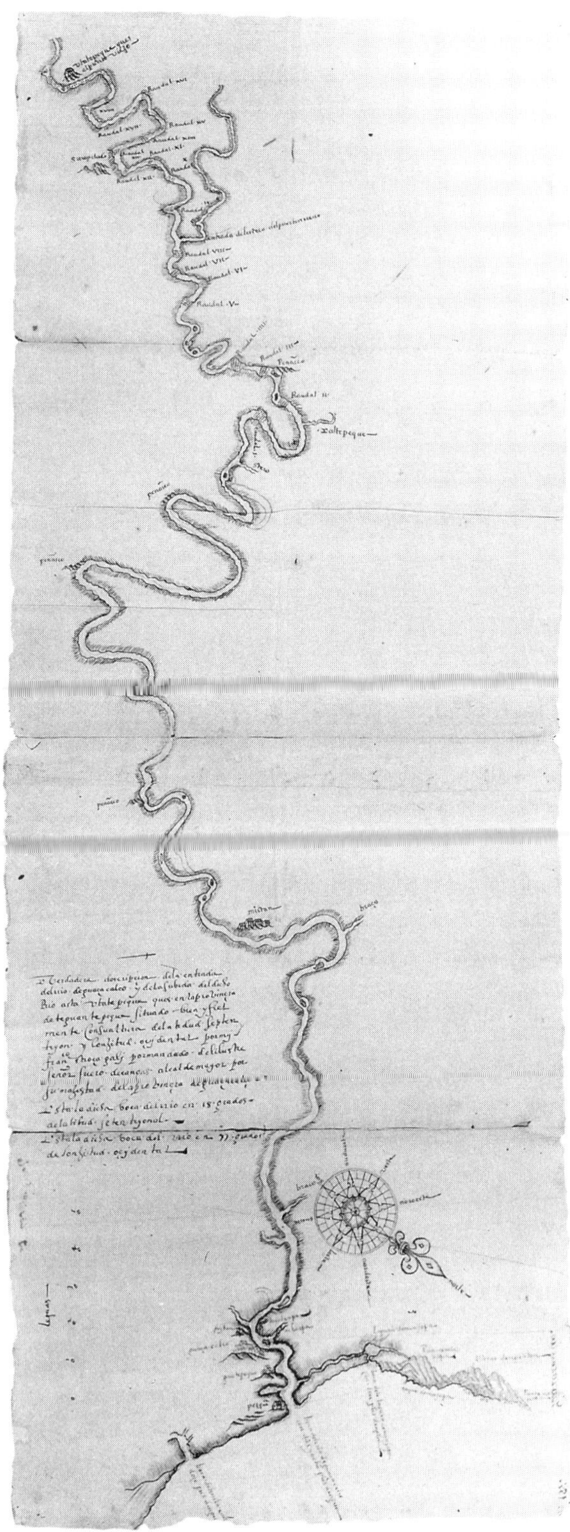

Figure 8. The Relación Geográfica map of Coatzocoalco or Villa de Espiritu Santo by Francisco Stroza Gali, 1580. The local alcalde mayor, responding to the Relación Geográfica questionnaire, called upon a Spanish mariner to create this accompanying map. Francisco Stroza Gali used nautical instruments to map the Coatzacoalcos river, a serpentine yet vital transcontinental passage. Size of the original: 31.5 x 84 cm. Photograph courtesy of the Benson Latin American Collection, The General Libraries, The University of Texas at Austin (JGI xxiv-2).

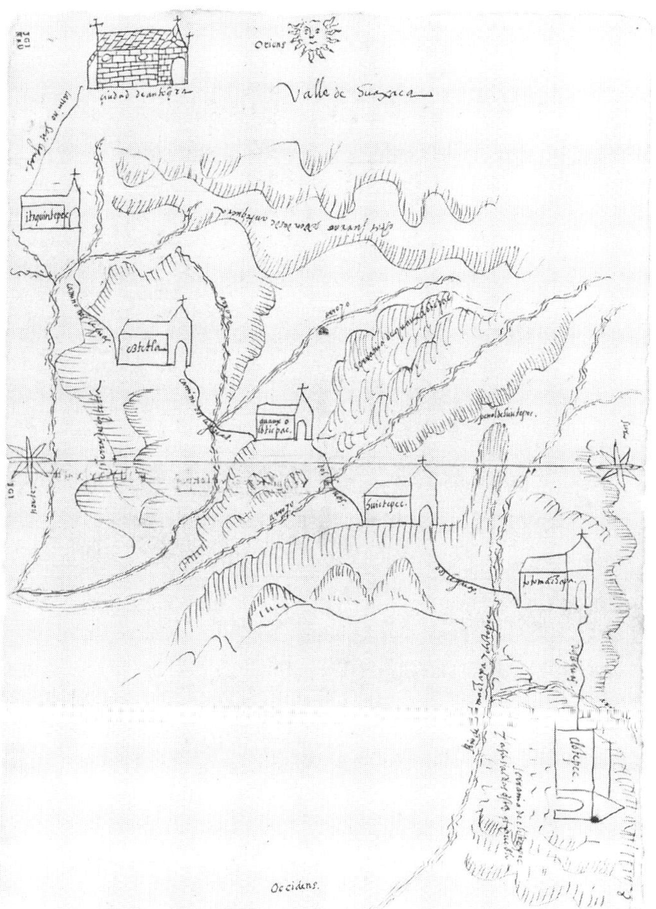

Figure 9. The Relación Geográfica map of Los Peñoles, attributed to Diosdado Treviño, 1579. This map seems to have been drawn by a local curate in the course of responding to the questionnaire. It is highly schematic, charting a rough itinerary of links between settlements. Size of the original: 43.5 x 31.5 cm. Photograph: B. Mundy; reproduced courtesy of the Benson Latin American Collection, The General Libraries, The University of Texas at Austin (JGI xxiv-15).

For years, López de Velasco waited in Spain, with hopes falling, for the trickle of responses from the New World. From Domínguez he heard nothing, his survey unfinished after over a decade (Edwards 1969: 21).[14] The maps resulting from the 1577 Relación Geográfica questionnaire that arrived from New Spain covered only a fraction of the landscape; drawn by Spaniards, Creoles, and indigenes, they differed widely (Robertson 1972a). Seen through the expectant eyes of López de Velasco, the map from Coatzocoalco or Villa de Espiritu Santo (Coatzacoalcos River; town near Tuzandepetl, Veracruz) was a nautical coastal chart (fig. 8) and would raise his hopes for yet more rational maps. Such hopes would no doubt be dimmed by a map like the one from Los Peñoles (Santa Catari-

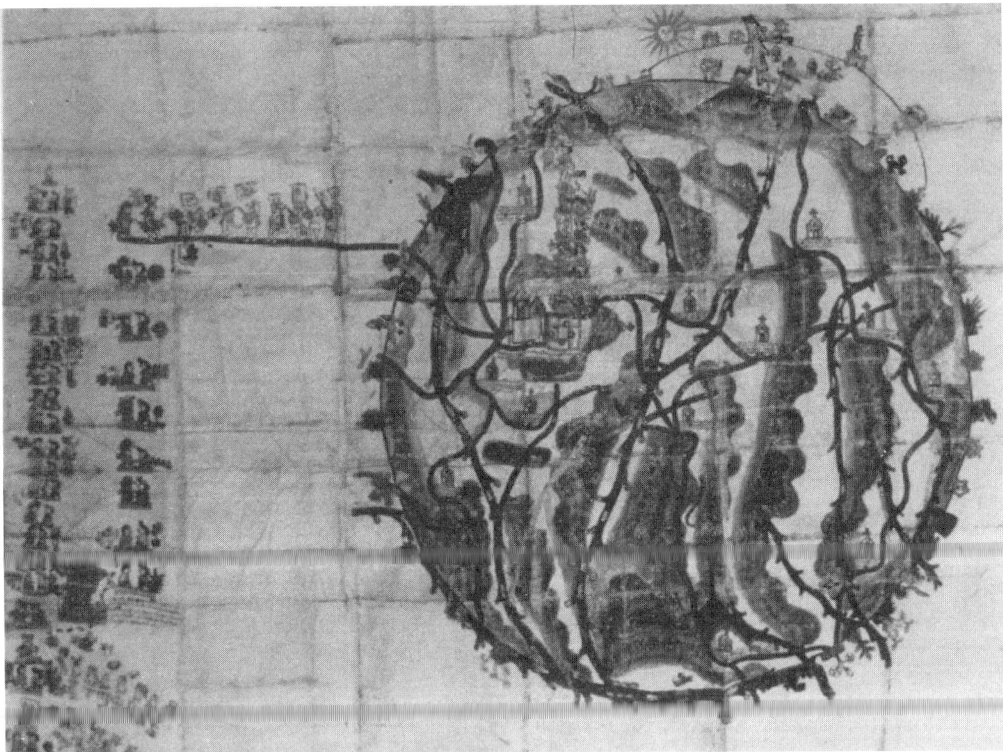

Figure 10. The Relación Geográfica map of Teozacoalco, 1580. (See also pl. 1.) This map of a remote town in a Mixtec-speaking region is undoubtedly the most famous map of the corpus. The figures on the left represent Mixtec royal genealogies, which were also the subject of some rare pre-Hispanic codices; the information on this colonial map and the pre-Hispanic books overlaps. The circular map on the right shows the region controlled by Teozacoalco; its boundaries are defined by pictographic toponyms and the interior maps the regional topography. Size of the original: 142 x 177 cm. Photograph courtesy of the Benson Latin American Collection, The General Libraries, The University of Texas at Austin (JGI xxv-3).

na Estetla, San Antonio Huitepec, Santa María Peñoles, Santiago Huajolotipac, San Juan Elotepec, San Pedro Totomachapan, Oaxaca), a rude sketch (fig. 9). The one from nearby Teozacoalco (San Pedro Teozacoalco, Oaxaca) was largely illegible to him, filled, as it was, with native symbols and signs (fig. 10; pl. 1). As López de Velasco tried to piece together a mosaic of New Spain, he would lack maps from the Pacific ports of Acapulco and Guatulco, as well as ones of three of the most important Spanish towns of the period: Mexico

City, Puebla de los Angeles, and Oaxaca. Compounding his predicament, no responses are known for the eclipse questionnaire of 1577 and 1578.

When López de Velasco compiled a terse catalogue of the texts and the maps of the Relaciones that had arrived to Spain by 1583 (Cline 1972d: 237–40), he must have been afflicted with an overriding sense of his project's failure. The next year he again sent out the questionnaire—with a slight amendment (Cline 1972d: 190)—as well as his fourth plea for eclipse observation (Edwards 1969: 18–9). Again, he was faced with dismal results.[15] Only four observations of the 1584 eclipse are known today, all coming from Mexico City, thereby defeating much of the purpose of numerous sightings (Edwards 1969: 21–2; Latorre y Setien 1916: 76). The survey Philip had planned of New Spain was also a fiasco. Domínguez responded to the eclipse instruction of 1584 from Mexico City, at which time his survey was still unfinished after thirteen years (Edwards 1969: 21). When the king dispatched another royal cosmographer, Jaime Juan, in 1583 to determine latitudes and longitudes, he was struck down by fever before completing his mission (Goodman 1988: 67–8). After 1584, another handful of written Relaciones Geográficas trickled in. Disillusioned, López de Velasco abandoned his duties as cosmographer-chronicler. Soon after, in 1588, members of the Council of the Indies asked of the king that López de Velasco's salary be suspended for neglecting his project (González Muñoz 1971: vi). López de Velasco aspired to map the New World no longer.

meztitlan
suchicoatlan
la ermita

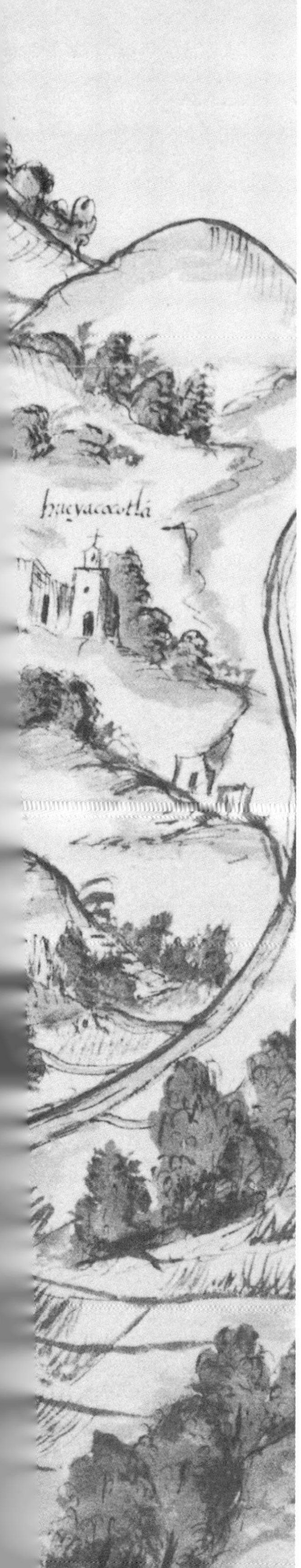

CHAPTER THREE

Colonial Spanish Officials and the Response to the Relación Geográfica Questionnaire

> 10. Describe the sites upon which each town is established. Is each upon a height, or low-lying, or on a plain? Make a map of the layout of the town, its streets, plazas and other features, noting the monasteries, as well as can be sketched easily on paper. On it show which part of the town faces south or north. . . .
>
> 42. What are the ports and landings along the coast? Make a map showing their shape and layout as can be drawn on a sheet of paper, in which form and proportion can be seen. . . .
>
> 47. What are the names of the islands along the coast? Why are they so named? Make a map, if possible, of their form and shape, showing their length, width, and lay of the land. Note the soil, pastures, trees, and resources they may have, their birds and animals, and the notable rivers and springs. . . .

When López de Velasco sent the Relación Geográfica questionnaire across the Atlantic, he entrusted its fate to New World colonists; those receiving the fifty-item questionnaire in the gobierno of New Spain were, on the whole, local administrators, the petty officials administering the crown's vast empire. Ninety-eight of them in this gobierno, that we know of, penned responses to the questionnaire, but more often than not they handed the task of drawing the map to native artists (table 1). Those fifteen Spaniards and Creoles who took it upon themselves to map their surroundings did so with maps that were indebted to European models. Most of these colonial maps are not marked by excellence in composition and execution. Instead, as a group they seem perfunctory, awkward, almost careless, sometimes in marked contrast to the texts the same colonists wrote, which are often discursive and thorough. In addition, most Spanish and Creole respondents dashed off their Relaciones Geográficas maps quickly, seeing nothing extraordinary in them, nothing that merited extra time or attention.

Why this indifference to maps? No doubt colonists' relationship to the image had a different nature—one perhaps closer to disdain than mere indifference—than López de Velasco's rosy faith. For in New Spain, unlike in Spain, images were the province of the "indios," as Spaniards called Amerindians; they were the texts of the conquered, who before the conquest had no alphabetic script and after the conquest still widely used "picture writing," that is, pictorial and logographic records. Alphabetic writing, on the other hand, was the realm of Spaniards; most Amerindians were not alphabetically literate. The Relación Geográfica corpus reflects the anti-image bias of the responding colonists: in the writing of the Relaciones Geográficas, colonial Spanish officials largely occupied themselves with texts and farmed out the maps to indigenous artists. Even when Spanish colonists did draw up the maps, they almost never took credit for their pictorial works, whereas they never failed to sign the accompanying texts, thereby celebrating their

Table 1 Relaciones Geográficas Maps and Artists in the Gobierno of New Spain

MAPS PRESUMED TO BE BY INDIGENOUS ARTISTS	
Artists	Maps
28 paint 1 map each	28
1 paints 2 maps (Cuzcatlan group)	2
1 paints 3 maps (Tlaxcala group)	3
1 paints 5 maps (Suchitepec group)	5
1 paints 7 maps (Gueytlalpa group)	7
32 artists (68 % of all artists) paint	45 maps (65% of all maps)
MAPS PRESUMED TO BE BY NON-INDIGENOUS ARTISTS	
Artists	Maps
10 paint 1 map each	10
1 paints 5 maps (Temazcaltepec group)	5
1 paints 3 maps (Francisco Stroza Gali)	3
3 paint 2 maps (Ixcatlan, Miranda, Yurirpúndaro)	6
15 artists (32% of all artists) paint	24 maps (35% of all maps)
TOTALS	
Artists	Maps
32 (68%) presumed indigenous artists	45 (65%)
15 (32%) presumed non-indigenous artists	24 (35%)
47 artists	69 (100%)

authorship and testifying to its authority. As Hernando de Cervantes, the *corregidor* (local governor) of Teozacoalco, wrote on his Relación, "I have faith in all that is herein contained . . . as witness to its truth, I sign it with my name" (RGS 3: 151).

This disavowal of authorship of the maps not only hints at the authors' stance towards pictures in general, and their own in particular, but also creates practical problems in the study of the Relación Geográfica corpus. If no one took credit, how do we know who created the maps? To address this problem, I have relied on old-fashioned connoisseurship—close scrutiny of individual drawing styles in order to correlate the author of the map to a particular scribe or signatory of the text—to reveal the identity of the Spanish colonists drawing the maps. This is an inexact science at best, but I have been able to designate the mapmakers in a few cases.

In New Spain, the views of respondent colonists, as expressed in the maps that they made, clashed harshly with the aspirations of López de Velasco. They certainly held the image in less esteem than did the Spanish cosmographer, who saw maps as the perfect complement to texts, spatial descriptions rounding out the temporal chronicles. The colonists' maps lacked the rational, repeatable principles that López de Velasco felt should underlie their—and all—maps. He, like other European cosmographers and mapmakers, saw maps as being based upon a regular, predictable framework, be it the graticule of lines of latitude and longitude or the diminishing parallels of Albertian perspective. Thus, to López de Velasco, the map showed something beyond the individual, or even the communal, experience of the landscape. Its rigid geometry owed its rules to absolute principles, being discovered but not constructed by man. The rational basis that López de Velasco saw as underpinning the map ruled out vagaries of individual choice and taste; in theory, if López de Velasco dispatched two mapmakers independently to map the same piece of territory, they would return having each made the same map.

But colonial officials in New Spain had not absorbed the rational ideals of López de Velasco or, if they had, wholly lacked the technology to execute them.[1] True chorography was beyond their reach, and given that most officials were interlopers on the local scene, so were the "communal" maps that Kagan has discussed for Spain (Kagan n.d.). As a result, the maps they made of New Spain were largely model-driven, by which I mean that colonists had clearly seen "rational" maps, and in response to the questionnaire, they modified and channeled their perceptions of the New World into pictorial forms they knew, wanting their maps to resemble the common printed maps of fifteenth- and sixteenth-century Europe. However, colonists lacked the technical know-how to apply to their surrounding landscape the rational principles that López de Velasco promoted.

Nonetheless, these maps, the part of the corpus painted by Spanish colonists, should not be dismissed lightly. They offer us a needed counterpoint to the indigenous maps in the corpus, and in showing us European cartographic practice in New Spain, they lay to rest the myth of colonial maps' untroubled progress toward technical perfection.

THE RESPONDENTS

The men who responded to the Relación Geográfica questionnaire when it arrived in New Spain in the late 1570s, and then again in the mid-1580s, were for the most part *alcaldes mayores* or corregidores (Cline 1972d: 191). These were local governors who had been appointed by the royal government to oversee the seventy large districts, called *alcaldías mayores,* or the two hundred smaller *corregimientos* (Gerhard 1986: 13–7) that the crown had carved out of lands that had once belonged to New Spain's indigenous peoples.[2] These same men may have also received López de Velasco's eclipse instrucción two years earlier; most of them, if not all, had ignored it. However, scores of them responded to the Relación Geográfica questionnaire, which asked for no complicated scientific observations, just questions about the history and the land of the regions to which they were assigned, questions to which they felt capable of responding.

Certainly it was López de Velasco's intent to have them fashion the local maps that he asked for in item 10. For him, the map was an essential counterpart to the textual replies, and I think he assumed that the king's representatives in New Spain would share his views: he envisaged the questionnaire circulating in a closed system. The written questionnaire did ask that local officials call upon local residents, often New World indigenes, to provide some of the local history, but their responses were to be filtered through the voice and the language of the corregidor or alcalde mayor before being set in the Spanish text that was to be sent back to the royal government. Nowhere did it ask anyone but the Spanish colonial respondent to draw maps. The Relación Geográfica questionnaire even assumed that maps were to be made exclusively by colonists. It directed items 1 to 10 only to pueblos de españoles, segregated towns for Spaniards and Creoles, and the last of these questions was the request for the map. The questionnaire never asked that indigenous towns be mapped or that indigenous painters map them.

Was López de Velasco right to entrust local colonial officials to make his maps? Was his assumption—of a closed circle of mutual understanding—flawed? Other than their names, we know little about these local officials, many of whom had been born in and probably never left the New World. The corregidores would rarely have been the most powerful or cosmopolitan of men, although they probably ranked as the colony's elite and certainly all were literate. On the whole they were Creoles, men of Spanish parents or grandparents, but who were born in the New World. A respondent like Gabriel de Chávez (1530–c. 1604) would be atypical only in that he was a slight cut above most of their number (RGS 7: 52–3). The son of a conquistador, Chávez was born in Mexico. The encomienda, or royal grant of indigenous labor, given to his father was not sufficient support, so Chávez entered the royal service and was sent from post to post across the face of New Spain. Chávez had an inquisitive and nimble mind and an easy proficiency in written Spanish. During the course of his early career, his lineage, his contacts, or his ability distinguished him, for when he answered his Relación Geográfica, he was the alcalde mayor of Meztitlan (Metztitlan, Hidalgo), a slightly better post than corregidor, for usually alcaldías mayores were larger and more populous than corregimientos. Along with answering the occasional questionnaire from the crown, Chávez's and other functionaries' duties included collecting tribute from their charges (from which these officials drew their salary), monitoring local litigation, and overseeing the activities of the local indigenous government (Haring 1975: 128–38; McAlister 1984: 187).

Since these local officials derived their salaries from whatever money they could extract in the form of tribute or otherwise glean from the native peoples in the region that they controlled, they were often exacting of, and in some cases abusive to, their charges. Their relationship with the community in which they lived was often antagonistic, but it never lasted for long: the crown kept them itinerant, usually posting them in a region for a three-year stint and then relocating them. Chávez is known to have held posts in different regions in 1562, 1576, 1579, and 1581 (RGS 7: 52–3). Another official, Francisco de Castañeda, a corregidor, wrote in part the Relación Geográfica of Tequizistlan (Tequisistlan, Mexico) in February of 1580; shortly thereafter, Casteñeda was reassigned to Teotitlan del Camino (Oaxaca), a post more than 200 kilometers away, and wrote its Relación Geográfica in September of 1581. His scribe, Francisco de Miranda, who accompanied him to both posts, seems to have drawn both maps, for close comparison shows that in both cases the hand of the writer was also the hand that drew the maps (RGS 7: 214–5).

The footloose local official was also isolated. For the many who oversaw districts with a population that was largely indigenous, their lives would have lacked the company of many resident Spaniards outside of their own households, although there was always the local curate for company. Few could speak the language of their indigenous charges. The isolation of the lives of Casteñeda and his fellow officials was heightened by royal law, which forbade royal officials to own property or conduct business in the districts in which they served; they were even banned from marrying locally.[3]

The isolation, itinerancy, and antagonism toward the land's occupants that many local officials experienced certainly affected their relationship to their temporary homes. Many knew little of their surroundings: once a corregidor or alcalde mayor arrived at a post, his experience of his larger district was usually cursory as he made a requisite tour of duty around the province. With so little time at each post, few would have known well the landscape that they were asked to describe and map in the Relaciones Geográficas. For instance, Juan López de Zárate, the corregidor of Los Peñoles, and Diosdado Treviño, its priest, replied to one question about local toponyms with the following:

> The land is all mountains and ranges, so harsh and terrible that we are unable to apply a particular name to any of them because they are countless. Even so, in the past (and even today), the Indians have given names to all of them, but they sound dissonant in our Spanish language. Since we deal with these names so infrequently, using them here [in the Relación and the map] seems out of place because these names are not important.[4] (RGS 3: 50)

The tenuousness of their connection to their immediate surroundings would certainly have been a factor in the carelessness that they displayed in their maps. They were not drawing landscapes that had great meaning to them; their relationship was strictly official. The corregidor's distant connection to the local landscape and community barred him from creating a map modeled on "communal" views of the city or village, those maps that Kagan had described as being metaphoric, showing how local inhabitants visualized their cities, rather than being strictly chorographic descriptions (Kagan n.d.). For instance, Casteñeda's official role determined how he would frame his maps. In the two maps that he had Miranda draw, the edges, or frame, coincided with the limits of the corregimiento, and there is little that distinguishes one locale from another. Treviño, who drew the map of Los Peñoles, did the same, and in doing so he, like Casteñeda, was framing the landscape according to the political limits of the domain, a response that had little to do with the request of the questionnaire, which only asked for a map of the main town.

MAPS AND MODELS

These men made maps that were not only colored by the distant, cursory relationship they had to their regions, but also determined by their ideas of what a map should be. Thus the maps colonists made mimic certain European maps, their most likely source being the pages of illustrated books, imported or printed in Mexico, or printed single-sheet maps.[5]

If we closely scrutinize López de Velasco's main map request of the questionnaire, item 10 (found at the beginning of this chapter), we see that any contemporary reader would have understood it as asking for a chorographic map, somewhat along the lines of van den

Wyngaerde's chorographic pictures of Spain (refer back to fig. 5), or perhaps, to offer an example better known at the time, the 1524 Cortés map (refer back to fig. 1). The Relación question asked for elements typical of chorography: a layout of streets, notable architecture, all of a single settlement. Since the question reined in the respondents so tightly, why then in their maps did they stray so wide of the mark?

No doubt the simultaneous demands made in other parts of the questionnaire for items of geographic information invited each local Spanish official to interpret the specific chorographic map item widely, following his creed of *obedezco pero no cumplo* (I obey but do not comply). For instance, other items asked for league distances to and names of nearby towns, and such items may have planted the seeds for maps that were closer to the genre of geographic than chorographic. Thus the local Spanish official in New Spain often drew up a map that contained the geographic information demanded by one or more of the other questions instead of the requested plans of a city or port.[6]

So although the crown, in the questionnaire, tried to delimit the kind of map it wanted (chorographic in item 10, nautical in items 42 and 47), local Spanish officials shook off such encumbrances. Instead, each colonist's understanding of maps acted as the driving force in the work he made, even though in following his own course he strayed from the explicit directives of the questionnaire. Although it might not have been exactly what the crown said it wanted, the pictorial response of a local Spanish official was often what he knew. He was drawn to familiar graphic models, like vine to trellis, to shape the geographic information for the crown.[7]

ITINERARIES

No doubt the European models colonists chose, either consciously or subconsciously, were those that dovetailed with their own experience of the landscape that they were asked to map. A ubiquitous experience was that of travel, as they journeyed from their homes to their outlying provinces and as they took their obligatory tours of duty. Treviño's map of Los Peñoles showed little but a series of towns and ranches connected by roads—perhaps the closest reflection of Treviño's experience of place, whose specifics, by his own admission, he found excessive to the point of tedium (figs. 9 and 11). On a journey, travelers see only the towns through which they pass, know only the roads upon which they travel, and experience all this in a very specific sequence. Treviño's map is based on a practical reckoning of the distances and directions between towns as one travels from Oaxaca, the nearest Spanish town, through the mountains to Elotepec.

Perhaps Treviño and other colonial respondents may have seen published itinerary maps in European books, for these maps in the form of rough sequences of towns, like links in a chain, are known from Roman times, and were used by European pilgrims and traders (Dilke 1987: 234–42, 254; Harvey 1987: 495–8). These printed maps would have simply confirmed to colonists that their primary experience of the landscape—travel—could be translated into a map form.

In addition to mirroring colonist's experience of their landscape, an itinerary map was perhaps the *least* pictorial of maps. Most maps offer a multitude of scenes. Itinerary maps, on the other hand, are linear, showing a progression of places in set order. They read like a text, and therefore offered to the respondent colonists—a group wary of images but certain of texts—a textual way of representing the world. The Los Peñoles map repeats what the Relación tells, both following the same linear progression. Compare figures 9 and 11 to the following passage:

> The first Peñol called Itzcuintepec, lies from the city of Antequera, bishopric of Oaxaca, where the cathedral is, six leagues, leaving the city [behind] to the east; in general with the rest of the Peñoles, they run, as I said, one after the next from north to south. The second, called Eztitla, lies from the aforementioned city nine leagues, the third, called Quauxoloticpac, 11 leagues, the fourth lies 12 leagues and is called Huiztepec;[8] the fifth, called Totomachapa, lies from the aforementioned city [of Oaxaca] 14 leagues and the sixth, called Elotepec, lies 17 leagues away (RGS 3: 45).

The parallels are unavoidable: Treviño's own experience of the landscape is like his textual description of that same landscape, which is like, in turn, his visual description, or map, of it. In fact, the first two alone could have served as inspiration for the map; while Treviño may have been familiar with printed European itinerary maps, he certainly could have drawn up his map without using one as a direct model.

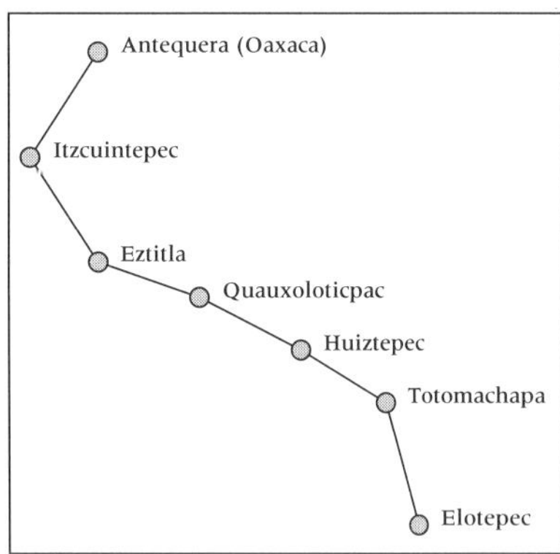

Figure 11. Schematic rendering of the Relación Geográfica map of Los Peñoles.

The itinerary map, both because of its alignment to a colonial official's experience of his region and because of its close kinship to a textual description, had a compelling presence in the corpus of colonists' maps. One pair of itinerary maps from the neighboring regions of Tecuicuilco (Teococuilco, Oaxaca; fig. 12) and Atlatlauca-Malinaltepec (San Juan Bautista Atatlahuca and Maninaltepec, Oaxaca; fig. 13) are close enough to

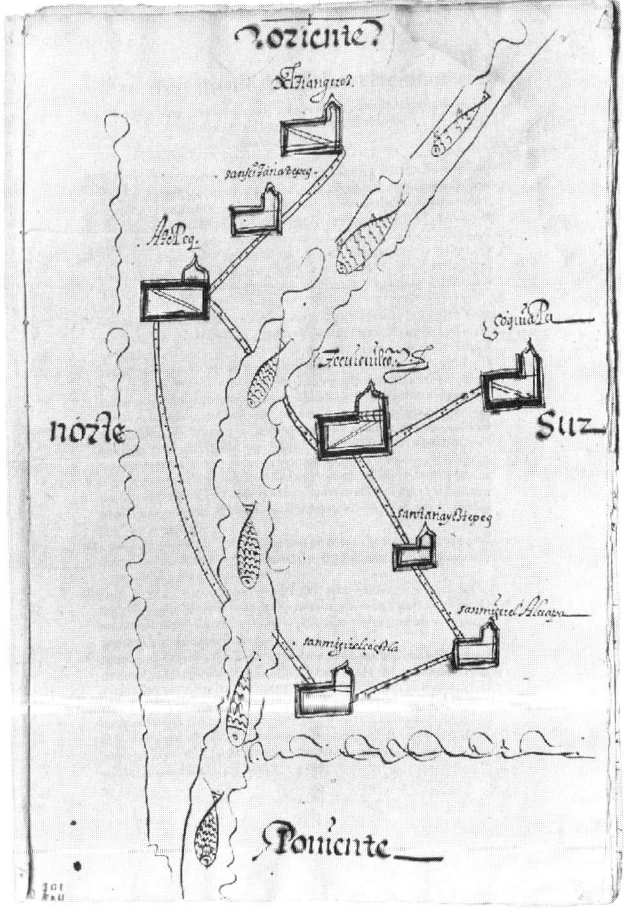

Figure 12. The Relación Geográfica map of Tecuicuilco, 1580. The corregimiento of Tecuicuilco is schematically represented; fish fill the river, which is Oaxaca's Río Grande. Size of the original: 31 x 21 cm. Photograph: B. Mundy; reproduced courtesy of the Benson Latin American Collection, The General Libraries, The University of Texas at Austin (JGI xxiv-19).

be considered virtual copies of each other, the Tecuicuilco map made in the European style, the Atlatlauca its native counterpart. The accompanying texts are similar as well, and it seems that the two corregidores penned their Relaciones in collusion (RGS 3: 85–6). Both maps show towns, identified by written glosses, connected by roads shown as short straight lines. These maps contradict the Tecuicuilco corregidor's account, which described this region as full of hills, its roads "rugged and mountainous, tortuous" (RGS 3: 88). These maps seem to have been shaped by each other and by the remembered experience of travel, rather than by a visual reckoning of the "tortuous" landscape.

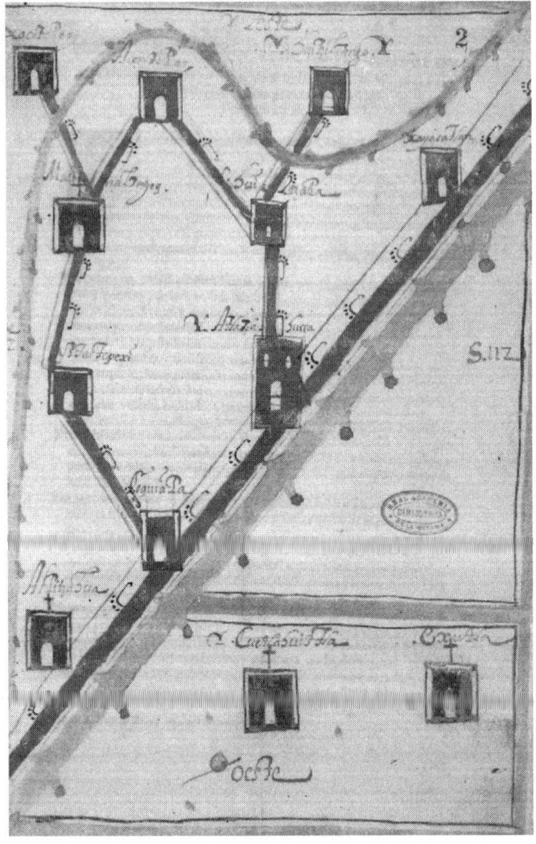

Figure 13. The Relación Geográfica map of Atlatlauca-Malinaltepec, 1580. This map shows the region to the north of Tecuicuilco, downstream along the Río Grande. It is the work of a native painter, who marked the rivers with indigenous water symbols. Size of the original: 32 x 21 cm. Photograph: Juan Jiménez Salmerón; reproduced courtesy of the Real Academia de la Historia, Madrid (9-25-4/4663-xxvi).

CHOROGRAPHY

While many colonists did model their maps on itineraries, charting schematic voyages with little reference to scale or to geographic barriers, others heeded López de Velasco's call for chorographic maps, which offered them means of registering the visual experience of their (often new) surroundings. Most would have been familiar with chorographic works. Although book inventories are few, we do know that a lively book trade kept colonists supplied with current volumes, including ones containing maps. For example, at least one edition of Ptolemy that reproduced chorographic views of towns was housed in the collection of the Franciscan library in Tlatelolco (Bagrow 1985: 13; Mathes 1985: 63). And some sale inventories register maps, most likely the innumerable printed single-sheet maps that offered chorographic views of towns; these spread like chaff across Europe and almost certainly reached New Spain (Leonard 1964: 199, 200).

Since in New Spain texts were the prerogative of Spanish colonists and pictures largely the province of indigenes, it comes as no surprise that the most sophisticated work of chorography was created by the alcalde mayor of Meztitlan, Gabriel de Chávez, a colonist who made little of the divide between colonist and indigene (fig. 14). By his own admission, Chávez had read, and held in high esteem native pictorial works, ending his own Relación with a statement of his debt to them: "and this was made clear to me by means

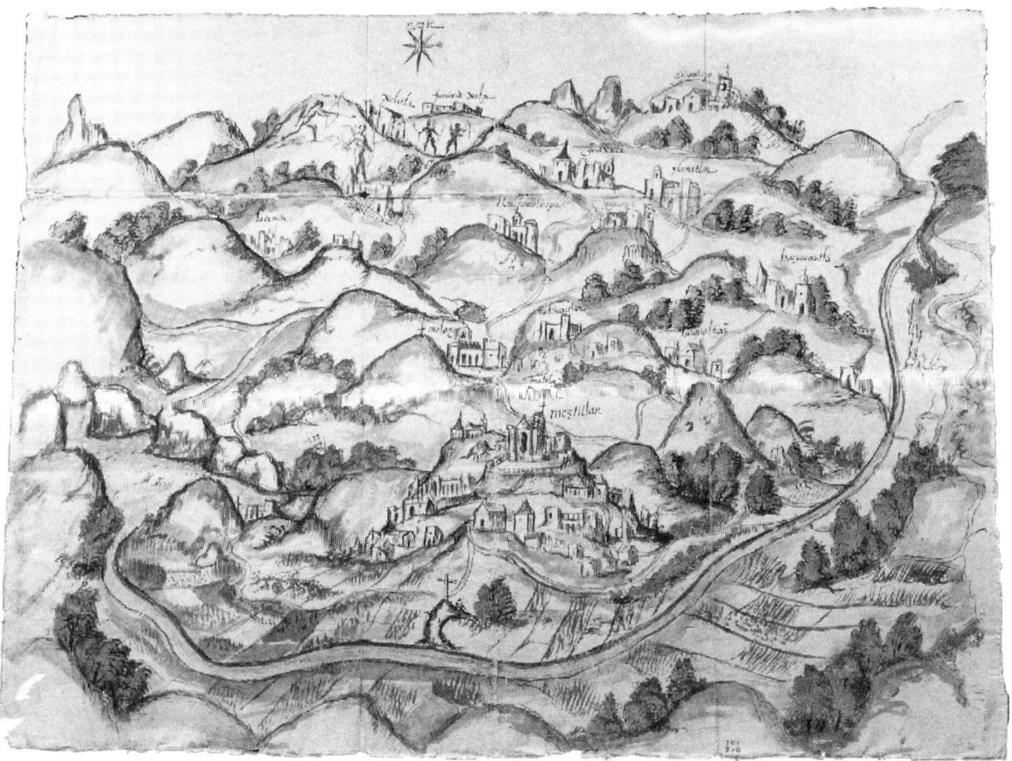

Figure 14. The Relación Geográfica map of Meztitlan by Gabriel de Chávez, 1579. The Creole alcalde mayor of Meztitlan used an oblique perspective to picture his large region, following the conventions of European cityscapes. He drew his map with ink and then shaded it with delicate washes of pigment. Chávez included emblematic architecture and figures—such as the Chichimec warriors fighting at top—to capture the particular character of the region. Size of the original: 42.5 x 58 cm. Photograph: B. Mundy; reproduced courtesy of the Benson Latin American Collection, The General Libraries, The University of Texas at Austin (JGI xxiv-12).

of ancient pictures that I saw, as well as by what I have surveyed with my own eyes" (RGS 7: 75). These "ancient pictures," along with "individual inquiries of the oldest Indian leaders," allowed him to write accounts of native life and the native calendar and feast days whose details make them exceptional among Relaciones Geográficas responses (RGS 7: 57; Kubler and Gibson 1951). Of all the colonists responding from New Spain, Chávez alone seems to have shared López de Velasco's view of the complementariness of image and text. He includes a diagram, set within the text and probably copied from an indige-

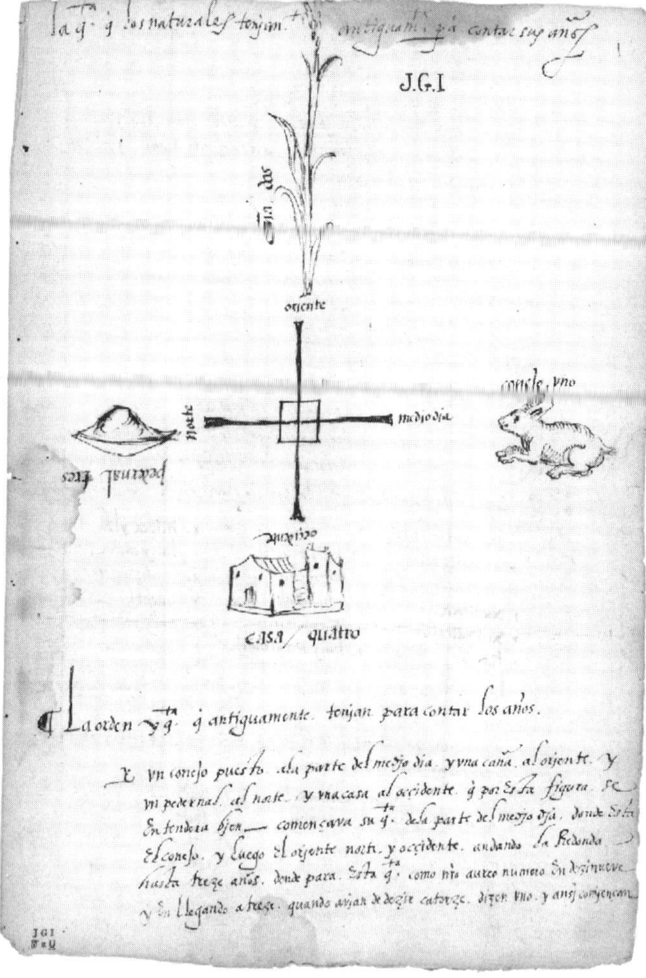

Figure 15. Native calendar from the Relación Geográfica text of Meztitlan by Gabriel de Chávez, 1579. Chávez included this schematic diagram of the central Mexican year names along with his Relación. The four years, read counter-clockwise—Reed, Flint Knife, House, and Rabbit—were also associated with the cardinal directions, as Chávez observed, by aligning Reed to east. Chávez's inclusion of this diagram shows his familiarity with native source material; it also reveals him to be the painter of the map, since the structures on the map and the symbol for the year "House" are quite similar. Photograph courtesy of the Benson Latin American Collection, The General Libraries, The University of Texas at Austin (JGI xxiv-12).

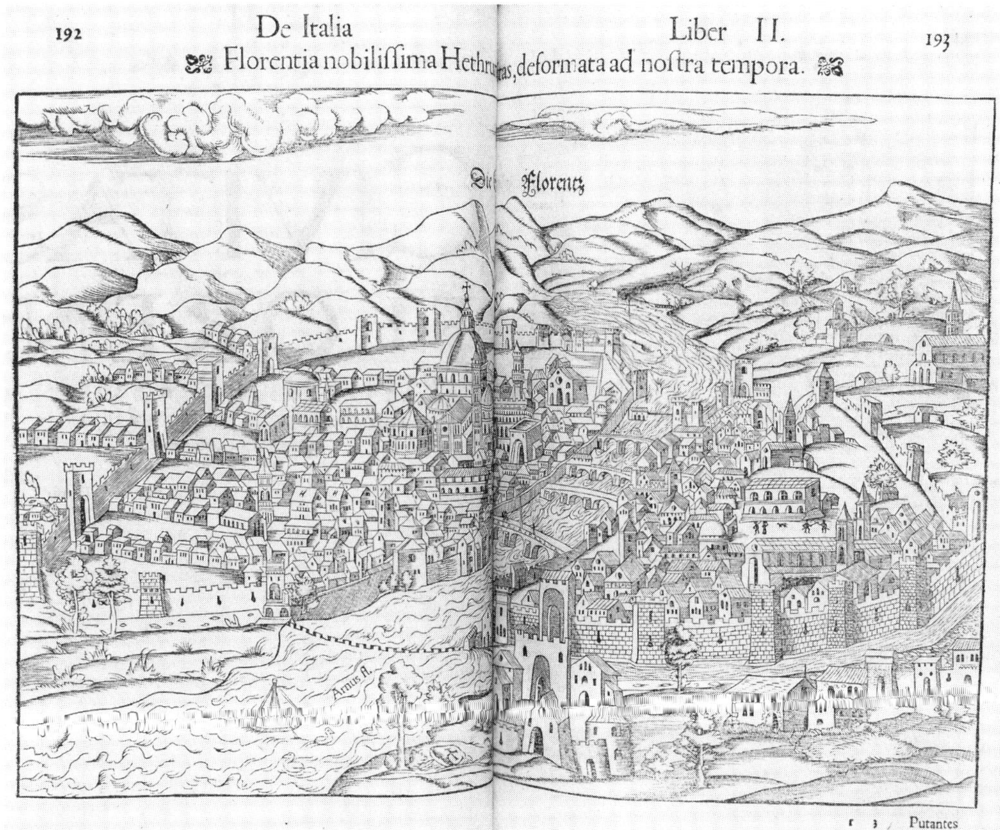

Figure 16. Map of Florence, 1550. This city plan was published in the 1550 edition of Sebastian Münster's widely popular *Cosmographiae universalis. . . .* (Basel, pp. 192–3). It is representative of the kinds of city plans that would have been circulating in New Spain at the time the Relaciones Geográficas maps were painted. Size of the original: 22 x 35 cm. Photograph courtesy of the Rare Books and Manuscripts Division, The New York Public Library, Astor, Lenox and Tilden Foundations.

nous source, to show the four years of the native calendar: Reed, Flint Knife, House, and Rabbit (fig. 15). Since Chávez's drawing of the year-sign House is identical in style to the houses appearing on the Relación map, we can harbor little doubt that this alcalde mayor was also the map's artist (RGS 7: 52).

Chávez's map shows its debt to European chorography; in both his map and a popular print like the view of Florence published in Münster's *Cosmographiae* (fig. 16), we see a hilly region from an oblique perspective, as if sketched from a towering mountain. In

Chávez's map, the town of Meztitlan is in the center foreground, its buildings individually discernible as they are in Münster's work. They are grouped around a prominent hill whose fields adjoin the river below. The region Chávez's picture embraces, as we see when we plot it on a modern map, fans out from Meztitlan towards the northeast (not north, as marked on the map), and his "view" spans about 100 kilometers out from Meztitlan (fig. 17). He does not show this expansive landscape with a receding perspective,

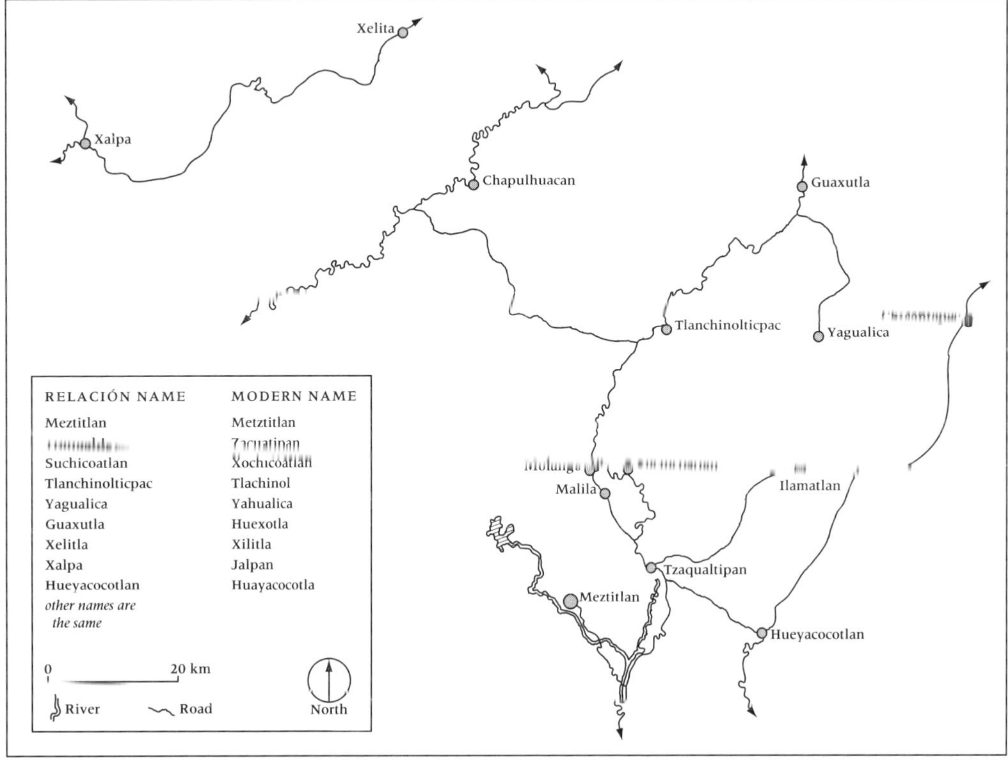

Figure 17. Region of Meztitlan. This modern topographical map shows the region covered by Gabriel de Chávez's Relación Geográfica map of Meztitlan.

Figure 18. Detail of the Relación Geográfica map of Meztitlan, 1579. On the top area of his map, Chávez included figures of Chichimec indians to show their occupation of the northern zone of the region. The bows and arrows they carry are also characteristic of their appearance in indigenous manuscripts. Photograph: B. Mundy; reproduced courtesy of the Benson Latin American Collection, The General Libraries, The University of Texas at Austin (JGI xxiv-12).

with distance signaled by diminution, as contemporary pictorial practice would dictate, but rather uses a kind of rising perspective also seen in contemporary prints: the farther a town is from the imaginary viewer on the towering hill, the higher up this town appears on the picture plane. Just as in chorographic illustrations found in versions of Ptolemy, these towns do not diminish in size if they are farther from the viewer, in order that all the towns in the region could be clearly described by the artist. As in Münster, prominent architecture receives its due and buildings are distinguishable in all towns. For instance, Chávez includes the "Fuerte de Xalpa" at the top of the map to mark a fortress-like presidio established at the edges of this untamed frontier.[9]

Perhaps Chávez's life at the outposts of the Spanish domain, in addition to the probing cast of his intellect, allowed him, even led him, to make a map that absorbed elements from a native pictorial tradition into the European chorographic one. On his map he draws, near the fort of Xalpa and the town of Xelitla, some battling warriors armed with bows and arrows; they are the only human figures Chávez depicts (fig. 18). Münster's print also included a few minuscule human figures to show the occupations of Florentines. Chávez, in contrast, means these fighters to distinguish this region as one inhabited by the feral Chichimecs, who, as he explains in his text, lived on the borders of Xelitla (RGS 7: 60). These native groups had been pushed north by Spanish expansion and were to remain on the periphery of Spanish rule into the eighteenth century (Gerhard 1986: 63).

Clearly Chávez drew on the native pictorials he boasted of having seen, and had used to draw up his native calendar, to depict the Chichimecs. These near-naked men fighting with bows and arrows are also found in some native pictorials, which show them dressed only in skins and carrying bows and arrows.[10] These same indigenous histories distinguished the Chichimecs from "civilized" natives who wore clothes of woven fabric and whose warriors used the *atlatl* and the *macana,* weaponry that Chávez describes in his text but does not portray. Chávez, like his indigenous neighbors, seems to have meant these half-naked fighters to stand for the uncivilized.[11]

PRINTED MAPS

Chávez's map of Meztitlan is a solitary island in the corpus, showing the possibilities open to colonial officials who held the image in equal regard to the text. When clearly Chávez drew inspiration from a print, he also knew of the artistic conventions of original artworks and aspired to create his own original rendering of a place. But many other colonists held the print in such high regard that they aspired to reproduce the print rather than the map. That is, they imitated, not the artistry of maps they knew, but the *printedness.*

What is printedness? Limited by their medium, printmakers developed techniques to make prints approximate the tone and shading of polychrome paintings, as did artists working in pen and ink. For example, they used hatching, closely set crossing lines, to convey different tones. These same engravers' techniques—hatching as opposed to shading—were used by colonists making the maps of the Relaciones Geográficas in order to make their drawings look like prints. We see that to these local colonial officials, the print, instead of the original drawing or painting, was the ideal measure that their works tried to match. If we imagine the visual experiences of these corregidores, we can understand such an attitude. Artistic drawings and paintings would infrequently come into the purview of local colonial officials, except perhaps for the ponderous religious paintings glimpsed in a cathedral.[12] Prints, however, were duplicated and published in books or sold by the sheet. If local colonial officials knew artistic landscapes, it would most likely

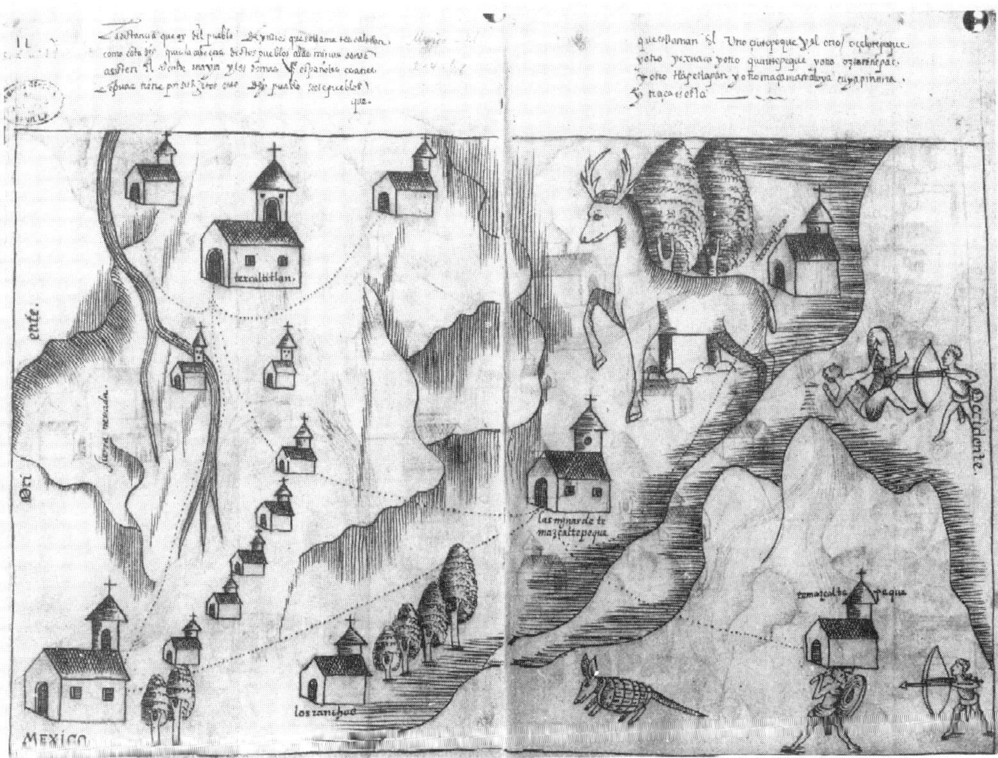

Figure 19. The Relación Geográfica map of Tescaltitlan, 1579–1580. This pen-and-ink drawing, one of five included in the Relación, depicts the background landscape in much the same way as did prints of the period. As in the Relación Geográfica map of Meztitlan, scantily clad warriors fighting with bows and arrows are depicted; the Relación text tells us that a pre-Hispanic fort once stood in this region. Size of the original: 24 x 39 cm. Photograph courtesy of the Archivo General de Indias, Seville (Mapas y Planos, 21).

be through prints of famous cities and regions: Rome, Florence, Tuscany. Owing to the subject matter of prints, to their wide dissemination both in Europe and in the New World, and to their presence in books, prints had a certain status and authority.

That the printedness rather than the artistry of printed maps was worthy of emulation shows up a number of times in the Relación Geográfica corpus, wherein colonial authors expended more energy making their work look printed than in trying (as true chorography did) to capture certain specific descriptive details of their region. For instance, Melchior Núñez de la Cerda, a Spanish scribe who was in the region of the Minas de Temazcaltepec (Real de Arriba, Mexico) in 1580, captured this region to the west of Mexico City with five fine-line maps in black ink as well as by recopying the five assembled Relaciones that had been created by others.[13] His maps and the Relaciones he copied covered Tuzantla (Michoacan) and the Minas de Temazcaltepec, the latter of which included accounts from the native towns of Tescaltitlan (Santiago Texcaltitlan, Mexico; fig. 19), Texupilco (Tejupilco de Hidalgo, Mexico), and Temazcaltepec (Temascaltepec, Mexico). It is the texts, which neatly border each map set within them, that offer the most specific description of the region. For instance, the Relación responds to a query on local flora by telling that the area around Tescaltitlan and Texupilco is filled with live oaks as well as *manzanilla de la tierra* (manchineel or *Hippomane mancinella?;* RGS 7: 148). To correspond, the map of Tescaltitlan is decorated with drawings of trees, although their stiff cones of

VOCABVLARIO
EN LENGVA MEXICANA Y CASTELLANA, COMpuesto por el muy Reuerendo Padre Fray Alonso de Molina, dela Orden del bienauenturado nuestro Padre sant Francisco.
DIRIGIDO AL MVY EXCELENTE SEÑOR
[illegible]
tuum Franciscum:
Signasti domine seruum
Signis redemptionis nostre.
¶ Indorum nimis te fecit prole parentem.
Qui genuit moriens, quos pater alme foues.
Confixus viuis, langues: cum mente reuoluis.
Vulnera, cum spectas, stigmata carne geris.
EN MEXICO,
En Casa de Antonio de Spinosa.
1571

Figure 20. Frontispiece of Alonso de Molina's *Vocabulario en lengua castellana y mexicana. . . ,* 1571. The background of this book page shows a typical printed representation of landscape. It was published in Mexico, appearing a decade or so before the Relaciones Geográficas. Size of original: 23 x 14.5 cm. Photograph courtesy of the Rare Books and Manuscripts Division, The New York Public Library, Astor, Lenox and Tilden Foundations (reprinted from García Icazbalceta 1885: fig. 60).

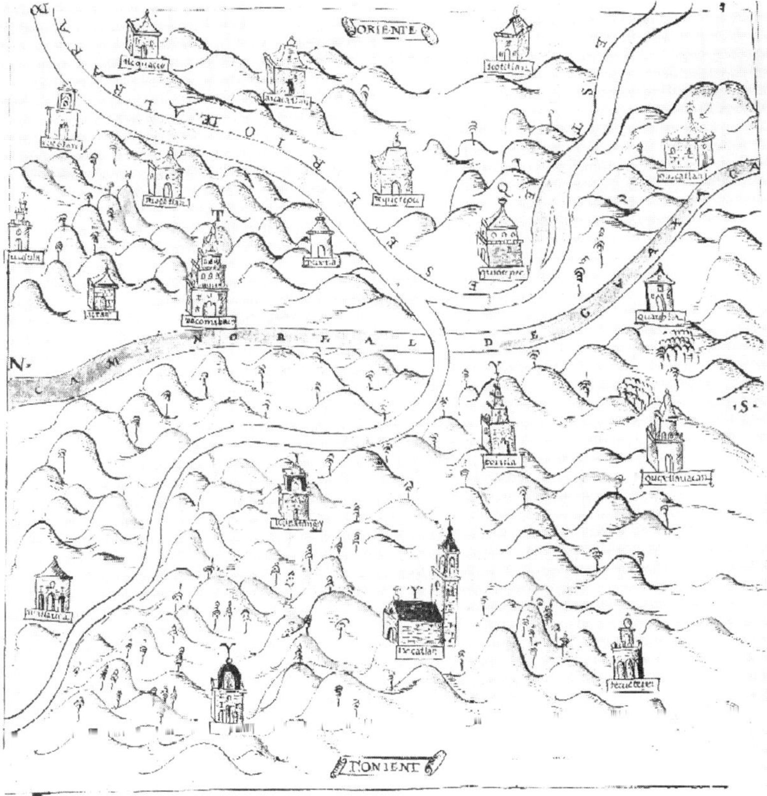

Figure 21. The Relación Geográfica map of Santa María Ixcatlan A, 1579. The maker of the map, probably the corregidor Gonzalo Velázquez de Lara, drew upon the conventions of prints when creating his two maps of Ixcatlan. He highlighted his pen-and-ink drawing with touches of paint to distinguish roads, rivers, and buildings. Size of the original: 32.5 x 31 cm. Photograph courtesy of the Benson Latin American Collection, The General Libraries, The University of Texas at Austin (JGI xxiv-7).

foliage are far less botanically specific than the text (fig. 19). The artist does include an armadillo at the bottom of the map, a New World creature that surprised and delighted Spaniards.

Núñez de la Cerda used carefully parallel hatching lines to throw his mountains into relief, to add shadow to the sides of leaves, just as an engraver would do. In addition, his settlements are all shown as identical houses or churches, another feature of a woodblock print. His sources could have been multifold, since such styles were ubiquitously used in the landscapes that filled the background of prints. For instance, a woodcut of Saint Francis that was used as the frontispiece of Alonso de Molina's *Vocabulario,* published in Mexico in 1571, shows the saint against a hilly backdrop that is similar to the Temascaltepec maps (fig. 20).

A pair of maps from Ixcatlan (Santa María Ixcatlan, Oaxaca) was most likely drawn by

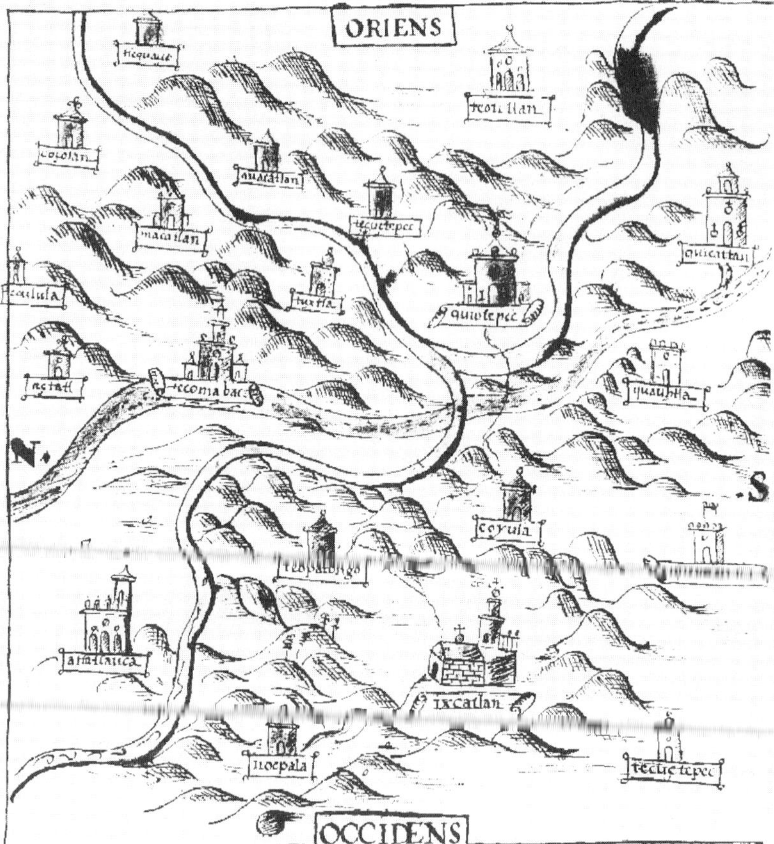

Figure 22. The Relación Geográfica map of Santa María Ixcatlan B, 1579. This map is nearly identical to the other map appearing in the Relación and seems to have been a trial run for the more polished version of Ixcatlan A. Size of the original: 31 x 21 cm. Photograph courtesy of the Benson Latin American Collection, The General Libraries, The University of Texas at Austin (JGI xxiv-7).

the town's corregidor, Gonzalo Velázquez de Lara, as was its Relación, and in making these maps of the region, Velázquez de Lara seems to have been trying to create drawings that could pass themselves off as prints (figs. 21 and 22). He drew with a fine black line and used hatching to add contour to the hills, a technique for shading that is characteristic of prints and is a marked contrast to Chávez's more painterly technique in shading with washes of color. The cartouches in the form of scrolls, unfurled to reveal the directions *"oriente"* and *"poniente"* seen at the top and bottom of Ixcatlan A (fig. 21), are ubiquitous in both manuscripts and prints of the period. Velázquez de Lara also included two nearly identical maps in his Relación, as if his hand-drawn works were impressions of an engraver's block.[14]

Velázquez de Lara shows the mountainous corregimiento as a network of human settlements, each symbolized by a church or other monumental building. These symbols were no doubt inspired by printed maps from Europe, and their specific contours are indebted to prints. With the exception of Ixcatlan's sober, dark-roofed, single-nave church and bell tower, which appear in the lower center of figure 21, the other churches on the map are fanciful confections. Tecomahuaca, for instance, is marked by a square church with a many-layered roof, the whole resembling a four-tiered wedding cake. The sources for these churches could have been the coat of arms or escutcheons that frequently decorated books published in Mexico and Europe and that often boasted architectural renderings. For example, the coat of arms of Archbishop Alonso de Montúfar, which appeared on the frontispiece of a Mexican book of 1566, shows two such buildings, one a church, the other a crenellated castle (fig. 23). Castles, symbolizing Spanish Castile, were also stamped onto almost every *real* and *maravedí* coin circulating in the colony (Nesmith 1955).

¶ REVERENDI PATRIS FRATRIS BARTHO-
lomæi à Ledeſma ordinis Prædicatorum et ſacræ Theologiæ profeſ
ſoris de ſeptem nouæ legis ſacramentis Summarium.
Cum indice locupletiſſimo.

PRO CHRISTO LE
GATIO
NE FVNGIMVR

Mcxici, excudebat Antonius de Eſpinoſa.
Cum priuilegio.
1566

Figure 23. The escutcheon of Archbishop Alonso de Montúfar, 1566. The castle and church seen in the center of this escutcheon are typical of printed renderings of architecture. This comes from Bartolomé de Ledesma's *Reverendi patris fratris . . .* , published in Mexico in 1566. Size of the original: 17 x 10 cm. Photograph courtesy of the Rare Books and Manuscripts Division, The New York Public Library, Astor, Lenox and Tilden Foundations.

NAUTICAL MAPS

Other than the work of sophisticated chorography that Chávez's map of Meztitlan offered, the maps colonists made would have proved disappointing to López de Velasco. On the one hand, there were sketchy itinerary maps which offered little new information about the landscape and lacked the rational principles that López de Velasco would have expected to order them. On the other hand, there were maps that were equally unscientific, but tried to hide their idiosyncrasies by dissembling as prints.

However, a few maps in the corpus were the kind of universal, rational maps that López de Velasco was seeking, created when three alcaldes mayores turned to a professional within their midst, a mariner passing through their ports, to draw their maps. They submitted the resulting nautical maps to the king. No doubt they felt these maps best fulfilled the request made by item 42 for a picture showing the "shape and layout" of "ports and landings," sensing that this item called for nautical charts. Such maps would have been familiar to the Spaniards who came to New Spain from the Iberian peninsula, for they would have seen such charts on their voyages. Nautical charts, or mariners' maps, while they did not include the chorographic elements asked for in item 10, were certainly paragons of the kind of rationally ordered maps to which López de Velasco aspired. Mariners produced the most carefully measured maps of the day because they depended upon them in the dangerous business of navigation. They also had at their disposal magnetic compasses and mariners' astrolabes, instruments whose technical precision was unequalled by those used by coeval land surveyors (Multhauf 1958: 400) and allowed their users to gauge astronomical positionings (Price 1964: 603–9; Taylor 1964: 547). For this reason, sixteenth-century world maps showed coastal outlines with a precision that was badly matched in many cases to the sketchiness of their continental interiors and placement of longitudinal lines. Yet not just any corregidor or alcalde mayor in New Spain would have been able to make such nautical maps himself, since they required a technical expertise out of his reach.

Thus the three alcaldes turned to Francisco Stroza Gali, an important captain and cosmographer who traveled along the transatlantic and transpacific trading routes that scored the Spanish empire.[15] In the first half of 1580, as his trail of maps reveals, he was making a trip between the Atlantic and Pacific via the Isthmus of Tehuantepec. On February 5, 1580, Stroza Gali finished his first map for Juan de Medina, the alcalde mayor of Tlacotalpa (Tlacotalpan, Veracruz; fig. 24), which lies on the Gulf. Medina added his own text to this map, proudly assigning its authorship to Stroza Gali, making this map one of the few in the corpus where authorship is specifically attributed. The captain-cosmographer then moved southeast down the coast along a well-traveled shipping route (McAlister 1984: 238), where the alcalde mayor of Coatzocoalco, Suero de Cangas y Quiñónes, commissioned him to make another Relación map (fig. 8). Like his fellow alcalde mayor up the coast, Cangas y Quiñónes attached the Stroza Gali map to his Relación, dated April 29, 1580. Stroza Gali signed this second map himself; his short text attests, and his signature confirms, its accuracy.

In making the maps from Tlacotalpa and Coatzocoalco, Stroza Gali's method was probably consistent with contemporary techniques used to make large-scale coastal maps, for the maps he made for the Relaciones are detailed enough for navigation. Stroza Gali, on

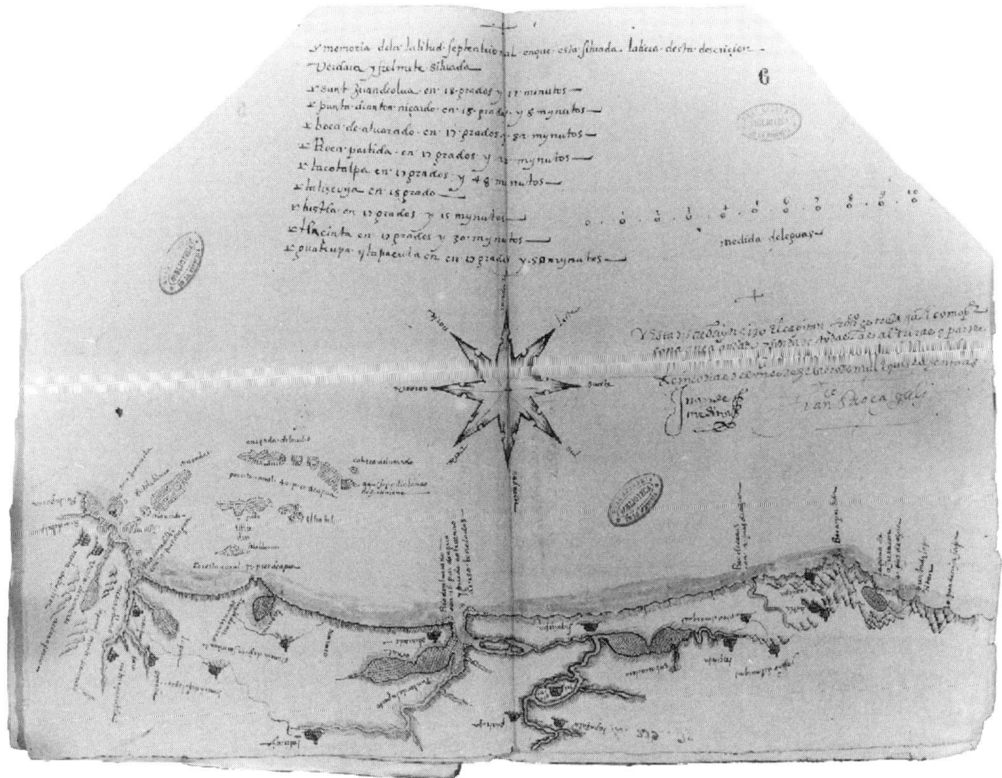

Figure 24. The Relación Geográfica map of Tlacotalpa by Francisco Stroza Gali, 1580. An itinerant Spanish mariner created this map of a portion of the Gulf coast at the behest of the local alcalde mayor. Shortly thereafter he drew up the map of the nearby Coatzacoalcos region that would also be included in its Relación. Size of the original: 31.5 x 43 cm. Photograph: Juan Jiménez Salmerón; reproduced courtesy of the Real Academia de la Historia, Madrid (9-25-4/ 4663-xxxvii).

the Tlacotalpa map, is described by the alcalde mayor as someone who "has traveled and taken all the latitudes and positions shown here." Stroza Gali may have begun with an existing nautical chart to get a pattern of the coast before adjusting and correcting it with measurements he took himself. In drawing the Tlacotalpa coastline, Stroza Gali used his compass, standard equipment for a seafaring cosmographer.[16] He visited some of the coastal sites on the map, according to the Relación, and probably used a mariner's astrolabe, another piece of common equipment, to measure the angle of elevation of the sun or Pole Star, an angle commensurate with the degrees of latitude (Price 1964: 603–9).[17] He also recorded soundings of the ocean's depth, measurements taken using a weight and

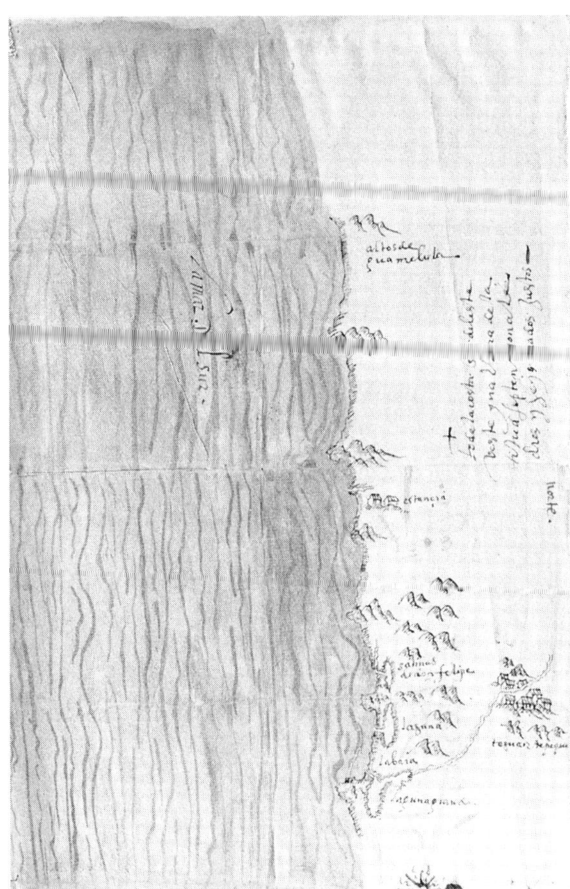

Figure 25. The Relación Geográfica map of Tehuantepec B, attributed to Francisco Stroza Gali, 1580. The Spanish mariner Stroza Gali seems to have created this third, unsigned, map after crossing into the Tehuantepec region via the Coatzacoalcos river. Size of the original: 21 x 31.5 cm. Photograph: B. Mundy; reproduced courtesy of the Benson Latin American Collection, The General Libraries, The University of Texas at Austin (JGI xxv-4).

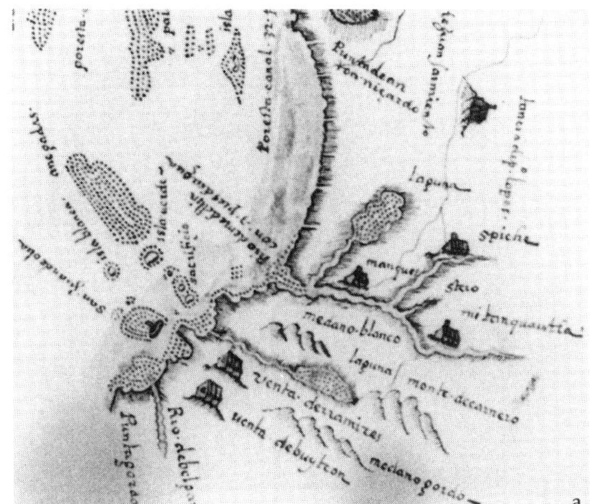

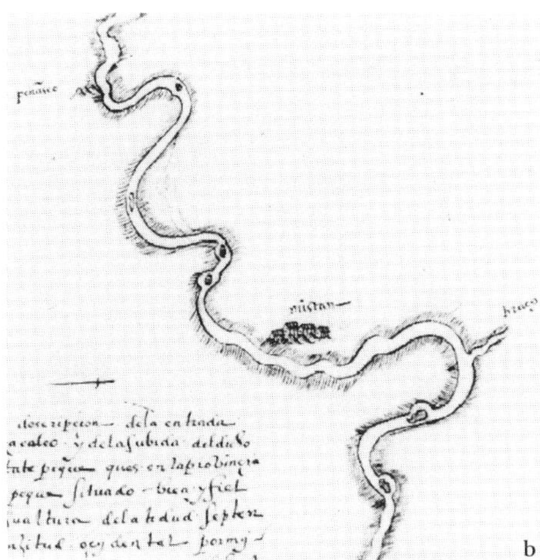

Figure 26. Details of three maps, 1580. Details drawn from the Relaciones Geográficas maps of (a) Tlacotalpa, (b) Coatzocoalco, and (c) the unattributed Tehuantepec B show both handwriting and building forms to be nearly identical. Thus, the author of Tehuantepec B seems to have been Francisco Stroza Gali, the author of the other two. Photographs: (a) Juan Jiménez Salmerón; reproduced courtesy of the Real Academia de la Historia, Madrid (9-25-4/4663–xxxvii); (b) the Benson Latin American Collection; (c) B. Mundy; (b) and (c) reproduced courtesy of the Benson Latin American Collection, The General Libraries, The University of Texas at Austin (JGI xxiv-2 and JGI xxv-4).

rope. In the Tlacotalpa region, Stroza Gali seems not to have measured the positions of the inland towns, for they are compressed along a northwest-southeast axis, that is, all towns crowd toward the shoreline.

While these two signed coastal charts were given pride of place in their respective Relaciones, a third, unsigned map also by Stroza Gali was attached without comment to the text of the Tehuantepec (Santo Domingo Tehuantepec, Oaxaca) Relación, which was written between September 20 and October 5, 1580 (fig. 25).[18] Unattributed until now, this sketch of Tehuantepec (Tehuantepec B in appendix A) bears an unmistakable resemblance to the two other works by the hand of Stroza Gali (fig. 26). In addition, chronolo-

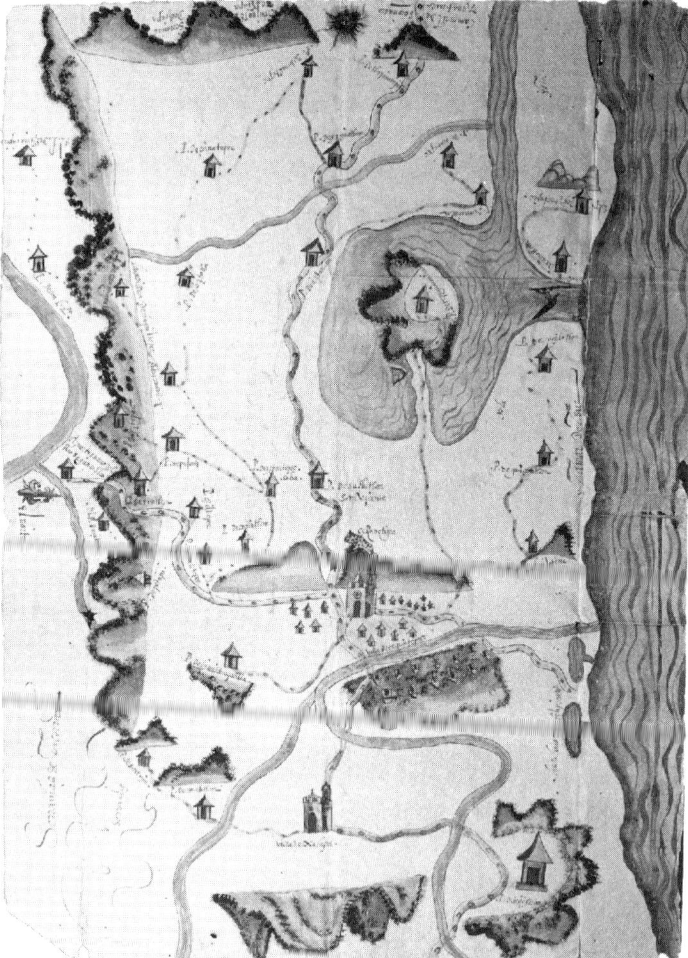

Figure 27. The Relación Geográfica map of Tehuantepec A, 1580. The alcalde mayor of Tehuantepec included this map with his finished Relación, along with a second map attributed to Francisco Stroza Gali. This map was created by an indigenous artist, who writes "Tehuantepec" pictographically. The name means "hill of the jaguar" in Nahuatl, and the artist drew a hill topped with a jaguar at the midcenter of the map, above Tehuantepec's church. Size of the original: 42.5 x 56 cm. Photograph: B. Mundy; reproduced courtesy of the Benson Latin American Collection, The General Libraries, The University of Texas at Austin (JGI xxv-4).

gy coincides: Stroza Gali was likely to pass overland through Tehuantepec to the Pacific ocean at just this time if he was making a trip to the Far East, as we know he often did.

Yet while the alcaldes mayores of the Tlacotalpa and Coatzocoalco regions were proud, even boastful, of Stroza Gali's contribution, the alcalde mayor who wrote the text of the Tehuantepec Relación never refers to this coastal sketch, discussing instead a second, more elaborate native picture of the region also included with the Relación text (Tehuantepec A; fig. 27). Perhaps Stroza Gali wanted little credit for his sketchy map, for it was clearly not the product of the same careful techniques of mensuration that he applied along the Gulf coast. In this map, Stroza Gali greatly diminishes the size of the enormous Tehuantepec lagoon and misplaces the mouth of the Tehuantepec River, which flows directly into the ocean, not the lagoon as shown on the map. There are no markings of scale or precise orientation as we find on the other maps. Nor is this Tehuantepec map signed, unlike the other two maps. I can only suppose that the mariner meant this map as a rough sketch, perhaps a preliminary drawing to be further refined by observation and mensuration, and never intended it as a finished map to be sent to the king, but the zealous alcalde mayor of Tehuantepec included it anyway.

MAP TECHNOLOGY AND ASTRONOMY IN NEW SPAIN

Apart from the maps that one professional, Stroza Gali, contributed to the corpus, there is little evidence of cartographic innovation—or even rational cartographic competence, as López de Velasco would have defined it—among the maps of New Spain made by Spanish officials. If we judge from the evidence that the Relación Geográfica corpus offers, we see it is implausible, as some have theorized, that Spanish colonial laymen would have been conduits for an advanced technology of mapping. Colonists' knowledge of maps and mapmaking was clearly limited, their interest in visually describing the world was meager, their estimation of images low. The results of two other cosmographic projects, the eclipse instrucción and Francisco Domínguez's survey, that along with the Relaciones Geográficas were meant to result in a new map of New Spain, show us that the "scientific" revolution that López de Velasco heralded had not yet visited colonists in the New World.

López de Velasco's project to map the New World depended upon the observations of local officials, first in making the group of eclipse observations collected by means of his eclipse instrucción and second in supplying the list of latitudes, Pole Star elevations, and zenith passages gathered through his Relación questionnaire. But few of the colonial officials answering the questionnaire knew the latitude or elevation of the Pole Star; the question about zenith passage was widely misinterpreted (Edwards 1969). When confronted with the question about latitude, the Pole Star elevation, and the zenith passage of the sun, the writer of the Relación of Gueguetlan (Santo Domingo Huehuetlan, Puebla) was, typically, at a loss, failing to understand the question. "I can find no one in this town who would know how to find the latitude because the town is in a gully and thus the sun rises late in this town, at eight o'clock" (RGS 5: 208). Tepeapulco's corregidor, also unable to answer the question, blamed his native informants: "they say that in this town of Tepeapulco [Hidalgo] the sun is measured midday as in the city of Mexico because it is at the same latitude and that they don't understand anything else [about the question]" (RGS 7: 174). The network of amateur cosmographers and mapmakers Juan López de Velasco hoped to find or to create in New Spain did not exist.

Ironically, with the entry of Spain into the New World such knowledge seems to have regressed, rather than progressed. The corregidor of Tepeapulco tried to blame his own confusion about latitude on the ignorance of his native informants (RGS 7: 174), but such knowledge was held by indigenous peoples before the conquest (Aveni 1980; Nuttall 1928). To them, the solstices, equinoxes, and zenith passages were of such importance they were celebrated with feasts and commemorated in architecture. For instance, the Aztec aligned the cleft in the twin temples of the Templo Mayor in their city of Tenochtitlan to the rising of the equinoctial sun during the feast of Tlacaxipehualiztli (Aveni 1980: 246–9; Maudslay 1913). During the early colonial period, alignment of buildings to astronomical events seems to have continued. At the native city of Cholula, for instance, the unusual orientation of the Franciscan monastery San Gabriel can be explained by its alignment with the rising sun on both the spring equinox and the feast day of San Gabriel, both of which fall in late March (Kubler 1985: 94). Had López de Velasco directed his astronomical questions to native sages, he might have had met with better results.

López de Velasco's other project, the instruction on the sighting of eclipses, was met with even less enthusiasm by local Spanish officials than the latitude, Pole Star and zenith passage questions of the Relación questionnaire. The four responses sent from New Spain were drawn up by the most learned men of the colony, but even so the margin of error in longitudes was huge (Edwards 1969: 21). One respondent, the cosmographer Jaime Juan, calculated the position of Mexico City from his observation of the 1584 lunar eclipse, and his reading was quite accurate for the time (Edwards 1969: 22). However, his previous effort, based on the eclipse of the previous year, resulted in calculations of the position of Mexico City that were about 14 degrees off (Menéndez-Pidal 1944: 18).

While colonists lacked cosmological knowledge of the order that most mariners would possess, they also lacked the practical knowledge of land surveying. At the same time that corregidores and alcaldes mayores across the country were answering their questionnaires, Francisco Domínguez was carrying out a country-wide survey. But few corregidores and alcaldes mayores seem to have known Domínguez's work, and the officials in New Spain who drew maps probably had only a vague notion of survey maps since few were ever published, owing to their tactical importance in warfare.[19] Little knowledge of this type of map shows up in the corpus of the Relaciones Geográficas.[20] While instructions on geometric surveying could be found in sixteenth-century mathematical textbooks (Taylor 1964: 539), there is little to suggest that such learning ever seeped down to local-level Spanish officials in New Spain. Thus it is also unlikely that local officials knew the city maps of sixth-century Roman land surveyors, a possibility suggested by Erwin W. Palm (1975). Indeed, even one of New Spain's most prominent monastic libraries held little in the way of such secular topics. Two cosmographies were among the books of the Franciscan monastery of Santiago Tlatelolco in the sixteenth century, along with Ptolemy's *Geography* (*Códice Mendieta* 1971: 255–6; Mathes 1985: 63),[21] but the cosmographies were sold by 1584 on account of being useless (*"se vendieron por inútiles"; Códice Mendieta* 1971: 262). Even in Europe, the skills to measure large tracts of territory and translate this information onto a map were in the hands of a very few, almost all of them professional cosmographers and cartographers (Taylor 1964: 537–44).

While cosmographers in Spain saw the map as a way to make visible their hegemony over a distant locale, their functionaries in New Spain did not hold images in great esteem. Nor did they feel much practical need for maps. In Europe, survey maps were often used to consecrate property, but Spanish officials in New Spain saw little need for detailed survey maps to record land ownership or decide regional boundaries.[22] At the time they responded to the questionnaire, colonists in most parts of New Spain outside the Valley of Mexico found plentiful amounts of land, owing to the precipitous drop in the native population over the course of the sixteenth century. In addition, in the early years of the colony, Spaniards were more concerned with native labor needed to exploit the land than with the land itself. Thus their territorial records were often little more than lists of population centers.

Compared to the methods Domínguez imported, the contemporary techniques of survey and mensuration in New Spain were rough and approximate. A typical survey in sixteenth-century New Spain was little more than a *vista de ojos,* literally "the eyes' view." Middling distances were commonly described as "more or less a musket shot," *un tiro de arcabuz, mas o menos,* which one source described as being one-sixth of a league (AGN Tierras, vol. 1715, exp. 6). Leagues were used to measure larger distances and modern historians convert them to about four and one-third kilometers (Haggard 1941: 78). In the sixteenth century, however, leagues could be "long leagues" or "short leagues" (see, for instance, item 8 in the Relación Geográfica questionnaire in appendix B). When a more detailed survey was called for, as when land disputes arose over fields in the well-populated Valley of Mexico, Spanish officials would be sent out to make measurements by pacing and using ropes, counting out *pasos,* steps, or using the *cordel,* a rope that might measure 50 pasos or *varas,* both units equivalent to about a yard (AGN Tierras 2683, exp. 11; Gibson 1964: 257; Haggard 1941: 84–5).[23] Another common measurement was the *braza,* whose length varied from place to place (Gibson 1964: 257). Orientation was reckoned by the position of the sunrise or by relationships to prominent features in the landscape, such as mountain peaks or lakeshores (AGN Tierras, vol. 2983, exp. 11).

Thus, from the moment López de Velasco's mapping projects arrived on the shores of New Spain, they were headed toward failure. While many colonists were interested in the history of the colony, most showed little interest in visually depicting it. They were hampered by their lack of technical ability, and life in the colony offered them few reasons to make maps. In most cases, texts would suffice. In land surveys carried out in New Spain, the measurements were often recorded as a written text, consistent with European practice before the sixteenth century (Harvey 1987: 464).

When colonists did map New Spain in the succeeding centuries, they did so piecemeal, limning the country as the need arose. As Stroza Gali's maps show, economically vital coasts and shipping routes were the first to enter the colonists' pictorial record by way of nautical charts. Following these were land surveys of contested and coveted regions, such as the Valley of Mexico. The recurrent floods that visited the Valley of Mexico meant that it was also mapped for various drainage projects, called *desagüe.*[24] López de Velasco's approach of culling maps from laymen in every region, while more systematic, never bore fruit because it was incommensurate with the abilities and interests of Spaniards in New Spain.

While López de Velasco may have anticipated some of the technical difficulties, he was blind to the particular valence of the image in New Spain. Calling upon colonists to respond to items 10 and 42 *"en pintura,"* he used the common broad term that was more or less equivalent to "painting," which could also be used to describe maps, landscape paintings, portraits, and written descriptions.[25] (A more specific term to guide colonists, such as *mapa,* was not in common use.[26]) But in using "pintura," López de Velasco unwittingly used the same term that colonists used to describe native pictographic writings, works that most, save the most curious antiquarian, regarded with a mixture of skepticism (for "pinturas" could and often did encode native religious practice and belief) and disdain.

Colonists in New Spain negated images because celebrating them would be ceding cultural power back into the hands of New Spain's indigenous peoples, who were well versed in the production and traffic of images millennia before the Spanish arrived. Ghettoizing the image allowed New Spain to maintain two distinct camps, one Spanish, the other native. Paradoxically, it was with the native communities of New Spain that López de Velasco could have found much of the cosmographic knowledge he desired. By default of the colonists, native painters did make most of the maps. Their way of representing their landscape would have been, in López de Velasco's eyes, as bereft of geometry as those of their colonial brethren. But when we look at these maps today, we see the way that the indigenous people of New Spain saw their world. When the colonists failed to, New Spain's indigenes pictured their world for the Spanish king.

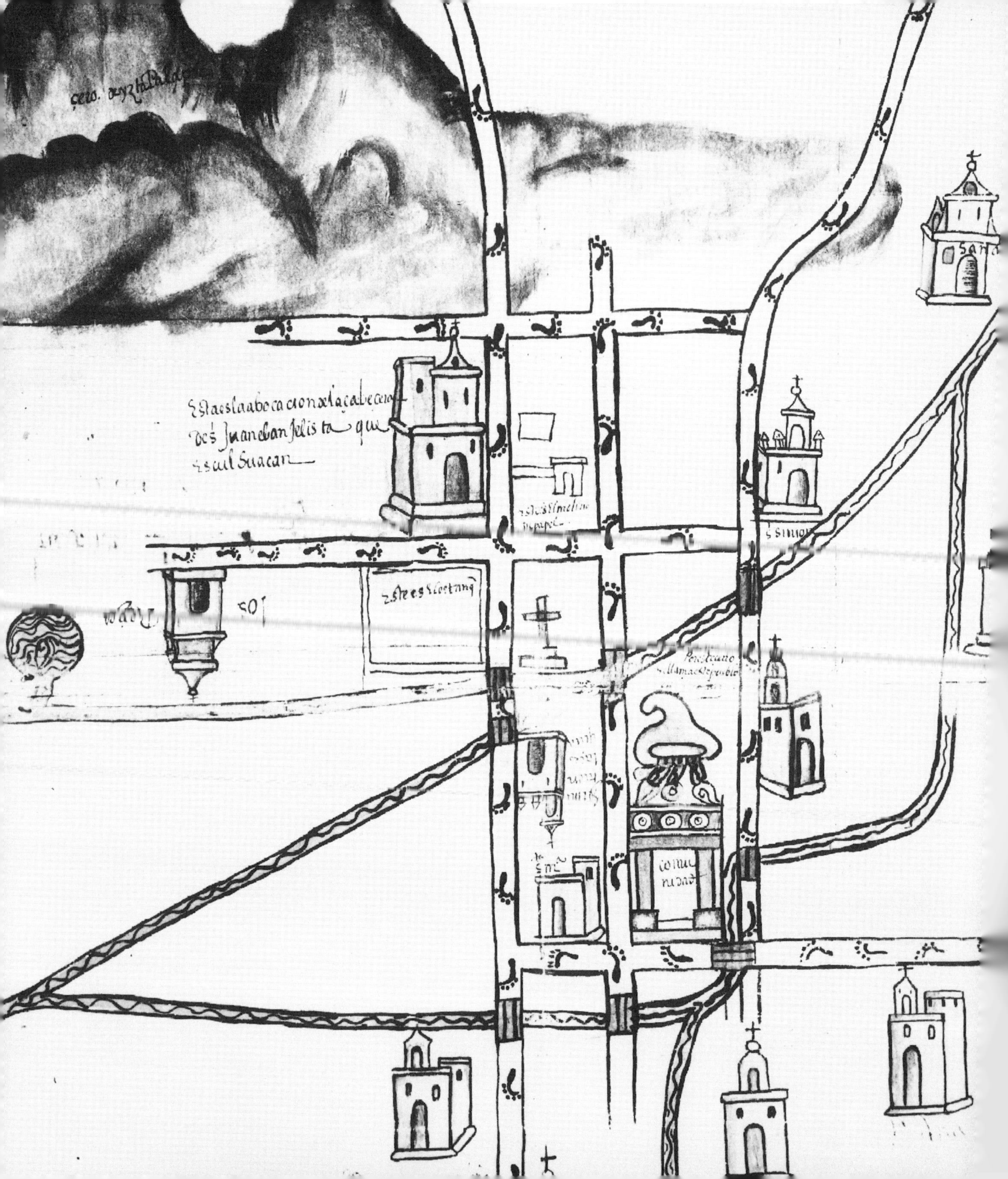

comu
nidad
s simon

CHAPTER FOUR

The Native Painters in the Colonial World

While some of the Relaciones Geográficas maps were drawn by the local corregidor or alcalde mayor, the majority were drawn by artists who counted themselves members of New Spain's *república de indios,* indigenous communities that were, in theory, segregated from the colony's newer inhabitants—such as Creoles, Spaniards, and African slaves.[1] Their participation in the corpus is the result of a misunderstanding: native artists drew maps of their communities in response to item 10, but the crown never intended native towns to answer this question; corregidores in native towns were clearly instructed to begin their reply with item 11, perhaps because the crown believed that no native painter could make a map. The crown was wrong: the corpus abounds with sophisticated maps marked with native iconography, bespeaking native authorship; in a few rare cases, a native painter is actually named.

This chapter discusses what we can reconstruct about the native painters of the corpus—their status, their education, their influences. Piecing together the shreds of evidence about native painters, we find that the indigenous artists of the Relación Geográfica corpus had a wholly different relationship both to pictorial images and to indigenous iconography than did the Spanish and Creole colonists who made other maps in the corpus. As we can clearly see from their works in the corpus, most native artists had a command of both the artistic conventions of their world as well as the dominant conventions of Europe; in fusing the two, these artists were inventing new ways to represent the landscapes of colonial New Spain. In short, they, not their Spanish or Creole counterparts, were the creative actors animating the corpus. These "indios," unlike Spaniards and Creoles, lived in a world where both ideology and practice pulled the image to center stage; as the world around them changed, they were compelled to find new ways to represent it. Thus it was not only the ethnic affiliations of these painters but also an entirely different orientation toward the iconography and role of the image that distinguishes the maps they made, and cleaves the corpus asunder, each half reflecting the two worlds, Spanish and indigenous, that maintained two distinct realities within sixteenth-century New Spain.

PAINTERS AND THE NATIVE ELITE

While the Spanish and Creole respondents writing the text of a Relación Geográfica often refer to accompanying maps or drawings, they almost never name the artist.[2] In a rare exception, each of the three Relaciones from the populous corregimiento of Mexicatzingo (Mexicaltzingo, Distrito Federal, Mexico) in the Valley of Mexico names the native painter of its map; in doing so, these three offer us some of the only attributed works by indigenes of sixteenth-century New Spain. Most other native painters remain hidden in shadows. In the Relación Geográfica of Ixtapalapa (Distrito Federal), its author, Francisco de Loya, reveals that the painter of its map (fig. 28) is Martín Cano, a resident of Ixtapalapa. While de Loya does not specify that Cano was a native painter, Cano tips his hand by including native iconography on his map; in addition, contemporary documents made as part of a land grant document identify Cano as an "indio." (AGN Tierras, vol. 2809, exp. 4). When the corregidor Gonzalo Gallegos and the friar Juan Núñez composed the Relación of nearby Culhuacan (Distrito Federal), they mention that the map (fig. 29) was done by a native artist, and the text on its reverse names its maker as Pedro de San Agustín. The text of the Mexicatzingo Relación Geográfica tells us that the maker of its map was "Domingo Bonifacio, indio pintor," but his map is now lost (RGS 7: 26–7; RGS 7: 47).

The Relaciones Geográficas texts offer us enough hints so that we can reconstruct the circumstances wherein native painters such as San Agustín, Cano, and Bonifacio made their maps. The genesis often began with the *cabildo,* or indigenous town council, whose members were gathered together in town after town in New Spain to answer the crown's queries in the Relación Geográfica questionnaire. We know this because a number of corregidores and alcaldes mayors, to show that they had faithfully followed the crown's directive to ask "people knowledgeable about the things of the land," mention the participation of indigenous men whom they named as *gobernadores* (governors), *alcaldes* (judges),

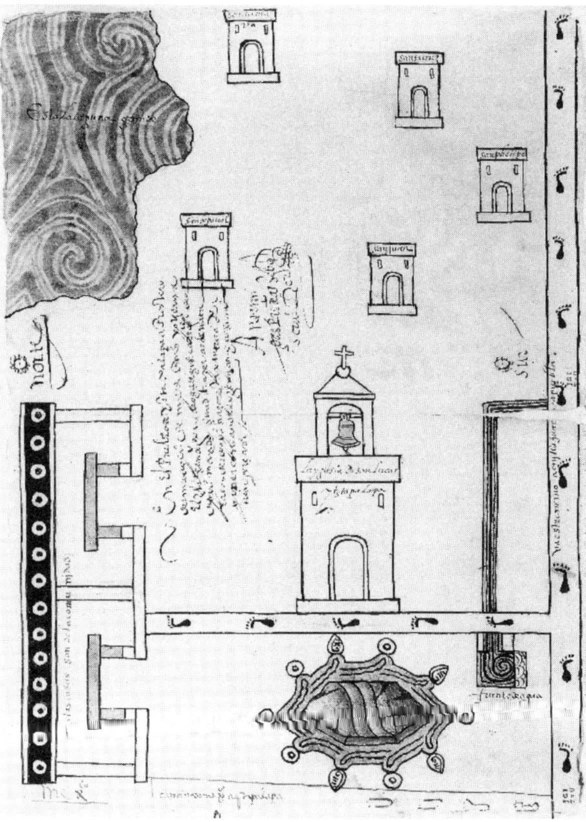

Figure 28. The Relación Geográfica map of Ixtapalapa by Martín Cano, 1580. This map of the indigenous town of Ixtapalapa in the Valley of Mexico was painted by a local artist, Martín Cano. The town name means "water near the flagstones" in Nahuatl, and Cano uses pictographs to write it. Below the church, a hexagonal flagstone is surrounded by a blue ribbon of water. Ixtapalapa's church dominates one axis of the composition, while its twin community buildings, at left, dominate the other. Size of the original: 43 x 31 cm. Photograph courtesy of the Benson Latin American Collection, The General Libraries, The University of Texas at Austin (JGI xxiv-8).

and sometimes *regidores* (councilmen) in the creation of the local Relación. The Relación of Cempoala (Zempoala, Hidalgo) is typical, opening with corregidor Luis Obregón's declaration that appearing before him are:

> Don Diego de Mendoza, *principal* of Cempoala, and don Francisco de Guzmán, governor of the town of Tzaquala, and don Pablo de Aquino, governor of [the town of] Tecpilpan, and Martín de Ircio, governor of [the town of] Tlaquilpa, and the alcaldes from the aforementioned towns and many other old and aged Indians of these same towns. . . . (RGS 6: 73)[3]

Men like don Francisco de Guzmán and don Pablo de Aquino were the indigenous elite, who borrowed their titles from municipal governments on the Iberian peninsula but usually derived their power from political hierarchies that were in place long before the

Spanish arrived (Gibson 1964: 154–5; Lockhart 1992). Many natives named as *principales,* like don Diego de Mendoza, also participated in the making of the Relaciones Geográficas texts. This title was not one borrowed from the Iberian model of town government, which never meshed perfectly with practices rising out of indigenous soil, but was particular to colonial America. The term *principales* designated native elites, likely the hereditary nobility without particular office in the cabildo, acting more like councillors-at-large.

Once gathered together, the native cabildo would offer answers to the Relaciones Geográficas questions posed by the Hispanic corregidor, often funneling them through an interpreter or series of interpreters. In the Zapotec town of Ixtepexic (Santa Catarina Ixte-

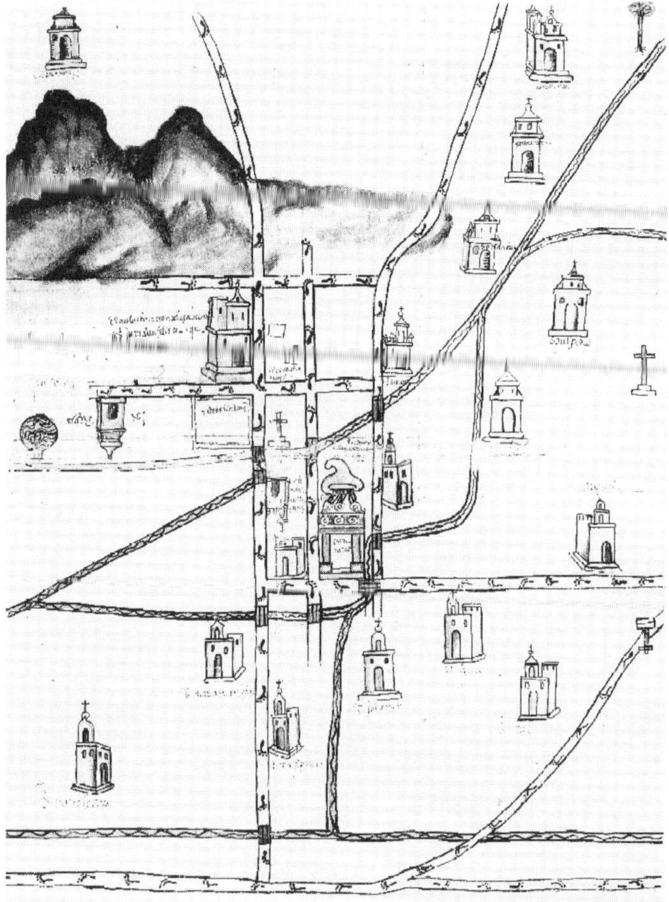

Figure [illegible] The Relación Geográfica map of Culhuacan by Pedro de San Agustín, 1580. This map, from a prominent town in the Valley of Mexico, was created by an indigenous resident of the town, who is named on the reverse. The town's Augustinian monastery, at upper left, is the largest building he portrays. Across the street from the monastery is a paper mill that probably made the paper used for this map. Size of the original: 71 x 54 cm. Photograph courtesy of the Benson Latin American Collection, The General Libraries, The University of Texas at Austin (JGI xxiii-14).

peji, Oaxaca), for instance, the scribe, Diego Ramírez de Castro, who spoke Spanish and Nahuatl, and Juan de Zárate, a resident who spoke Spanish, Nahuatl, and Zapotec, used Nahuatl as the bridge between the corregidor's Spanish questions and the Zapotec replies that local cabildo members offered (RGS 2: 248). In some cases, perhaps days elapsed as the cabildo transmitted a wealth of information about native culture; the length of the response seems directly proportional to the interest of the Hispanic official responding. The quizzical alcalde mayor of Meztitlan, Gabriel de Chávez, filled twenty-three pages with handwritten responses; his consultations with native elders and ancient manuscripts were certainly more than a day's labor. But in most cases the writing of the Relación seems to have taken no more than a day.

When the cabildo members reached item 10, they presumably chose a painter to carry out the request for a map. The elected painter might have been a cabildo member: Pedro de San Agustín, the native painter who rendered the Culhuacan Relación map, was in 1585 a member of the Culhuacan cabildo. As an alcalde, he held an office second only to the native gobernador (Cline 1986: 41). Since few complete records of cabildo members from the hundreds of Indian towns across New Spain exist, and even more rarely are the names of painters available, we cannot say how many Relaciones Geográficas painters were cabildo members, but most likely most (if not all) were members of the native elite, since painting was an elite occupation. In the preconquest period only the children of elites were trained in the arts of writing and painting (Reyes-Valerio 1989). Bartolomé de Las Casas, the renowned Dominican, wrote that to be a native manuscript painter "was a highly esteemed career" (Las Casas 1967, vol. 2: 505); the mestizo don Juan Bautista de Pomar wrote how it was the nobles who took up painting (RGS 8: 86). As we will see below, the friars who educated colonial natives continued to favor elites, teaching them to write and paint.

After a native painter like San Agustín was handed his commission, he seems to have carried it out somewhere apart from the caucus at work on the Relación text. The evidence for this lies in the physical objects of the texts and maps: most of the native maps are separate from the texts, being bound in with the texts after they were painted or at some later date, perhaps after their arrival in Spain. Most of them are painted on European paper with the same watermarks as the texts, suggesting that the Spanish respondent supplied the artists with their media. In a few cases, the map is painted on different paper stock or native fig-bark paper or hide and may have existed in the hands of the community even before the Relación Geográfica questionnaire arrived. In Culhuacan, San Agustín used locally made paper that bore no European watermark, and perhaps acquired it from the town's paper mill that is pictured on the map.[4] In the Relación Geográfica from the native town of Misquiahuala (Mixquihuala, Hidalgo), the corregidor, Juan de Padilla, writes that the native map "was remitted to me," suggesting that the map was in existence before he wrote his text (RGS 6: 31 and fn. 6). But in most cases, the native painter seems to have completed his (or her) map at the time the texts were being written and submitted it to the corregidor or alcalde mayor, who then sent the map and text back to the viceregal seat.[5]

From the Relaciones Geográficas texts coming from native towns and telling us about them, we can also picture the local worlds wherein the painters and their fellow elites moved, and throughout we sense how powerfully elites were able to shape their immediate surroundings. We glimpse their political importance within their local communities. Within the texts, the Hispanic writers often dismissed the native masses, called *macehualtin* (singular: *macehualli*) as brutish and stupid. The Creole alcalde mayor, Constantino Bravo de Lagunas, who penned the Relación from Xalapa de la Vera Cruz (Jalapa Enríquez, Veracruz), for example, complained that "all the Indians have no greater intelligence than eight-year-old Spanish children" (RGS 5: 340–1, 344). Nonetheless, to write his text, he gathered together "indios principales" and listened carefully to the answers these leaders supplied. Hispanic administrators such as Bravo de Lagunas stood on different footing with this native elite because they depended upon the elite's compliance to run their regions. Juan González, the corregidor of Xonotla (Jonotla, Puebla) and Tetela (Tetela de Ocampo, Puebla), revealed the true nature of this traffic when he wrote that the native governor of the town, don Antonio de Luna, held permanent stewardship because he was a "natural lord" as well as because the viceregal government, and its representative González, supported his rule (RGS 5: 381–2). In Xonotla, as in other places, the relationship was interdependent: Don Antonio de Luna looked to corregidor González to confirm the authority he exercised among his native subjects, and in return, González depended upon de Luna, and his father before him, to extract taxes and tribute from the macehualtin (RGS 5: 383). The testimony the Relación Geográfica corpus offers as to the political importance that elites had within the colonial state is confirmed by other historical material. Hispanic corregidores, alcaldes mayores, and encomenderos needed their native counterparts, the gobernadores, alcaldes, regidores, and principales, to keep large and scattered populations orderly and to run the tribute machine that fed the Spanish state, just as they often depended upon native elites for answers to the king's questions (Gibson 1955; 1964; Spores 1967; Lockhart 1992).

The relationship was not simply one of Spanish exploitation, for both sides stood to profit. For the native elites such as those supplying the answers to the Relación, their association with European friars and Hispanic officials was their link to power and prosperity. The closer they came to the tribute flowing into both Spanish coffers and their own community treasury, the greater their chances to control its uses, or in some cases to skim from the till (Pita Moreda 1987: 223–9). The native elite's role in the church as *cantores* and *fiscales* gave them added prestige in the community (Ricard 1982: 97–8). In short, many of the same social and economic interests bound elite natives and Europeans together into the colonial state.

But the native elite's role was not just about getting and spending. As the linchpin of the Spanish colonial state, standing between, on the one hand, Spanish government officials, private encomenderos, and resident priests, and, on the other, a large body of native commoners, the native elite had an instrumental role in shaping colonial reality (Gruzinski 1993; Haskett 1987; Haskett 1988; Haskett 1991; Lockhart 1992). Externally, they negotiated with the colonial powers outside of their communities; internally, within their own towns and villages, they forged ideologies that would be acceptable to their communities and would be aligned with the new political and economic order of the colonial state.

The Relaciones Geográficas, both picture and text, give examples of how elites were actively shaping a new reality, a new way of understanding the order of things. When in Meztitlan, Gabriel de Chávez asked elites about their history, giving them questions to guide their replies, and they in turn tailored their histories to meet his questions. That is, the native elites in Meztitlan and elsewhere may not have thought of their own histories as divided neatly into the categories the Relación questionnaire provided—"warfare," "costume," and so on—but these elites recast their history along the lines of the king's directive. At the same time, though, Meztitlan elites were able to broadcast their view of themselves as a historically civilized nation, the kind of civility Spaniards would recognize. They had, as they stated in their replies to Chávez, a hierarchy of rulers, a system of laws, and a prescription of punishments. In emphasizing that they and their communities were a civilized people as the Spanish would define it, Meztitlan elites were actively shoring up their authority (as civilized leaders) within the framework of the Spanish colonial state.

It is in the native maps, however, that we see most clearly the shaping hands of the elites. In the written responses, it was the questionnaire that gave form and shape to the replies, and in addition, the answers were translated out of native tongues and passed through the filter of the Spanish-speaking respondent and scribe before being written down on the page. The maps certainly had their constraints: the questionnaire asked for "the layout of the town, its streets, plazas and other features, noting the monasteries. . . ," echoing López de Velasco's conception of space as being defined by architecture. But it was not as if an indigenous map was recast and copied by the corregidor; these maps were drawn entirely by members of the native elite, and sometimes they, and sometimes the scribe, added the glosses. Given their contexts, and above all the Spanish commission that led to their creation, these maps oscillate between how elite painters envisioned and depicted their own communities and how they saw fit to present them to the larger colonial world.

MENDICANT INFLUENCES

While the Relaciones Geográficas texts are revealing about the political networks within which elites, and elite painters, moved, they are less than telling about the role of the Catholic church, which not only had a profound influence on the spiritual life of native communities, but also was probably the sphere of colonial life wherein Europeans and Creoles had the greatest contact with and influence upon the native painters at work on the maps. The texts, written as they were by secular government officials, only touch on

Figure 30. The Relación Geográfica map of Guaxtepec, 1580. (See also pl. 2.) Created by an indigenous artist, this map presents a richly hued portrayal of the region. At center is the monastic church, and below it is the indigenous place-name of the town, which translates as "hill of the huaxin tree." Size of the original: 62 x 83 cm. Photograph courtesy of the Benson Latin American Collection, The General Libraries, The University of Texas at Austin (JGI xxiv-3).

Figure 31. The Relación Geográfica map of Acapistla, 1580. Like the mapmaker in nearby Guaxtepec, the indigenous mapmaker of Acapistla placed the monastic church with its large courtyard at the center of the map. The surrounding landscape is vividly colored, and outlying settlements are often named with indigenous toponyms. Size of the original: 84 x 61.5 cm. Photograph courtesy of the Benson Latin American Collection, The General Libraries, The University of Texas at Austin (JGI xxiii-8).

churchly matters, but many of the native painters made maps where the image of the Catholic church dominates, as in the Relaciones Geográficas maps from Guaxtepec (Oaxtepec, Morelos; fig. 30; pl. 2), Acapistla (Yecapixtla, Morelos; fig. 31), Cuzcatlan (Coxcatlan Puebla; figs. 32 and 33), Cholula (Cholula de Rivadabia, Puebla; fig. 34; pl. 3), Chimalhuacan Atengo (Santa María Chimalhuacan, Mexico; fig. 35), and Coatepec Chalco (Coatepec, Mexico). On these maps, communities are organized by and around a central religion, and they reveal this by their iconography, in which communities are symbolized by churches (Leibsohn n.d.: 1). The colorful Acapistla map (fig. 31) is typical. At the center of the map, dominating the composition, is the Augustinian monastery (San Juan Bautista Yecapixtla) that was founded in 1535 (Ricard 1982: 74); radiating out from it are the town's subject villages. Like the monastic church, they are painted a vivid salmon pink. These smaller chapels are all cast from the same mold: large arched doorway, pointed roof, and a crowning cross. In order to show the monastic church as central, the artist has had to compress the arrangement of subject towns along the north-south axis and elongate those along the east-west one. In the Acapixtla map, as in others, the artist has drawn the build-

Figure 32. The Relación Geográfica map of Cuzcatlan A, 1580. This map concentrates on the regional network of roads and rivers and uses churches to symbolize human settlements, as was typical of coeval European maps. However, the indigenous artist also lines the edges of the map with many pictorial symbols that almost certainly are toponyms of regional boundaries. Size of the original: 43. 5 x 31.5 cm. Photograph courtesy of the Archivo General de Indias, Seville (Mapas y Planos, 19).

ings in a three-dimensional perspective; their appearances often have little in common with the building portrayed. These church-centered maps, then, serve as a corrective to the texts, showing the central role the church played within native communities.

Or do they? We are bound to question the role of audience. This was, after all, a corpus of native maps painted for Spaniards. Was showing communities organized around churches how native elites wanted to present their community to the Catholic king's eyes? Or was this how they saw it themselves? Most of the maps, I believe, reflect in part how

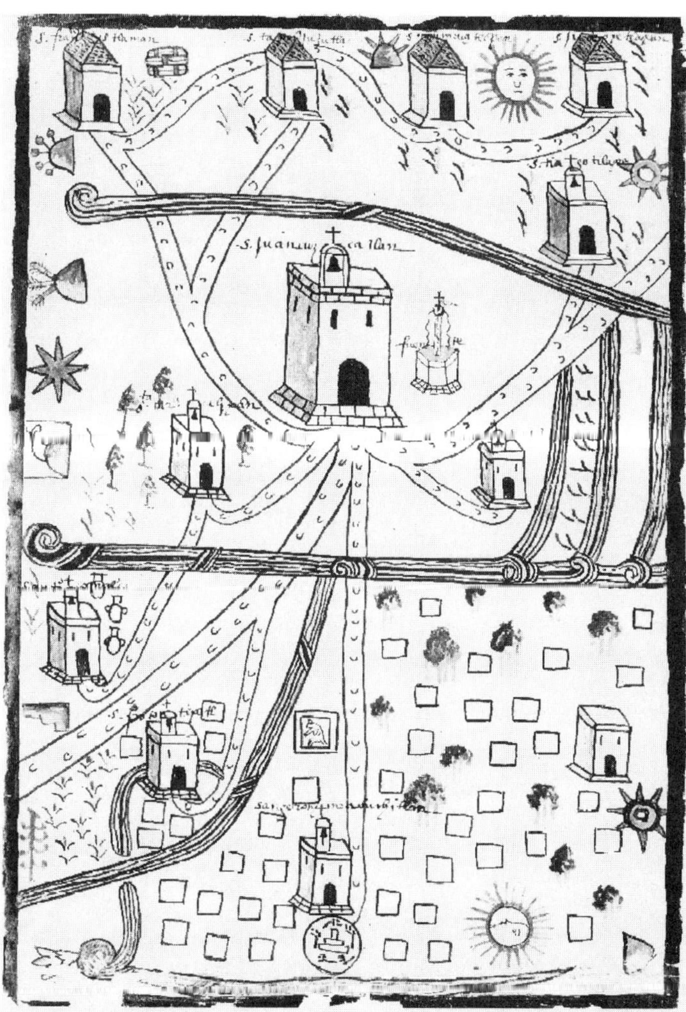

Figure 33. The Relación Geográfica map of Cuzcatlan B, 1580. This map largely duplicates the map of Cuzcatlan A, except it includes fewer pictorial symbols. Size of the original: 43. 5 x 31.5 cm. Photograph courtesy of the Benson Latin American Collection, University of Texas at Austin (JGI xxiii-15).

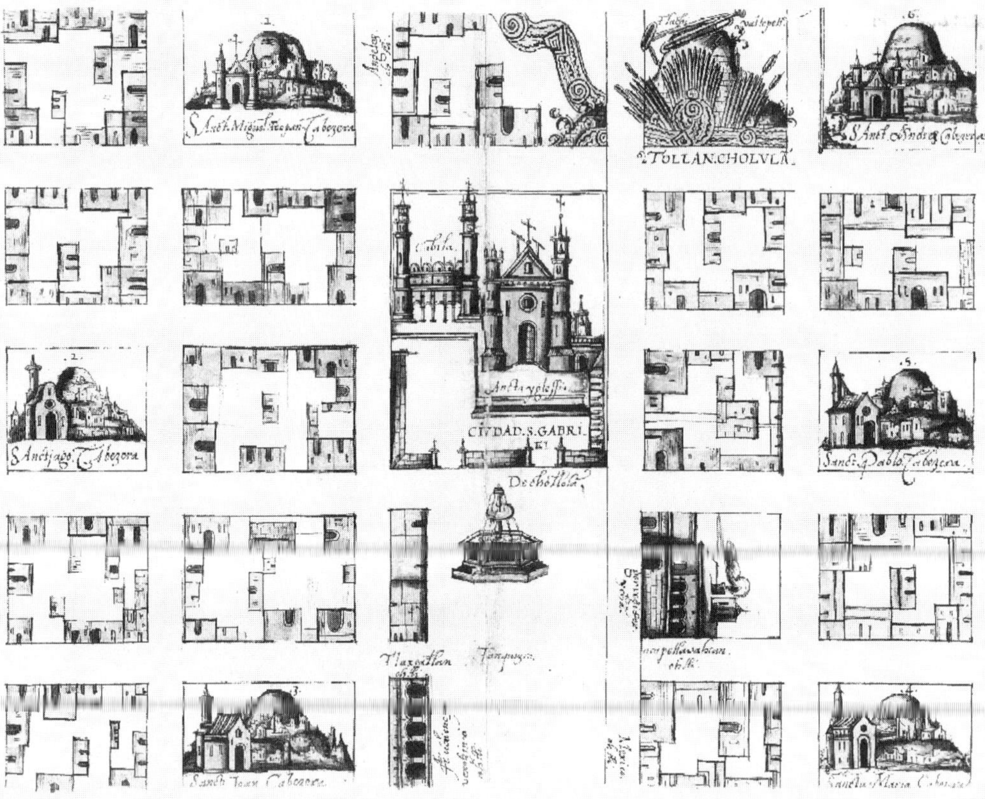

Figure 34. The Relación Geográfica map of Cholula, 1581. (See also pl. 3.) This extraordinary indigenous map from the city of Cholula shows it to have been laid out along a strict grid, a pattern followed today. While the Franciscan monastic complex is central, Cholula's large pre-Hispanic pyramid, the *tlachihual tepetl*, is pictured at upper right and named as "Tollan Cholulã." Size of the original: 31 x 44 cm. Photograph courtesy of the Benson Latin American Collection, The General Libraries, The University of Texas at Austin (JGI xxiv-1).

the native community saw itself and contain some projections of what native elites thought a Spanish audience wanted. In short, the native colonial artist's work was colored by his (or her) "double-consciousness," as he (or she) painted for the local community as well as for a shadowy, but powerful Spanish patron. This double-consciousness manifests itself in the disjunction between the nuclear communities, shown on this group of maps by means of churches, and the reality of sociopolitical arrangements.

The mapmakers' use of the church to symbolize communities was, I think, an autonomous choice. European map conventions dictated its use, as we see on the maps of Ixcatlan (figs. 21 and 22). In addition, the church was certainly the most important European institution to enter native communities in the sixteenth century. It filled a spiritual and ritual chasm left by the banning of indigenous religious practice. More specifically, the church was both the place that many of the painters at work on the corpus would have been educated and the institution that defined the status of the image in colonial New Spain.

Most of these church-centered maps came from towns hosting mendicant friars; the Franciscans, Augustinians, and Dominicans had been in charge of converting New Spain's indigenes to Catholicism since soon after the Conquest (see table 2). It was the priest, not the corregidor, who had the closest day-to-day contact with the native elite. Again, elites were intermediaries in this colonial interchange. Friars, whose aims were

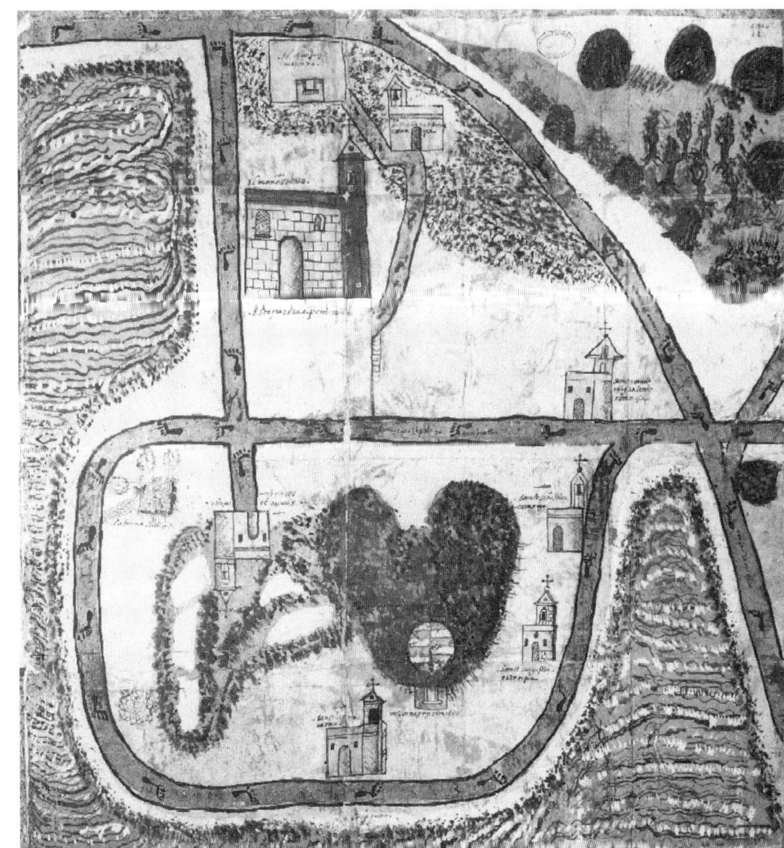

Figure 35. The Relación Geográfica map of Chimalhuacan Atengo, 1579. Chimalhuacan Atengo lies in the Valley of Mexico, on the edge of Lake Tetzcoco, which fills the bottom and left side of this map. The first name of the town means "place of the shield possessors," and a striped circular shield is depicted within the bell-shaped hill symbol, which appears upside-down at bottom. Size of the original: 55 x 52 cm. Photograph courtesy of the Archivo General de Indias, Seville (Mapas y Planos, 11).

otherworldly, needed elites to lead the community towards Christianity and often directed their mission to the upper ranks of society (Gómez Canedo 1977; Gómez Canedo 1982). While elites often dealt with colonial administrators through translators, they could often speak comfortably with priests in their native tongues, since many priests had learned the local indigenous language, be it Nahuatl, Mixtec, Zapotec, Otomi, or another of the dozens of languages spoken across New Spain at the time (Burkhart 1989; Harvey 1972; Ricard 1982). In the town of Teozacoalco, which produced the largest and most important map in the corpus, it was the resident priest, Juan Ruiz Zuazo, who questioned native elders because, as the Relación put it, he "understands very well the language of the aforementioned natives" (RGS 3: 151). In larger towns, friars were close at hand in the local monastery. Their monastic schools were the only institutions throughout New Spain dedicated to native education. All these factors helped the churches they established to find themselves in the center of community life and as symbols for the community on the Relaciones Geográficas maps.

However, the hierarchy shown by the church symbols—large main church versus smaller subject churches—may not have been an automatic choice for the native mapmaker. Instead, this hierarchy of nucleated towns seen in the Acapistla map accurately reflected how Spaniards envisioned town life, and can also be found on the European-style Relación maps, such as the Relación Geográfica of Acambaro (Guanajuato; fig. 36). Such nucleation and hierarchy were sometimes at odds with traditional native arrangements. Many Nahuatl-speaking polities, for example, had settlements scattered throughout the area of their domain. Each of these settlements, though, to the resident

Table 2 Relationships between the Production of Relaciones Geográficas Maps and the Presence of Secular and Monastic (Regular) Religious Establishments

Map Production in Secular and Monastic Towns and Regions	Number	Percent
Maps painted in secular towns/regions	41	59
Maps painted in monastic towns/regions	28	41
Total maps	69	100
Within Monastic Towns and Regions		
Native maps	20	71
Non-native maps	8	29
Total maps from monastic regions	28	100
Within Secular Towns and Regions		
Native maps	25	61
Non-native maps	16	43
Total maps from secular regions	41	100

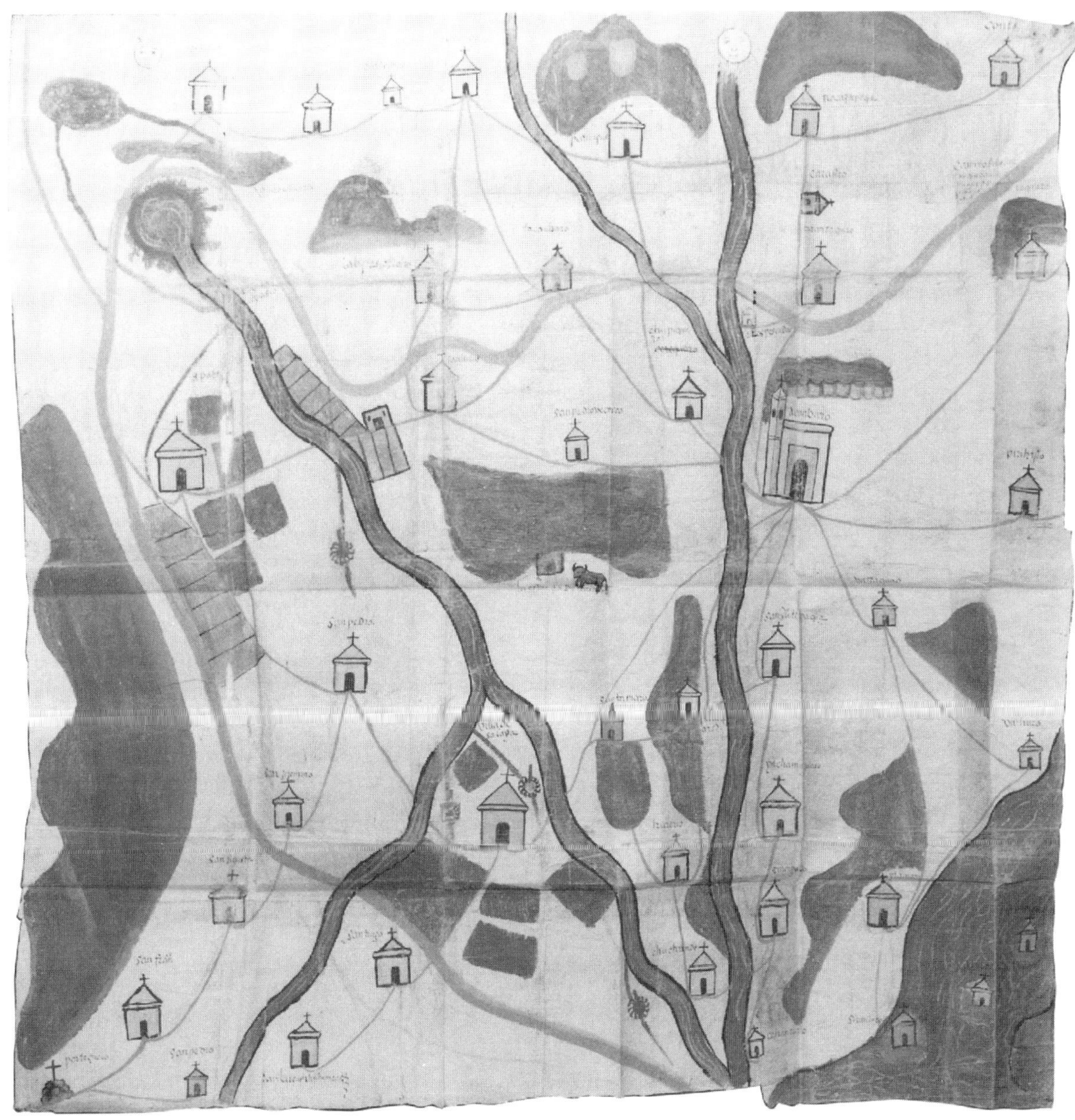

Figure 36. The Relación Geográfica map of Acambaro, 1580. The artist of this map used churches to symbolize towns and villages, as was consistent with European practice. The main town of Acambaro is symbolized by the larger monastic church on the right. Size of the original: 114.5 x 115 cm. Photograph: Juan Jiménez Salmerón; reproduced courtesy of the Real Academia de la Historia, Madrid (9-25-4/4663-x).

Nahua, was an equal member of the polity; one did not predominate over the others (Lockhart 1992). However, among the Mixtec, who lived to the south of the Nahua in what is now the state of Oaxaca, the nucleated, hierarchical model seems to have been close to sociopolitical arrangements in pre-Hispanic times (Byland and Pohl 1994; Spores 1967: 94–6). In the model we see on these Relaciones Geográficas maps, with settlements nucleated and ranked in a hierarchy of *cabecera* ("head town") and subjects, native artists made visible their current political and religious arrangements, which were often ones Spanish administrators and friars had imposed upon them. These maps showed an order the colonial government was eager to see.

But while native artists did undoubtedly tailor their self-presentation in these maps to a Spanish audience, they were not painting only what community outsiders wanted to see. Sometimes they welcomed the hierarchy the church and state imposed. In Nahua regions, leaders of the settlement that was elevated by church and crown officials to be the cabecera of the entire polity often stood to benefit from their new status, since they could command the corvée labor from other settlements that were assigned to be their subjects.

MONASTIC SCHOOLS

Many of the native artists of the Relación Geográfica corpus were trained at monastic schools. Evidence for this shows up internally in the corpus: some of its most polished painting, marked by clear forms, confident line, and skillfully applied colors, comes from towns with large and active monasteries. The painting on a map is often on a par with the mendicant architecture of the town itself. Guaxtepec is the site of an impressive Dominican monastery established in 1528, one of the earliest of the Dominican parishes in New Spain (Ricard 1982: 69). From it comes an exquisitely drawn and colored map (fig. 30; pl. 2). The map from Cholula (fig. 34; pl. 3) is equally fine; its artist displays the fine-line precision of a well-trained hand and was perhaps the product of Cholula's monastery school, which at the time the Relaciones Geográficas were written, was staffed by as many as twenty friars (RGS 5: 144). The map of Culhuacan also comes from an important mendicant town (fig. 29). Other correlations between mendicant presence and the production of Relaciones Geográficas maps are tabulated in table 2.

A primary role of mendicants in these native towns was to run schools. As Diego Muñoz Camargo, the mestizo author of the Relación of Tlaxcala (Tlaxcala) wrote, native education "is the absolute principal work that the friars do in both this land [of Tlaxcala] and in all of New Spain" (RGS 4: 52). One important monastic school, San José de los Naturales in Mexico City, created by fray Pedro de Gante, gives us some idea of what the mendicant curriculum was like. At San José, native boys learned not only the rudiments of reading, writing, and numbers, but also trades instrumental in decorating churches: painting, sculpture, jewelry making, and embroidery (Torquemada 1986, vol. 3: bk. 17, ch. 2). As one Franciscan wrote "the pomp and adornment of the churches is absolutely necessary to elevate [the Indians'] spirits and move them towards the things of God" (*Códice Franciscano* 1941: 58). The main Franciscan academy at Tlatelolco in the Valley of Mexico had a cadre of native painters whom fray Bernardino de Sahagún employed to paint the illustrations of the encyclopedic Florentine Codex (1979).

Not all natives were welcome in these schools. Rather, the friars concentrated on elites. While Franciscans held educating the native population to be one of their primary goals (*Códice Franciscano* 1941: 55), theirs was a selective process. The children of the commoners were only to be taught the basic elements of Christian doctrine so as to remain "in the simple state of their ancestors" (*Códice Franciscano* 1941: 55), as were daughters of the elite. But the sons of the elite were taught not only Christian doctrine, but also to read, write, sing, and assist the friars (Gómez Canedo 1982). These elite sons were expected to return to their homes to assume positions as local leaders (*Códice Franciscano* 1941: 63), to practice and disseminate this doctrine and these arts. Sons of elites were taken from many regions of New Spain to train at Tlatelolco and San José. For instance, don Francisco de Salinas, the son of a Cuicatec lord, graduated from Tlatelolco then returned to his native town in Oaxaca to eventually become its gobernador (Hunt 1972: 219); the Relación Geográfica from Cuicatlan (San Juan Bautista Cuicatlan, Oaxaca) cites his participation and names him as a principal at the time of its composition (RGS 2: 165). Through educating and evangelizing the elites, the friars hoped their doctrine would trickle down to all sectors of native society.

Within these schools, those elite native students chosen to be artists undoubtedly learned the pragmatics of painting, such as the application of ink in straight, well-regulated lines, the techniques for gathering, grinding, mixing, and applying pigments, and the conventions of proper composition. But perhaps the lessons that registered the greatest impact would have come from copying images from prints in the possession of the friars (Torquemada 1986, vol. 3: bk. 17, ch. 1). Such prints were the main conduit of European artistic style and convention into the colony. By teaching their charges to copy them, the friars promoted an art that arose from previous works of art, rather than from any direct visual experience. This idea of art was well-received by native painters, whose preconquest tradition taught an art that grew out of the preexisting model rather than from direct observation. Within prints, at the same time, native artists would also have encountered illusionism, the idea that painting should mimic visual experience.

SYMBOLS AND ILLUSIONISM

Illusionism, in the New World, could have meant artistic revolution, since it asked the indigenous artist to rethink his or her precepts about the relationship of the image to what was represented. On the whole, native painting favored a conventionalized symbolic vocabulary and a rendering that clearly articulated important parts of an object or figure. Illusionism encouraged native painters to abandon all this and draw images that mimicked visual experience by using shading, foreshortening, and perspectival rendering.

In the Relaciones Geográficas maps, however, the forces of conservatism largely prevailed, and not because native artists were unable to master European convention. The Franciscan Motolinía (fray Toribio de Benavente) noted with admiration how native painters were so skilled that "they make as good images as any in Flanders" (Motolinía 1971: 240). It seems as if their conservatism was a conscious choice, and their use of symbols and signs to represent objects and places gained added authority because of the parallel system of symbols commonly used to represent places on European and Spanish colonial maps. While native artists often used perspectival rendering, an outgrowth of illusionism, they used it to revamp their symbolic vocabulary rather than to recreate what their eyes saw. We see perspectival drawing in the native map of Guaxtepec (fig. 30; pl. 2), where the *casa de justicia* (courthouse) at the bottom center is painted according to its dictates, as is, to a certain extent, the central monastery. Nonetheless, the artist presents a monastery that is a symbol of "church" or "monastery" rather than an illusionistic rendering, perhaps drawing the image of the Guaxtepec monastery from a print. As pictured on the map, the monastery is shaded in tones of rose; its facade is carefully delineated with quarry stones; its roof is domed, with an elaborate pediment framing a cherub. The whole concoction bears little resemblance to the actual monastery in the town, which has a rather severe church facade, with a long, low arcade attached to its right side (fig. 37). The town's church does have a high, round window but lacks the sloping pediment, the central belfry, and the dome shown in the Relación map. The artist seems to have been trying to draw a symbol or conventionalized representation of a monastery, and probably turned to a European printed illustration for inspiration, an image such as the frontispiece of Bartolomé de Ledesma's *Reverendi patris fratris . . .* , a book published in New Spain in 1566 (refer back to fig. 23; García Icazbalceta 1886; Robertson 1974: 155). The artist used similar conventionalized representations in painting the outlying subject churches. These seven smaller churches, which look nearly identical, are made of pink blocks and are dominated by arched doors and topped with a central belfry, like smaller versions of the Guaxtepec monastery. Their shading gives the illusion that they are receding into space,

Figure 37. Facade of the Monastery of Guaxtepec (Oaxtepec, Morelos). Photograph: B. Mundy.

but their repetitiveness suggests that the native artist of this map meant them to read as a symbol for "church," rather than to visually replicate individual churches.

In the Guaxtepec map, the use of "church" symbols correlates closely with symbols used in European maps, but we cannot assume that the idea was drawn from European models because other maps show how the use of symbols was rooted in native representation and was merely buttressed by European map conventions. In making the Relación Geográfica map of Texupa (Santiago Tejúpan, Oaxaca; fig. 38; pl. 4), the artist was pulled toward the idea of a European illusionistic landscape, perhaps introduced at the hands of the Dominican friars who settled in Texupa in 1571, whose monastery appears on the map flanked by a large enclosed garden.[7] Whatever he (or perhaps she) may have learned within the monastery, the artist was unwilling to abandon native symbols. All around the town, the painter overlapped individual native hill symbols into ranges in an attempt to depict the landscape surrounding Texupa. This new idea of an illusionistic landscape, the map shows, was not powerful enough to override the fundamentals of native depiction, which demanded that each hill symbol be shown individually. The Texupa artist reached almost the same conclusion as did the artist of the Codex Kingsborough from the Valley of Mexico (fig. 39). In both the Texupa map and the Codex Kingsborough, the distinctness of almost every hill is maintained in native fashion, but the hills are grouped together to give the impression of the visual unity that the landscape presents to the eye, in the European manner. Thus, no matter what the friars may have taught native painters, native painters clearly had choices about how they would, in the end, represent their surrounding world, and often the choices were conservative ones.

The blending of native conventions with European illusionism that shows up in the Texupa, Guaxtepec, and other maps of the corpus can be perhaps traced to the painters'

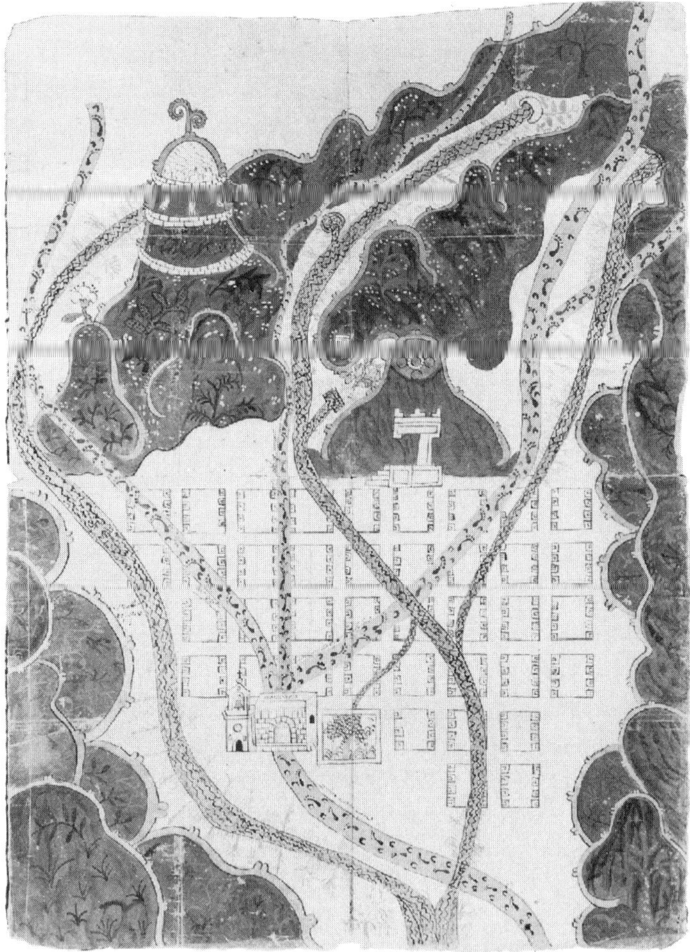

Figure 38. The Relación Geográfica map of Texupa, 1579. Size of the original: 56.5 x 43 cm. Photograph: Juan Jiménez Salmerón; reproduced courtesy of the Real Academia de la Historia, Madrid (9-25-4/4663-xvii).

Figure 39. Codex Kingsborough, fol. 204r, c. 1555. A ledger of complaints brought by the native community of Tepetlaoztoc against their Spanish encomendero, the Codex Kingsborough also includes this beautifully painted map of the Tepetlaoztoc region. In it, the artist fuses indigenous symbols with European landscape representation. Size of the original: 29.8 x 21.5 cm. Photograph courtesy of the British Museum, London (Museum of Mankind, Add. Ms. 13964).

training within monasteries. For this training, although overseen by friars, must have been carried out by other native painters. Sixteenth-century documentary sources, written by friars, are not always clear, perhaps purposely so, on who actually trained native disciples to paint (e.g., Mendieta 1971: 403–10; Motolinía 1971: 240). We do know, however, that friars were few in number, and they oversaw enormous native congregations. For instance, the twenty friars at Cholula oversaw a region of about 9,000 tributaries (Gerhard 1986: 117). Friars were not always of an artistic bent, and within the ranks of the secular Spanish community there were scant numbers of artists. In order to train new painters, friars were forced to turn to older native painters. Thus, the education carried out in monasteries was in some respects a conduit whereby native painters taught novices and continued a native painting tradition.[8]

Once it had trained native painters, the church then employed them. A common task was making the large "church cloths," canvases that were filled with scenes the friars used in imparting Christian doctrine to natives. These cloths could be easily transported as friars traveled to small chapels in the countryside, where the church cloths could immediately transform empty walls into didactic panoramas (Burgoa 1934, vol. 1: 288). But perhaps the most honored work for these painters was to decorate the monastery itself. From descriptions by sixteenth-century writers and from the fresco fragments that exist today, it appears that lively fresco cycles carried out in the 1560s and 1570s covered most of the larger sixteenth-century monasteries, both inside and out (Peterson 1993).[9]

Most murals were probably done by native painters, and these painters formed the pool from which the artists of the Relaciones Geográficas maps were drawn. The maps were commissioned slightly later than the murals, between 1579 and 1584, but both projects would nonetheless fall within the life span of a native painter. The Relaciones Geográficas maps that come from monastic towns with known fresco programs are summarized in table 3. These maps—those of Guaxtepec (fig. 30; pl. 2), Cholula (fig. 34; pl. 3), and Culhuacan (fig. 29)—are marked by an even hand, trained line, and careful coloration, along with an admixture of illusionism, all qualities that bespeak some training. For most native artists in communities responding to the Relación Geográfica questionnaire, there were few places that they could be trained except the monastery.[10] In addition, the monasteries on the Guaxtepec, Cholula, Acapistla maps are in central, dominant positions, indicating that their artists envisioned each of the monasteries to be the *axis mundi* of the town. The name of San Agustín, Culhuacan's painter, certainly suggests a connection to Culhuacan's Augustinian friars. Although the monastic church that San Agustín pictured at the center left of his map is a ruin today, the cloister survives, and its lower walls are filled with unusual landscape paintings of the mid-1570s showing scenes

Table 3 Towns Housing Monastic Establishments Known to Have Produced Both Monastic Frescos and Relaciones Geográficas Maps

Augustinian	Culhuacan (Distrito Federal, Mexico)
	Acapistla (Yecapixtla, Morelos)
	Epazoyuca (Epazoyucan, Hidalgo)
	Tezontepec (covered by Misquiahuala map, now in Hidalgo)
Franciscan	Cholula (Puebla)
	Cempoala (Zempoala, Hidalgo)
Dominican	Guaxtepec (Oaxtepec, Morelos)

Sources: Edwards 1966: 97-8; Moyssén 1965; Peterson 1993; Reyes-Valerio 1989: 118; Ricard 1982: 214.

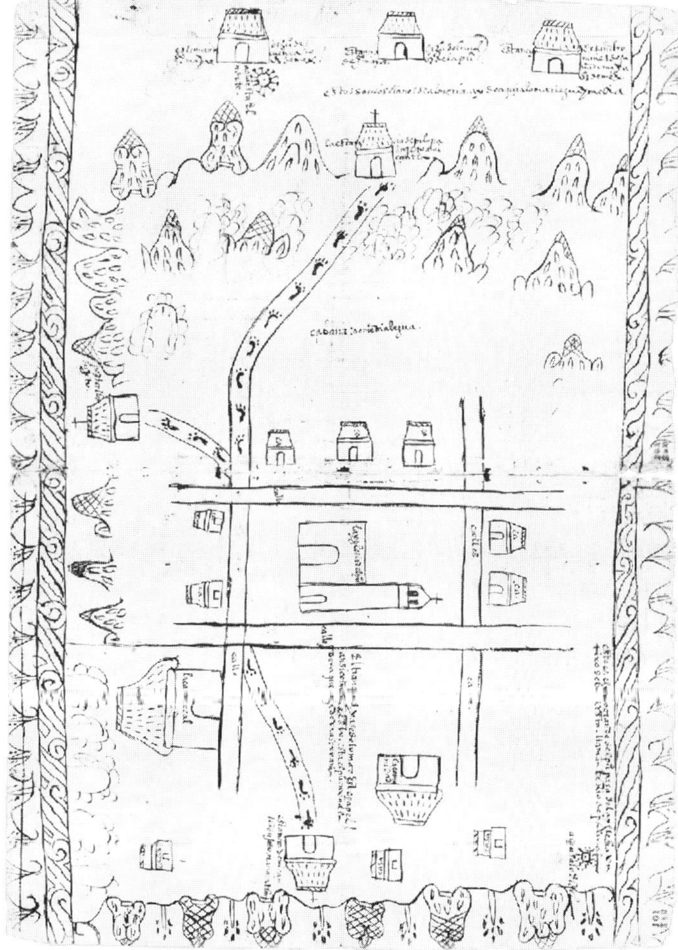

Figure 40. The Relación Geográfica map of Mizantla, 1579. This map is notable for its sloppy execution and wavering lines. It comes from an indigenous region controlled by secular clergy. Size of the original: 43.5 x 31 cm. Photograph courtesy of the Benson Latin American Collection, The General Libraries, The University of Texas at Austin (JGI xxiv-13).

of the eremitic life of Augustinian friars (Peterson 1985: 343–4; Peterson 1993: plate 12). These landscape paintings deploy the same varying washes of earth tones to convey an impression of recession in space as does San Agustín on the hill on the upper left corner of the Culhuacan map, and it is likely that he not only knew these fresco paintings in the local monastery, but helped paint them as well.[11]

The impact of mendicant training on the corpus can also be gauged by contrasting maps from native towns under the control of mendicants to those inhabited by secular clergy. In contrast to mendicants, who were committed to poverty and indigenous evangelization, secular priests often ministered solely to Spanish communities, up until the slow reforms that began in the late 1570s. Their goals were frequently as worldly as those of their parishioners: comfortable working conditions and a good income (Schwaller 1987: 129). Seculars took no vow of poverty and spent their careers moving from post to post, jockeying for positions at mines and ports because these wealthy sites offered greater chances for enrichment (Schwaller 1987: 78, 102, 139). There is little to indicate that sixteenth-century seculars were dedicated to native congregations or to improving the lot of these charges through extensive schooling; as a result they had little impact on native populations. For example, from Cuzcatlan came two copies of a Relación; in one copy, the transient secular *beneficiado* is mentioned but not by name, while in the other, all mention of him is omitted (Schwaller 1987: 105; RGS 5: 94). There is no mention of his works. The Cuzcatlan maps (figs. 32 and 33), like many of the other native maps from secular parishes, seem to exhibit little evidence that the native artist underwent any extensive formal training in the European tradition. Other maps from secular parishes were also made by barely trained native artists, such as, for example, the maps of Cuahuitlan (Cahuitan, Oaxaca), Ixtepexic, and Mizantla (Misantla, Veracruz, fig. 40). The correlations in the corpus are too few to prove any point definitively, but they do suggest that native towns with monastic establishments yielded trained native painters, while those with secular priests did not (see table 2).

MENDICANTS AND THE IMAGE

In New Spain, the mendicant friars celebrated the image as the way to reach and to teach New Spain's native peoples. Such a conduit was a natural one to choose, for pre-Hispanic societies were the creators of complex and sophisticated imagery and were more conversant with logographic, rather than alphabetic, writing both before the conquest and after. In short, cultural communication was dominated by images. The church promoted the teaching of painting to indigenous artists, and its use of European prints introduced European stylistic conventions and notions of representation to the hundreds of painters trained under its auspices. But to concentrate just on training is to miss the main point: the church widely promoted images in New Spain at the same time that it nurtured indigenous use of images, not only because images were a way to reach indigenes, but also because it gave the church a way of maintaining political and ideological control. By governing the traffic of images, church fathers could shape the indigenous conception of their world. Yet mendicants alone could not do this, and again they turned to the native elite as partners in the enterprise. The native elite, in turn, fashioned the images that both the indigenous community and outside actors of the colonial world saw; in these pictures they could carve out a new visual reality for all the inhabitants of New Spain. It is in this arena that many of the native painters of the Relación Geográfica corpus were working.

In promoting images, friars were perhaps not thinking of expedience alone, at least initially. For the regular friars were well educated, the European intellectuals of New Spain, and as such they probably stood closer to Philip II's view of the complementarity of image and text. But once in New Spain, friars soon realized the central role that the image—most particularly didactic painting—played in native culture. No matter how highly the friars may have held images when in Europe, when in New Spain, images—and their control—became essential, as friars sought to replace not the image with the text but the indigenous image with the Christian one (Reyes-Valerio 1989: 68–72). They destroyed native religious manuscripts so that Christian images would prevail. Friars used pictures to teach native peoples the rudiments of Christian doctrine, because they saw that images had the potential to be the bridge between their literate Spanish-speaking world and the world of their charges. The writer of the *Códice Franciscano* insists that "for the Indians, the best method [to teach] is with paintings" (1941: 60); another Franciscan, Jacobo de Testera, is credited with inventing peculiar catechisms constructed of images alone (Normann 1985: 11). Their battle for the minds and souls of New Spain's indigenes soon became a war fought with images.

For all the friars' grand claims about the efficacy of images and their efforts to control them, we must remember that it was largely the indigenous elite who were charged with painting the pictures that were to promote the church's evangelical program, just as in the Relación Geográfica project, the majority of painters were indigenes. The Franciscan Jerónimo de Mendieta said it bluntly: "of practically all the good and skillful works which are made in any of the arts and crafts in this land of the Indies (at least in New Spain), the Indians are those who execute and make them" (Mendieta 1971: 410). In depending upon them, the friars made members of the elite their partners in the evangelizing process. Thus the creation of images, particularly Christian images, was a collaborative process. We see this in the murals of the Augustinian monastery at Malinalco, Mexico (Peterson 1993). The friars seem to have chosen the overall millefleurs schema that covered the walls of the lower cloister, but the native painters who carried out the project incorporated elements of native iconography into the preexisting millefleurs scheme. The climbing vines now bloomed with *yolloxochimeh* (heart flowers) and *xiloxochimeh* (corn-silk flowers), twining with European roses and pomegranates. The painters included both native and European species because of the particular ideological valences they had within each of the two cultures. Thus when the Relación Geográfica questionnaire arrived in New Spain, elites were bound to see its maps in the same vein as they did other contemporary painting projects: a channel whereby they could broadcast their views of the order and the meaning of the surrounding world.

While some friars may have been uneasy about the degree of native collaboration that went on at Malinalco and other monasteries, they squelched their misgivings because they found, perhaps accidentally, that giving native elites ample opportunities to cultivate painting was a means of reaching another monastic goal: a separate *república de indios.* They cultivated the separation of this republic from the more worldly culture of sixteenth-century colonists, which, not incidentally, held the text in high regard. From the time the first twelve Franciscans disembarked in Veracruz in 1524, they harbored utopian ambitions for the indigenes of the New World, their hopes centering on maintaining a separate, more pure república. While the early conquistadores acted out a program of elite miscegenation by marrying indigenous noblewomen, and the crown itself pulled for a policy of hispanization, vocal and visible friars such as Sahagún and Mendieta directed the new colony towards a different path by continually seeking to promote and preserve the separateness of indigenous communities. In 1552, the Franciscans in New Spain described the colony, perhaps wishfully, as "these two republics, Spanish and Indian" (Motolinía 1971: 467). The friars maintained that they could create utopian Christian communities among the freshly converted people of New Spain, one condition being that indigenes be isolated from the deleterious effects resulting from contact with worldly Spanish colonists (Baudot 1983; Phelan 1970). Mendieta, for instance, cautioned against the vices to which the company of Spaniards exposed the indigenes and proposed barring Spaniards from native towns (1971: 502). While this strain of mendicant isolationism was on the wane by the time the Relaciones Geográficas were painted, this particular emphasis on the image would have a lasting effect in New Spain.

The mendicants' promotion of a distinct and isolate *república de indios,* and their wide tolerance of indigenous participation in the creation of an image-centered culture, proved worrisome to the viceregal government. To keep native painters in check, the viceroys insisted on keeping a watchful eye over them. Within San José de los Naturales was an examination center for native painters (Gómez Canedo 1983: 43; Ricard 1982: 208–9, 212–213), and in 1552 Viceroy don Luis de Velasco commanded that no native be allowed to paint religious images without first being examined at San José (Gómez Canedo 1982: 84). In fact, Martín Cano, the painter of the Relación Geográfica map of Ixtapalapa, is identified as an *"oficial de pintor,"* perhaps an indication that he had successfully completed his examination. Despite the viceregal attempts to control them, images were established as a way for elites to create and maintain an indigenous colonial identity. Thus, it is no wonder that so many native painters jumped aboard the Relación Geográfica project, lavishing care and attention on the maps they painted.

STYLE, ICONOGRAPHY, AND AUTHORSHIP

When the "indio pintor" San Agustín made his map of Culhuacan, he drew at its center the conventional native place-name of the town (fig. 29); this pictograph was a bell-shaped hill-symbol with a curved top representing *coloa,* the Nahuatl term for "something curved," which stood as a homonym for Culhuacan. Such elements of native iconography as these appear time and again throughout the corpus. What should we make of such inclusions of native iconography? Anything? Does native iconography, and its attendant native style, have meaning?

For the native artists, such inclusions must have had meaning. San Agustín, who clearly had been exposed to European models, probably through prints, would have easily known, like other indigenous painters, that such native iconography did not appear in European art; it was wholly indigenous, in this case referring to a name in Nahuatl, an indigenous language and San Agustín's native tongue. And if a painter like San Agustín ever forgot the parameters of indigenous representation, he (or she) could always turn to the "ancient pictures" (presumably pre-Hispanic or early colonial indigenous works) that we find as frequently cited sources in the Relaciones Geográficas texts. For example, the Relaciones Geográficas texts from Coatepec Chalco and Chicoalapa (Chicoloapan de Juarez, Mexico) refer time and again to the "ancient pictures" wherein their histories were kept.

This connection between style and ethnicity is bolstered by negative evidence. We have almost no evidence of a Spaniard or European in the early sixteenth century ever working in the native tradition or composing native-style manuscripts complete with native symbols unless he was copying a preexisting native manuscript.[12] In fact, I have seen only a few clear instances of a colonial European artist borrowing any iconographic element whatsoever from the native traditions.[13] The traffic between European mapping and native mapping was on a one-way street. In addition, all the Relaciones Geográficas texts which specify (or name) an indigenous author for their maps contain either native iconography or native stylistic elements or both.

The idea that style carries meaning harkens back to Donald Robertson's original categorization of the Relaciones Geográficas maps according to a tripartite schema of "native," "mixed," or "European" (Robertson 1972a). If the map showed an absence of stylistic features particular to native manuscripts,[14] Robertson classified it as "European." If it showed the presence of both native and European stylistic features, he classified it as "mixed." If it showed only native features and an absence of European style, he labeled it "native." Along with his schema, Robertson issued a disclaimer: he wrote that these labels "make no reference to the artist *per se,* to his racial antecedents, or to the cultural ambient in which he lived" (1972a: 255).[15] But in the image-saturated world of indigenous New Spain, style and iconography *did* have meaning, meaning that arose out of the difference between indigenous and European systems. Many painters at work on the corpus had to be aware, by virtue of their training, of those differences. The stylistic and more primarily iconographic elements that they employed (and that Robertson singled out) call out to be interpreted as nothing less than a self-conscious brand of native authorship.

In arguing that indigenous style was as meaningful, if not more, at the time it was painted as it is today, I mean to suggest that the artists used styles consciously, not reflexively or automatically, but I do not mean that they saw themselves as part of our general category of "indigenous." It is more likely that they identified themselves most strongly as members of local communities. The república de indios, promoted by the friars, may not have made great inroads into elite consciousness, but even so, the indigenous elite would have been aware of themselves, and of their community, as a separate and distinct sphere from the Spanish colonial culture of New Spain. They were differentiated, after all, by their language, their customs, and their dress. In having a sense of identity that was local rather than national, indigenous elites would have set their particular community in opposition to other communities as well as to the norms of the Hispanic state, be they manifest by the friars or the habits of the local corregidor. Their sense of identity plays out in the maps in that the artists were likely to use indigenous style, which they probably saw as local style, to show communal symbols or institutions. It is no coincidence that in both the Ixtapalapa and Culhuacan maps (figs. 28 and 29), Cano and San Agustín show the "casas de la comunidad," those community buildings where the native cabildo would meet, with their overextended lintels and decorated cornices, in a fashion similar to that of pre-Hispanic manuscripts. These community houses were local, indigenous spaces, and their architecture and their representation followed local convention. On the town plaza, they would have stood in counterpoint to the church, whose architecture was generally adapted from European models. To depict their community houses, these two artists depended upon local indigenous convention, in marked contrast to the more European forms that they used for churches; here, local style is matched to local institutions.

In employing indigenous style, each of these artists was also marking himself as a local. This is one reason, I believe, we find that place-names, supplying the name of the locale, are so frequently drawn according to indigenous convention. I think that we are called to read the artist's proclamation of localness as a very self-conscious act. The meaning of this, of course, has changed over time. When he painted, Cano was proclaiming specifically, I am of Ixtapalapa, and San Agustín, I am of Culhuacan; today we see them saying more generally, I am indigenous.

While native elements in the Relaciones Geográficas maps must have signaled local authorship to their contemporary audience and convey native authorship to us today, the schema developed by Robertson out of this neglects the ability and consciousness of the artists at work. "Mixed" appears as something derivative, as if the authors were aiming at a purely "native" style but their work was polluted by introductions from Europe; or as if they aspired to reproduce European maps, yet were unable to suppress all native elements.[16] But we find that in making their maps, the artists often used different styles, indigenous (or local) or European, selectively. That is, they were fully aware that local style and iconography was different than European, and they used this indigenous style to mark the names and institutions that defined their community, as well as to show themselves—through their knowledge and manipulation of local style—as members of this distinct community.

Thus, as these elite painters indulged in a play of style and iconography, they reveal a surprising self-consciousness about their own position and its representation in a colonial state. Although they might not have called it such, these artists lived in a world that embraced corregidores as well as principales, one where "ancient pictures" coexisted with prints of the Christian saints. The church had nurtured their talents, and provided a climate wherein images thrived. As did other members of elite groups who produced painted works, these artists, in their maps, were attempting to synthesize a new vision of the world around them that was consonant with this new reality. They manipulated styles to proclaim their local affiliations as well as to show the pull that colonial institutions, especially the church, had upon them. Seen in this light, the maps show us not only the spatial reality of these Amerindians—the plazas and monasteries that the king requested—but also an ongoing discovery of identity among their artists and the communities from which they came.

CHAPTER FIVE

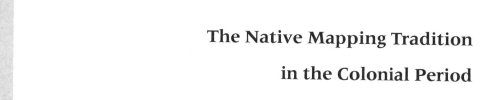

The Native Mapping Tradition in the Colonial Period

When Hispanic corregidores turned first to indigenous cabildos and then to painters in response to item 10, the request for a map was relayed as if over frayed wires, with signals delayed and scrambled, as often occurs in cross-cultural communication. The Spanish questionnaire clearly demanded a chorographic description of a town, showing the "the layout of the town" along with notable buildings. But for the native cabildos, and their attendant native painters, the idea of "town" as an architectural entity that defined a particular human community had little meaning. This was because in much of New Spain settlements were not nuclear towns, but were looser and more scattered, and because native respondents thought of themselves, to use the closest Spanish word, as "pueblos," as both land and its people, or as communities (Lockhart 1992: 15). Thus the Spanish commission to paint a map of the town was quickly (and somewhat mistakenly) translated by indigenous painters into a bid to paint a map of the community.

Thus understood, native painters could take up the commission with a confidence born of knowledge: the creation of maps of community was a Central Mexican tradition reaching back to pre-Hispanic times. Around the time of the Spanish conquest, community maps dominated the horizon of indigenous cartography.[1] Given the widespread and long-lived tradition of indigenous community maps, it was these maps (and their underlying conception of community) that guided many of the painters in constructing their Relaciones Geográficas maps (Glass 1975b; Glass and Robertson 1975). While none of the artists sought to recreate pre-Hispanic community maps, since these were "ancient pictures," of a different era from the present one, many artists were certainly, if quietly, influenced by indigenous community maps.

This chapter examines the two kinds of indigenous community maps that most directly influenced the native artists of the Relaciones Geográficas. To do so it asks, What did it mean to map a community in indigenous New Spain? How in some cases were the modes of mapping, or representing space, in indigenous community maps continued in the colonial corpus of maps at hand?

THE SURVIVAL OF THE PRE-HISPANIC TRADITION

Soon after the conquest, Spanish conquistadores and Catholic priests engaged in the wholesale destruction of indigenous art and manuscripts, with Archbishop Juan de Zumárraga setting entire libraries aflame. However, it also is clear that both the memory of earlier history, including the pre-Hispanic past and the documents associated with it, survived in many of the thousands of indigenous communities across New Spain (Gruzinski 1993). In the Relación Geográfica corpus, many of the reports coming from the cabildos either mention their cache of "ancient pictures" or recount histories in much the way that they were graphically encoded in the pre-Hispanic period, thus indirectly revealing the presence of an indigenous pictorial source. For instance, Gabriel de Chávez in his Relación describes "six painted figures" of gods as if he had seen them in a manuscript dealing with pre-Hispanic deities (RGS 7: 61–2). The Relación of Cempoala details the tributes the region paid to the Aztec emperors, the kind of information that was commonly encoded in pictorial form (Boone 1992a; Pasztory 1983: 179–208; RGS 6: 75–6; Robertson 1959). While few of the indigenous cabildos enumerated the contents of their caches of "ancient pictures," they probably included community maps. Unlike, say, religious almanacs, community maps were recognized as secular documents by Europeans and never specifically earmarked for destruction. In addition, we know that many native communities held sixteenth-century community maps through the eighteenth century and hold them even to the present (Oettinger 1983). Artists commissioned to paint the Relaciones Geográficas may have turned to the community maps in the hands of the cabildo and within these found specific inspiration both for the structure and content of their maps and specific items of pre-Hispanic iconography to draw upon.

While the native cabildos' own claims that they possessed old papers are the best evidence that indigenous manuscripts were still known and used within native communi-

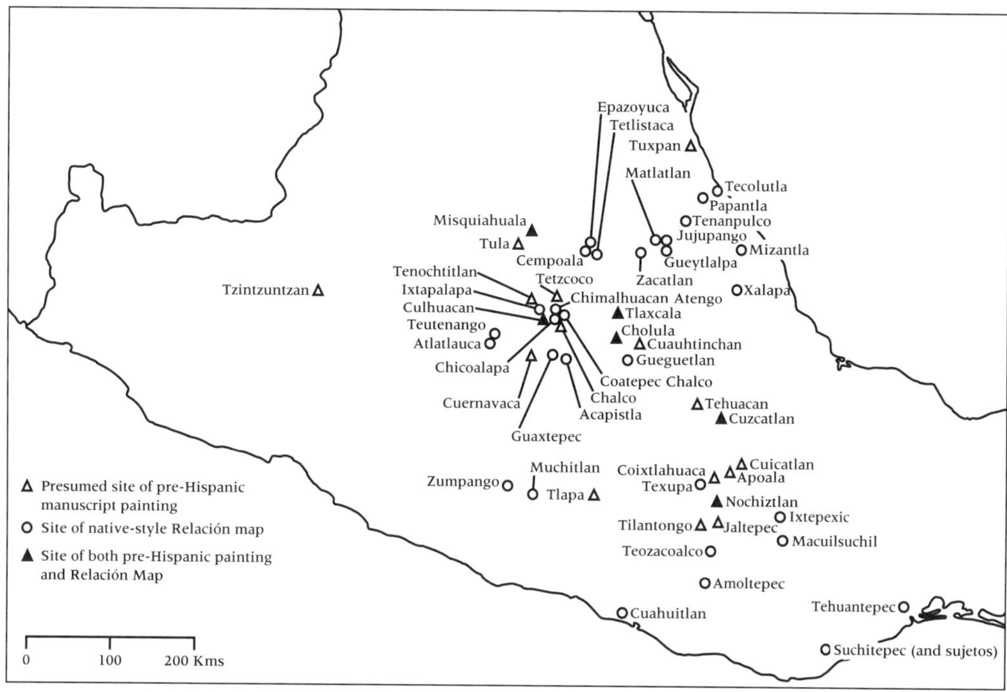

Figure 41. Map of New Spain. This map shows the general overlap, both in regions and in specific towns, between towns that produced native-style Relaciones Geográficas maps and pre-Hispanic centers of painting.

ties, the geographic distribution of the corpus is another, perhaps more indirect meter. When we look at a map of central Mexico that shows both where pre-Hispanic indigenous groups had a tradition of manuscript painting[2] and where indigenous maps were made to accompany the corpus, we find that the two largely overlap (fig. 41). The seeds of mapping had long been rooted in these soils and continued to bear fruit under Spanish domination.

In some areas, a fecund pre-Hispanic tradition led to a large harvest of Relaciones Geográficas maps. In the Valley of Mexico, where the pre-Hispanic Aztecs once had libraries filled with painted manuscripts, five different native painters each made a map for the Relación Geográfica corpus, this region establishing the high-water mark of the rate of mapped responses in the corpus. Of course, there were more administrative districts in the heavily populated Valley than anywhere else, and thus more corregidores and alcaldes mayores to call upon to answer the questionnaire. Because they dwelt so close to the viceregal capital, officials were more likely to receive the printed sheet than their counterparts in the hinterland and were likely to see their responses arrive safely in the capital, and from there to Spain. Nonetheless, that such a high proportion of the Valley Relaciones Geográficas were accompanied by maps and that all but one of these were painted by natives indicate that there was a large pool of trained native painters in the Valley to get the job done.

The artistic quality of some maps is certainly indebted to pre-Hispanic inheritance. Within the Valley itself, particularly noted for their manuscripts in colonial chronicles were the aristocratic Acolhua, whose capital of Tetzcoco, home to a line of poet-kings, lay on the east bank of Lake Tetzcoco, and was first rival, then ally to Tenochtitlan. Acolhua manuscript painters cultivated a certain "delicate" line, identified as a diagnostic of the "Te[tz]coco school" by Donald Robertson, that is clearly present in works from the Tetzcoco region such as the exquisite Codex Xolotl, Mapa Quinantzin, Mapa Tlotzin, and the Codex en Cruz (Robertson 1959a: 134–54).[3] The same delicate line turns up in the Relaciones Geográficas maps of Cempoala and Tetlistaca (San Tomás, Hidalgo), which are perhaps the most finely drawn works in the corpus (figs. 42 [pl. 5] and 43). With these maps, from provincial regions once under Tetzcoco's control, we can trace the Tetzcoco School pedigree late into the sixteenth century. The Cempoala Relación declares its region to have been once "of the lord of Tetzcoco, Nezahualcoyotzin" (RGS 6: 75) while neighboring Tetlistaca also declared its allegiance to Tetzcoco (RGS 6: 92).

But the colonial presence wore at the ties between indigenous communities and their pre-Hispanic pasts. In a telling episode, the maps originating in Tetzcoco itself—once the artistic epicenter of the Valley—were sketches made by the barely trained and itinerant colonial scribe Francisco de Miranda. Here the ties had frayed and broken, while in outlying Cempoala and Tetlistaca, exquisite indigenous-style maps were being made. In other words, in central Tetzcoco, indigenous painting was displaced, only to resurface—fresh and clear—on its margins.[4]

The displacement of native painting from the great preconquest centers towards the peripheries of their control was not true of all towns with historic painting traditions, nor were the ties to the pre-Hispanic tradition under equal stress throughout New Spain. Both Texupa and Cholula had deeply rooted preconquest traditions of native manuscript painting and continued to spawn manuscripts in the postconquest years, including two fine Relaciones Geográficas maps (figs. 34 [pl. 3] and 38 [pl. 4]). Cholula may have been the

Figure 42 *(opposite)*. The Relación Geográfica map of Cempoala, 1580. (See also pl. 5.) At the upper center of this extraordinary map is the indigenous place-name of Cempoala. To the left is the local Franciscan monastery. The landscape is thickly sown with indigenous toponyms and images of native leaders. Size of the original: 81 x 66.5 cm. Photograph courtesy of the Benson Latin American Collection, The General Libraries, The University of Texas at Austin (JGI xxv-10).

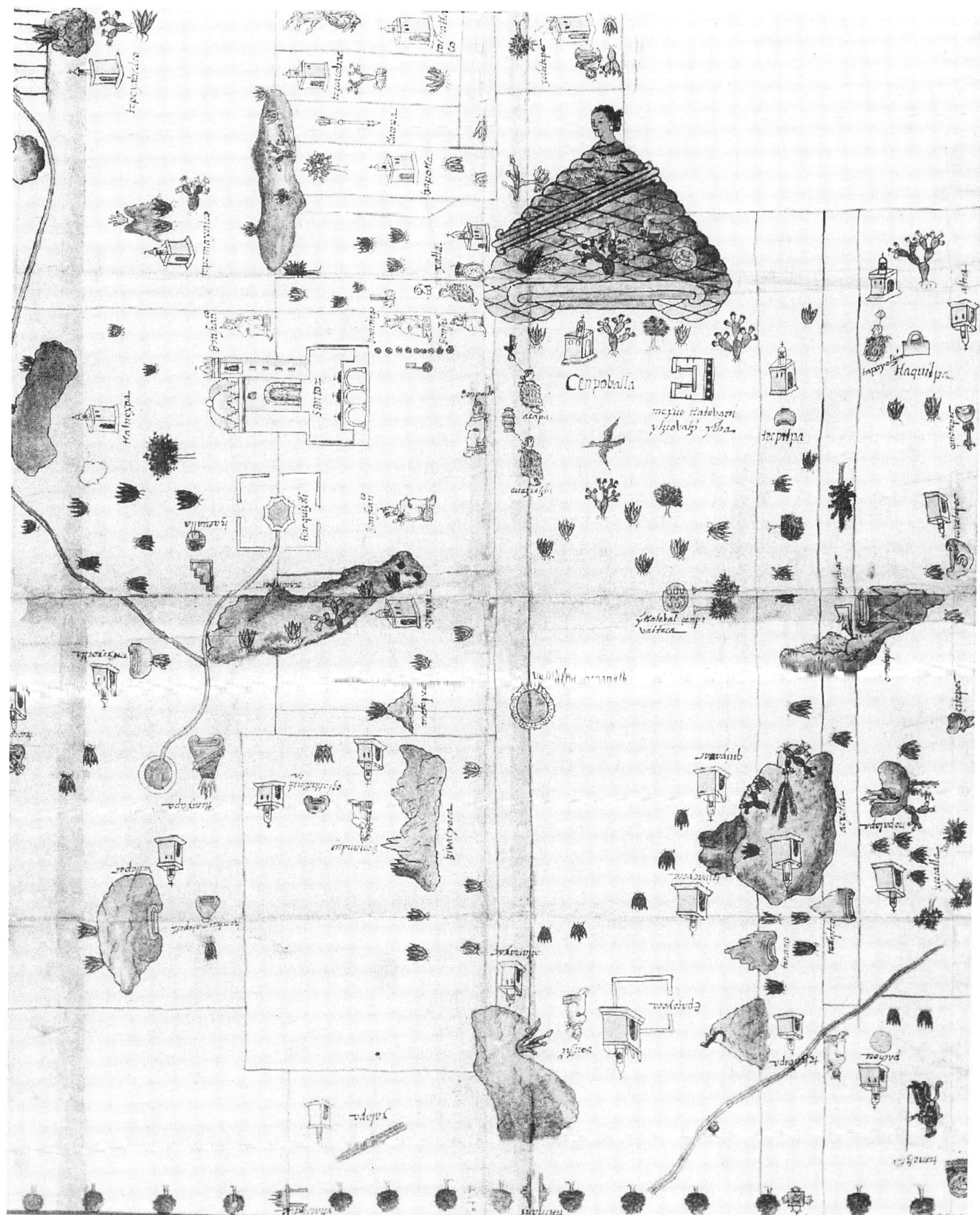

Cenpoballa
acapa
tecpilpa
tlaquilpa

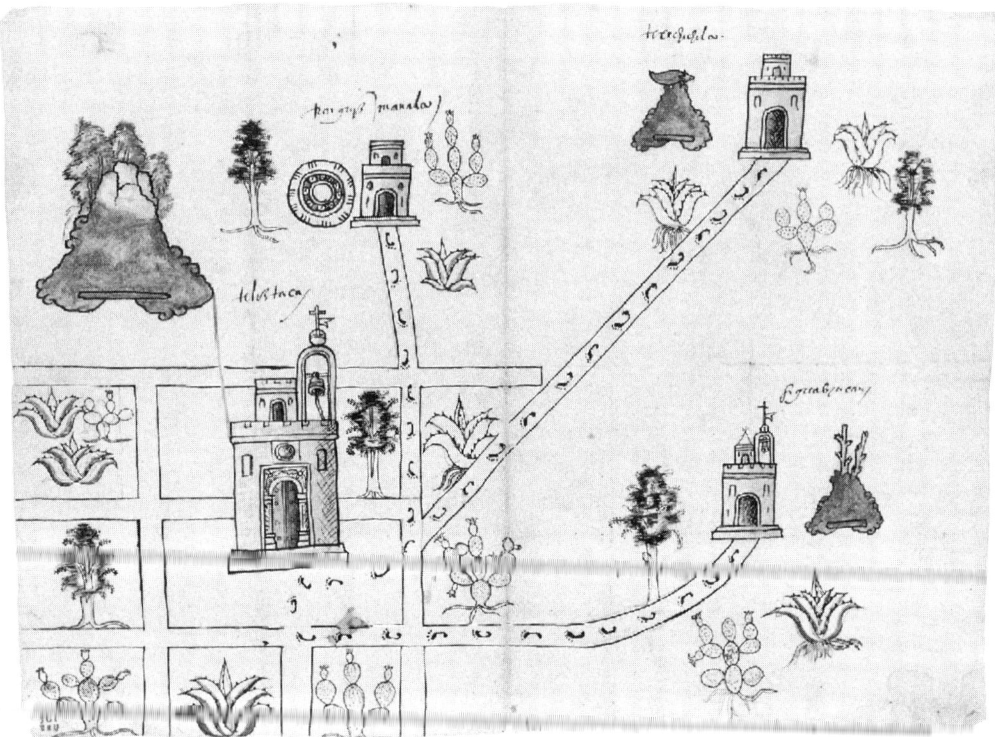

Figure 43. The Relación Geográfica map of Tetlistaca, 1581. This elegant fine-line map comes from a region once controlled by Tetzcoco, a city reknown for its fine manuscript painting. Each of the churches, symbolizing settlements, is named both alphabetically and with pictorial toponyms. The artist included roots in the depictions of plants, following indigenous practice. Size of the original: 31 x 43.5 cm. Photograph: B. Mundy; reproduced courtesy of the Benson Latin American Collection, The General Libraries, The University of Texas at Austin (JGI xxv-12).

birthplace of the Codex Borgia, a ritual manuscript that ranks as one of the masterpieces of pre-Hispanic painting;[5] after the conquest in the Cholula region, in nearby Cuauhtinchan, the native-style maps of Cuauhtinchan were painted (Simons 1968), as well as the logographic- alphabetic history, the Historia Tolteca-Chichimeca (Kirchhoff et al. 1976). Well to the south of Cholula, in mountainous Oaxaca, Texupa figured prominently in preconquest-style manuscripts from the Mixtec region (Smith 1973a) and was the birthplace of the Codex Sierra of 1550–1564 (Bailey 1972). So for a host of factors that varied from region to region—the strength of a pre-Hispanic painting tradition, the degree of a colonial presence, the survival of the indigenous elite during a period of demographic collapse—connections to and the presence of pre-Hispanic traditions varied across New Spain, covering that land like a cloak whose fabric was thin and tattered in some spots but strong and true in others.

THE INHERITANCE OF THE PRE-HISPANIC MAP

For indigenous colonial painters working about 1580, what were the tangible aftereffects of a pre-Hispanic tradition? As I see it, such tradition meant, first, that communities had adequate numbers of native painters to train others and to keep their art alive, often under the auspices of the church, as discussed in chapter 4. Second, it meant that artists in these communities were able to see painted manuscripts, perhaps ones kept in the storerooms of the cabildo, and from these could learn the conventions and symbol systems of indigenous manuscripts. Third, it meant that communities were still illuminated by an image culture, so that painting remained a central means of expression. This experience of being trained within a community, of being exposed to pre-existing manuscripts and of playing an important cultural role, linked colonial artists to their pre-Hispanic predecessors. Thus when asked to paint maps—to picture the community to itself—colonial artists had a tradition to invoke. As San Agustín painted his map of Culhuacan, his idea of the image of Culhuacan was shaped by past images as well as by his community's sense of itself and his sense of it (refer back to fig. 29). Thus many of the native works of the Relación Geográfica corpus have an almost accretive quality about them, shaped as a community's sense of self, always changing and growing, was given visual expression by one generation of artists after another. Because it was a work in an ongoing tradition, the Relación Geográfica map that was sent back to Spain was like the papery skin of an onion, showing all the contours and striations of the living bulb from which it had been separated.

Many Relaciones Geográficas maps, the Culhuacan map among them, bear a visible connection to those stockpiled documents, kept within the native cabildo, by incorporating elements of native iconography that originated in the pre-Hispanic era. In addition to being a self-conscious reference to local authorship, artists' use of pre-Hispanic imagery in the Relaciones Geográficas maps, as in other native colonial maps, was also convenient, because the pre-Hispanic repertory offered rich stores of symbols of topographical elements with which to describe space: hills, rivers, caves, plains, and gullies, to name but a few. On the map of Minas de Zumpango (Zumpango del Rió, Guerrero), for instance, the artist used indigenous convention to represent the spring below the town; it radiates with symbols standing for drops of jade and tiny shells, conventions drawn from pre-Hispanic iconography (fig. 44). In the map from Culhuacan, San Agustín marks his roads with footprints (fig.

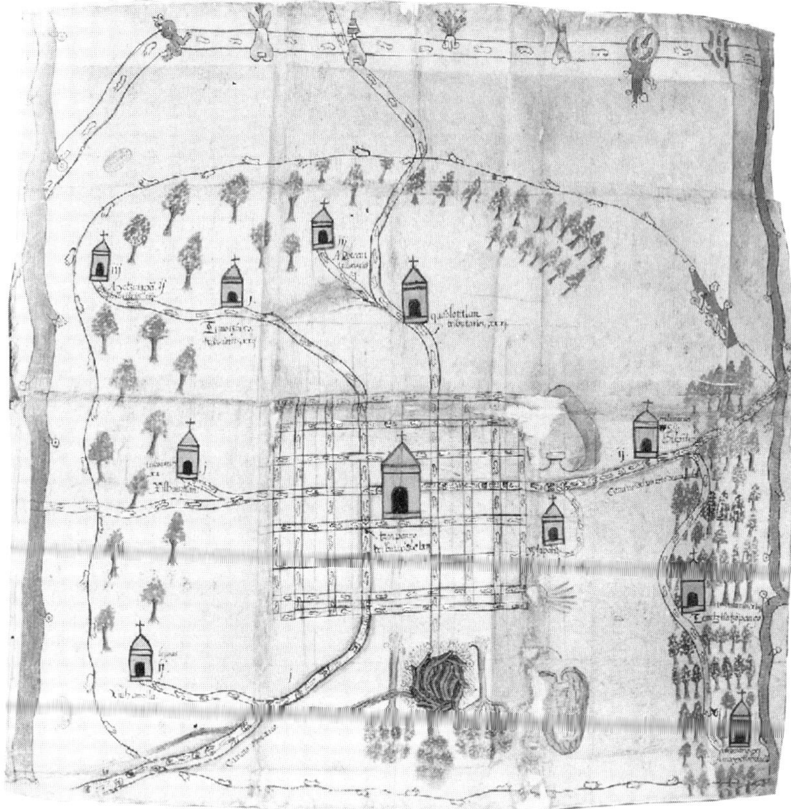

Figure 44. The Relación Geográfica map of Minas de Zumpango, 1582. Native iconography, such as the bell-shaped hill symbols at bottom and to the right of the town, as well as the line of pictographic toponyms at top, distinguish this map as the work of a native artist. Size of the original: 71 x 72.5 cm. Photograph: Juan Jiménez Salmerón; reproduced courtesy of the Real Academia de la Historia, Madrid (9-25-4/4663-xxxvi).

29). The Texupa artist must be credited with most inventive use of native iconography, for he (or perhaps she) combines indigenous hill symbols into a visually mimetic landscape (fig. 38; pl. 4). In a few cases, the artists portrayed architecture following indigenous convention. Some elements of pre-Hispanic iconography are listed in table 4.

The indigenous iconography on these maps was unremarkable and turns up in scores of colonial maps made by native painters. To the native artist as well as to the colonial viewer, such referents to landscape seemed untainted by the idolatrous associations ascribed to religious manuscripts that made this genre the target of destruction in Catholic New Spain, among both converted indigenes and zealous colonists. For in destroying religious manuscripts European colonists had, unwittingly perhaps, severed the visual associations between their iconography and the iconography of maps. By the time native

Table 4 Indigenous Symbols of Topographical Features

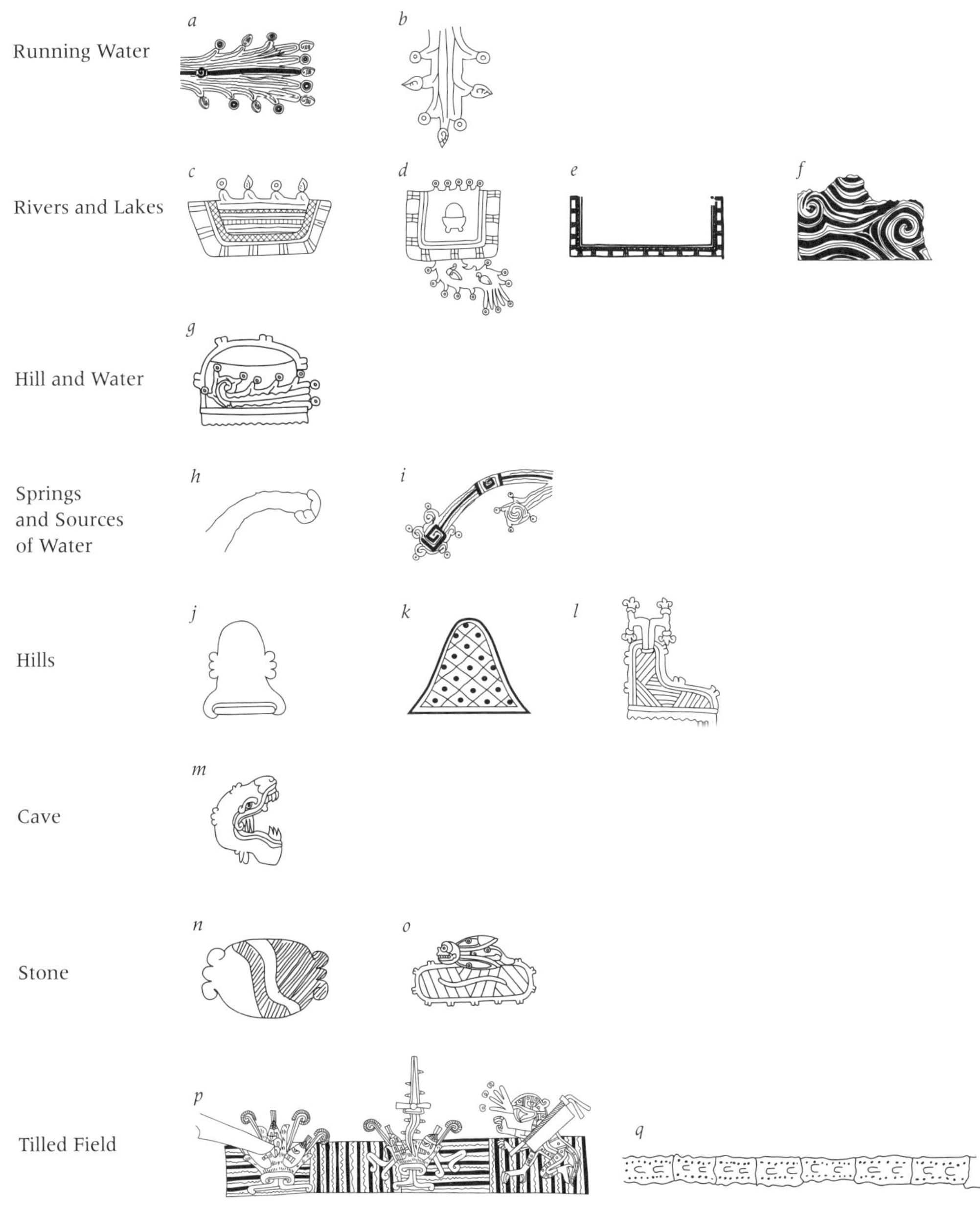

Sources: (a) after Codex Borbonicus, p. 5; (b) after Codex Mendoza, fol. 28r; (c) after Codex Mendoza, fol. 16v; (d) after Codex Zouche-Nuttall, p. 51; (e) after Codex Zouche-Nuttall, p. 75; (f) after Relación Geográfica map of Ixtapalapa; (g:) after Codex Selden, 5-III; (h) after Codex Xolotl, p. 6; (i) after Relación Geográfica map of Guaxtepec; (j) after Codex Mendoza, fol. 31r; (k) after Codex Xolotl, p. 6; (l) after Codex Selden, 10-II; (m) after Codex Mendoza, fol. 18r; (n) after Codex Mendoza, fol. 18r; (o) after Codex Vienna, p. 49; (p) after Codex Borgia, p. 20; (q) after the Humboldt Fragment II. Drawing reprinted from David Woodward and G. Malcolm Lewis, eds., *The History of Cartography,* vol. 2, book 3 (Chicago: University of Chicago Press, forthcoming). Prepared by the University of Wisconsin Cartographic Laboratory, Madison.

Table 5 Uses of Pre-Hispanic Iconography in the Forty-five Relaciones Geográficas Maps

Name of Relación	Prehispanic Iconography Used					
	In Main Toponym	In Subsidiary Toponym	To Depict People	To Name Boundaries	To Show Topography	To Depict Architecture
Acapistla	•	•			•	
Amoltepec	•	•	•	•	•	•
Atlatlauca and Malinaltepec					•	
Atlatlauca and Suchiaca					•	
Cempoala	•	•	•	•		•
Chicoalapa	•				•	
Chimalhuacan Atengo	•				•	•
Cholula	•				•	
Coatepec Chalco					•	
Cuahuitlan					•	
Culhuacan	•				•	•
Cuzcatlan A		•		•	•	•
Cuzcatlan B		•		•	•	•
Epazoyuca					•	
Guaxtepec	•	•			•	
Gueguetlan					•	
Gueytlalpa	•	•				
Ixtapalapa	•				•	•
Ixtepexic		•				
Jujupango	•	•				
Macuilsuchil	•				•	
Macupilco, San Miguel			•		•	
Matlatlan and Chila	•					•
Misquiahuala	•	•	•	•	•	
Mizantla					•	•
Muchitlan	•	•	•		•	
Nochiztlan					•	•
Papantla	•					
Suchitepec	•	•	•		•	
Tamagazcatepec, S. Bartolomé			•			
Tecolutla					•	
Tehuantepec A	•					
Tenanpulco and Matlactonatico	•					
Teozacoalco	•	•	•	•	•	
Tetlistaca	•	•			•	
Teutenango	•					•
Texupa	•	•			•	•
Tlacotepec	•		•			
Tlaxcala A			•			•
Tlaxcala B						•
Tlaxcala C						
Xalapa de la Vera Cruz						
Zacatlan	•				•	
Zozopastepec			•		•	
Zumpango, Minas de		•		•	•	

painters made their Relaciones Geográficas maps, most had either forgotten or ignored the wide meaning of their symbolic repertory. For instance, the jade and shell symbols that mark the water on the Zumpango map were also attributes of Chalchiuhtlicue, a central Mexican water goddess, and the hill symbol was once associated with the crocodilian earth-monster, but there is little to indicate that viewers still made these associations at the time the Relaciones Geográficas were created.

The shaking off of religious and cosmic associations was just one way in which indigenous iconography was evolving in these colonial maps. The use of native iconography to visually mimic the features of the landscape, seen so clearly in the Texupa map, was a departure from pre-Hispanic practice; in most of the Relaciones Geográficas maps, in fact, elements of native iconography are used as they were on maps of the pre-Hispanic period: to name things (table 5). That is, symbols were used to describe elements of the landscape, not by depicting them, but by naming them. In the way of naming, they followed pre-Hispanic practice which was to create rebus-like logographs, often called "picture writing." For example, the town of Xilotepec, or "hill of green ears of maize" in Nahuatl,[6] is represented by ears of maize, *xilotl* or *xilo* (the absolutive ending *-tl* is dropped) on top of the bell-shaped hill symbol, *tepetl* or *tepe* (fig. 45).[7] A locative suffix, -c or -co, meaning "place of," is also added, but, as is common, is not represented. In the map of Texupa, for example, the tallest hill symbol, sitting to the left of the town, looks like a visually mimetic picture of a hill but it is also, or perhaps exclusively, the logographic place-name of a hill (fig. 38; pl. 1). The Relación gives the hill's name in Nahuatl as Miahualtepec, "hill of the maize tassel," and this name is represented on the map by the double curlicue—a maize tassel, or "miyahuatl"—at the top of the conical hill symbol, or "tepetl."[8]

Figure 45. The indigenous place-name of Xilotepec (redrawn from Codex Mendoza, fol. 31r).

The way pre-Hispanic manuscripts used the name rather than the image to define and depict the landscape can perhaps be better seen in a page from the Codex Zouche-Nuttall (figs. 46 and 47; Codex Nuttall 1977; Byland and Pohl 1994; Smith 1973a). This manuscript comes from a Mixtec-speaking region in the modern state of Oaxaca and most scholars agree it was painted before the Spanish conquest. It is a mainly historical work, tracking various ruling families in the region around the town of Tilantongo, and most of its pages are lined with figures, places and dates arranged in three horizontal registers that were read in sequence.[9] But page 36 is arranged markedly differently from most of the rest of the manuscript. It is not broken into registers; instead, the whole page

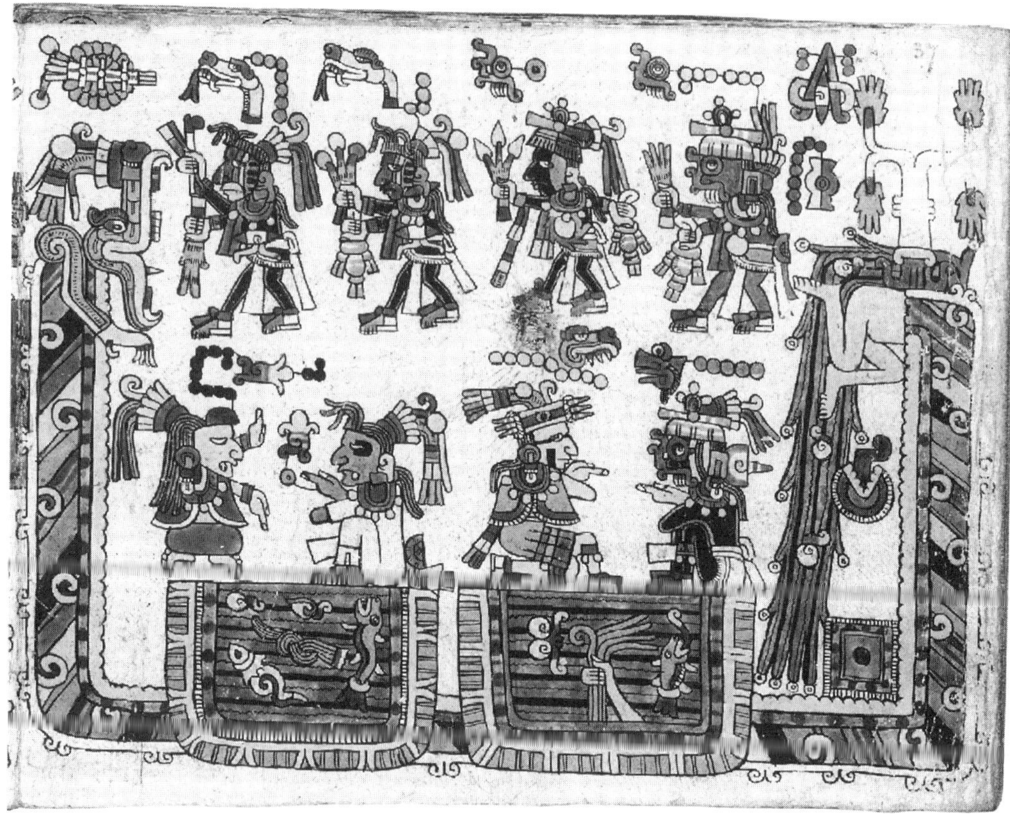

Figure 46. Codex Zouche-Nuttall, page 36. This page, part of a screenfold history, is one of the earliest known pre-Hispanic maps. Largely through pictographic toponyms, it represents the Apoala valley, in the modern state of Oaxaca. Size of the original: approx. 19 x 25.5 cm. Photograph courtesy of the British Museum, London (Museum of Mankind, Add. Ms. 39671, fol. 37).

is taken up with a schematic map of the Apoala Valley, where most of the ruling lineages of the Mixtec people originated (Jansen 1979). Figures 47 and 48 compare a topographical sketch of the Apoala Valley to a simplified drawing of page 36 of the Codex Zouche-Nuttall to show the correlations scholars have made between many of the logographic toponyms and geographic features in the Apoala Valley. The open-mouthed serpent at the left edge of page 36 of the Codex Zouche-Nuttall is a logographic symbol standing for the Mixtec toponym *yahui coo maa,* "deep cave of the serpent," the name of a spring on the northeast edge of the Valley. The central symbols within each of the two

U-shaped rivers pictured on the Valley floor are also toponyms. On the left, a hank of knotted grass within the river represents the Mixtec name *yuta ndua nama,* "river of the barranca of the soap plant." In the river to the right, a hand grabbing a bunch of feathers is used to show *yuta tnuhu,* the name of the river. This name means "river of the lineages" but is represented on page 36 of the Codex Zouche-Nuttall by a near homonym, *tnoho,* meaning "to pluck," as birds, and is shown by the hand with the bunch of feathers that appear to be plucked from a bird (Smith 1973a: 75). These two rivers do, in fact, run across the floor of the Valley. The figure of the lower half of a human may represent the

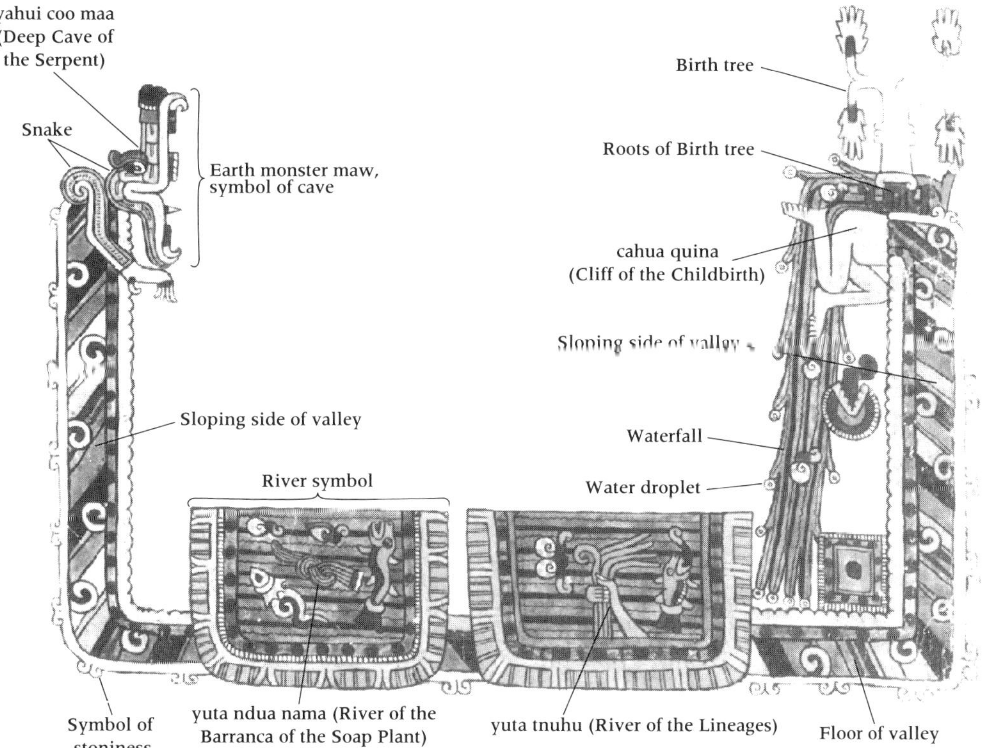

Figure 47. Drawing of Codex Zouche-Nuttall, page 36, with annotations. Drawing reprinted from David Woodward and G. Malcolm Lewis, eds., *The History of Cartography,* vol. 2, bk. 3 (Chicago: University of Chicago Press, 1998). Prepared by the University of Wisconsin Cartographic Laboratory, Madison.

drop-off between the upper and lower plains of the Apoala Valley, or it may be a logographic rendering of the name "cliff of the childbirth," as this precipice is known today (Anders and Jansen 1988: 173; Jansen 1979). A comparison to a schematic map of the Apoala Valley confirms that the rendering on page 36 of the Codex Zouche-Nuttall maps the Valley. The frame of the map, the large U-shaped enclosure is, in the most general sense, visually mimetic as it suggests the enclosing form—floor and sides—of a valley, but the topographical elements that specifically define the Apoala Valley—the rivers, the cliff, the cave—are represented in the Codex Zouche-Nuttall by their names.

Thus, for indigenous native artists who came to work on the Relación corpus around 1580, traditions supplied them with a flexible vocabulary of symbols and logographs to represent their surroundings. Many, if not most, artists would have had first-hand knowledge of symbols and logographs from pre-Hispanic and early colonial manuscripts kept by their communities. They may not have seen a sharp divide between pre-Hispanic and early colonial manuscripts; however, they may have seen more European imagery appearing on more recent community documents and paintings and recognized it as being different from local imagery. In addition, their understanding of the meaning of indigenous imagery, particularly its cosmic or religious associations, would have differed from their counterparts of three or four generations earlier. Nonetheless, an essential part of the pre-Hispanic tradition would have reached them unscathed: the use of names, rebus-like logographs, or "picture writing" to define and represent their surrounding environment.

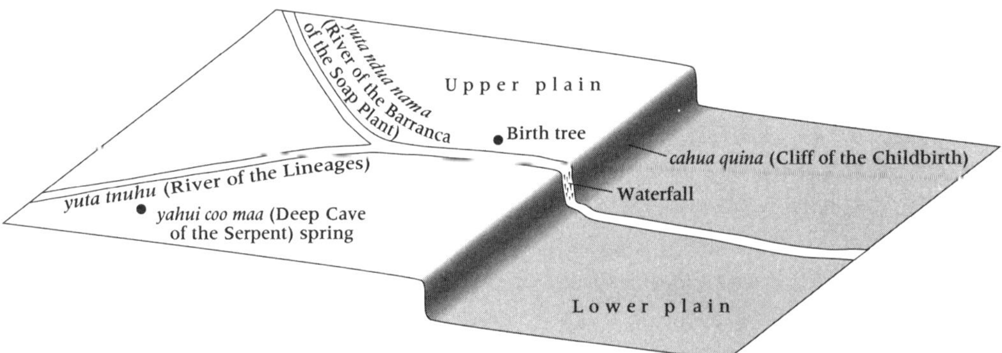

Figure 48. Schematic map of the Apoala Valley. The toponymns here are those found in the Codex Zouche-Nuttall, page 36. Map reprinted from David Woodward and G. Malcolm Lewis, eds., *The History of Cartography*, vol. 2, bk. 3 (Chicago: University of Chicago Press, 1998). Prepared by the University of Wisconsin Cartographic Laboratory, Madison.

COMMUNITY IN CENTRAL MEXICO

For the Western viewer of Amerindian maps, the logographic place-names, or "picture writing," used to define the landscape leap out as evidence of the maps' indigenous roots. Equally distinctive among these maps is their focus on communities as their subject, rather than cities or topography. Indigenous maps present us with the ways in which communities envisioned and pictured themselves. But before we look the modes of self-presentation, it is perhaps useful to understand a little better what, in Central Mexico, "community" entailed, to know the general traits of what the indigenous artist was being called upon to represent.

In recent years, historians and anthropologists have shed new light on the social and political layout of communities in pre-Hispanic and early colonial indigenous Mexico. Around the Valley of Mexico, where Nahuatl speakers clustered, the main communal unit was the *altepetl,* called by one historian "an ethnic state. . .an organization of people holding sway over a given territory" (Lockhart 1992: 14). Politically, altepetl were somewhat like Russian nesting dolls, holding within them smaller and smaller subunits; most comprised numerous calpolli, each with its own leader, which in turn comprised family-centered households (Lockhart 1992: 16–20). Physically, the shape of the altepetl is less clear. Each calpolli laid claim to shares of the larger altepetl's territory, but the altepetl's territory may not always have been contiguous. Altepetl were like city-states and could range in size from a few thousand yards with a few hundred people to hundreds of square miles with a population of thousands. No matter what the size, or that members of most altepetl ate, talked, and lived similarly to their neighbors who might live but paces away, all were fiercely individualistic, as "each altepetl imagined itself a radically separate people" (Lockhart 1992: 15). Thus the Culhua-Mexica, the altepetl that led the Aztec empire, distinguished itself sharply from its neighbors and allies; like all other altepetl, the Culhua-Mexica had a different origin myth and a different ruling family. At times, though, altepetl organized into larger groupings of complex altepetl (in much the same that calpolli organized themselves into altepetl).

Altepetl was the term Nahuatl speakers used for this basic communal unit; in the Mixtec region, the units were much the same and were called *ñuu,* "town, place where something exists" (M. E. Smith: personal communication, 1995). After the conquest, the Spanish called such communities *cacicazgos* after the *caciques,* the indigenous lords who ruled them (Spores 1984: 74–80). In much of the Mixtec region, geography added a further defining force: the Mixteca is furrowed by mountains, and each community kingdom tended to cluster within one valley, with the mountains protecting it and defining it from its neighbors. Like their Nahua counterparts, Mixtec kingdoms varied in size and density. Texupa, for example, was populous, having nearly 20,000 people in the postclassic era (1200–1500), but covered a small area; other kingdoms, like Teozacoalco, were certainly smaller in population but larger in area (Spores 1984: 96). Zapotecs, too, seem to have created ethnic affiliations within communities, but less is known about their pre-Hispanic arrangements in the postclassic era (Spores 1967; Whitecotton 1977).

Whatever their size or name, communities across Mesoamerica established their singularity and encouraged the solidarity of their members by invoking a common history. For instance, the Culhua-Mexica of Tenochtitlan claimed descent from a handful of peripatetic founders (Boone 1991). Communities often had their own deities, who were worshipped in a central temple. In some regions each community occupied a clearly defined block of territory that was separated from its neighbors by marked boundaries. This powerful sense of communal identity was ever present in all aspects of life as community members worshipped their local deities, bartered and traded in regional markets, or offered up tribute for altepetl leaders. But it is only in maps that we specifically see how communities across New Spain expressed their sense of self in relation to the space they occupied.

COMMUNITY MAPS

Given that altepetl or community kingdoms were the basic nodes of social affiliation in pre-Hispanic Mexico, it comes as little surprise that these communities were also the dominant subjects of maps. But we must recognize that there is a basic difference between the way in which we can reconstruct the social, political, and economic life of the altepetl or indigenous community and the way in which the community envisioned and represented itself. My aim here is not a reconstruction of how these communities were structured or operated from day to day, but rather an exegesis of how they represented themselves in maps: how did they envision community and territory? Such conceptions play a crucial role in shaping the native maps of the Relaciones Geográficas.

Broadly speaking, the extant material evidence, that is, the maps made in pre-Hispanic and early colonial times, reveals two ways in which maps made both the community and its spatial substrate visible and shaped viewers' perceptions of the community. Both ways are alike in confirming that Central Mexicans did not envision a community as only an expanse of topography or architecture, but then they diverge, presenting two alternative ways to present the community in maps. The first, the cartographic history,[10] concerns itself with the establishment of the community, specifically historical events such as conquest or the foundation of a ruling lineage, and typically defines the territory of the community by its boundaries. The second is the social settlement map, which tries to show how the groups that make up the community populace have organized both themselves and the space they inhabit. Without a doubt, there is some blending and interchange between these two kinds of maps, perhaps because all communities in Mesoamerica saw both their common history and their composite social groups, along with territory, as the ties that bound them into polities. However, these two modes of mapping were distinct and powerful enough to find discrete expression on a number of the Relaciones maps of the late sixteenth century.

NATIVE COMMUNITY MAPS: CARTOGRAPHIC HISTORIES

Page 36 of the Codex Zouche-Nuttall, introduced above, is one of the earliest known versions of the cartographic history. It may have been an innovation growing out of screenfold histories, since it introduces, within the regular rhythms of its lined pages, a new kind of image (fig. 46). As seen on page 36 of the Codex Zouche-Nuttall, the community not only shares the territorial expanse of the Apoala Valley, it also shares a ruling lineage of semi-divine origin. For set within the valley on the banks of the rivers are four human figures. Their names are the names of their birthdays in the Mesoamerican calendar; by this same system, we might be named Sunday the 4th or Tuesday the 25th. At left, the deity 13 Flower faces her husband, the deity 1 Flower. At right is their daughter 9 Lizard, who faces her husband 5 Wind, who will found Apoala's ruling dynasty.[11] As this map shows it, and also as the Apoala community once believed, this dynastic foundation breathed life into Apoala, bringing it into being. While Apoala's territory is not defined here by a set of boundary markers, it is specifically named and depicted in the two rivers, the spring, and the cliff discussed above.

While page 36 of the Codex Zouche-Nuttall is one of the few surviving pre-Hispanic prototypes of the cartographic history, many examples of cartographic histories made in the colonial period survive to the present; each represents the community by showing its common bounded territory and its shared history. This community-defining history often centered on ruling elites, detailing their lineages or acts of historic conquest, no doubt because painters numbered among the elite, who in turn chose themselves as the subjects of historical narratives.[12]

The Lienzo of Zacatepec 1, also from a Mixtec-speaking region, like the Codex Zouche-Nuttall, offers a slightly more elaborate example of how the community used the map to define itself both as shared territory as well as shared history of lineage (fig. 49).

This large cloth sheet is called a *lienzo* (the Spanish word for canvas), a term also used for many other community histories. It was made c. 1540–1560 and comes from the Mixtec town of Zacatepec in the modern state of Oaxaca.[13] Logographic place-names arranged along the edge of this large sheet name the boundary markers separating the lands of the community of Zacatepec from those of its neighbors. Zacatepec's own place-name lies in the middle of the boundary rectangle, and the place-names of neighboring towns are shown outside this rectangle. Within this cartographic framework lies a historical narrative. The genealogy of the rulers of Zacatepec wanders above and within the boundary map established by the rectangle of place-names;[14] the Lienzo shows three generations of Zacatepec's rulers entwined in a historical tale that begins in A.D. 1068 and results in the foundation of the town of Zacatepec and the establishment of its territory.

Just as the double name implies, cartographic histories focused as much on a community's sense of the bonds of history and leadership that held it together as they did on the territory that the community claimed as its own. The cartographic histories that offer a more extended context than the single-sheet format of, say, the Lienzo of Zacatepec seem to suggest that the genre grew out of historical narratives. For example, page 36 of the Codex Zouche-Nuttall is but one page in a larger historical narrative dealing with Mixtec kingly lineages. Another formative group of cartographic histories is found in the Historia Tolteca-Chichimeca of 1547–1560. This mixed logographic-alphabetic book recounts the migrations of a few Nahuatl-speaking clans into the Cholula region and their eventual establishment of the altepetl or community-kingdom of Cuauhtinchan. In this larger historical narrative, Cuauhtinchan presents a cartographic history of itself (on fols. 32v–33r) showing the events that led to Nahua rule (fig. 50). The logographic place-name of Cuauhtinchan, meaning "home of the eagle," lies in the center of the paper sheet, where an eagle stands inside a cave mouth. Place-names, most of them comprising the hill-symbol, line the edges of the map, representing Cuauhtinchan's boundaries with neighboring altepetl. Within the frame of place-names that defines Cuauhtinchan's territory, a historical narrative is projected. The group of recently arrived Nahuatl speakers is shown conquering the indigenous population of the region; each unfortunate local leader, shown flanking the place-name of his town, has been sacrificed. Eight at right and center have arrows piercing their necks; two others, at lower left and upper right, are stretched on frames and shot through with arrows. In the midst of their conquests, victorious leaders surveyed their boundaries, their actions shown by the string of footprints marking their path.

Figure 49 *(opposite)*. The Lienzo of Zacatepec 1, c. 1540–1560. Painted on a large sheet of cloth, the Lienzo of Zacatepec is an early colonial example of a cartographic history. It shows the boundaries of the town of Zacatepec, in the present-day state of Oaxaca, as well as a brief genealogy of its ruling family. Size of the original: 325 x 225 cm. Photograph reprinted from Antonio Peñafiel 1900. (Museo Nacional de Antropología, Mexico, No. 35-63.)

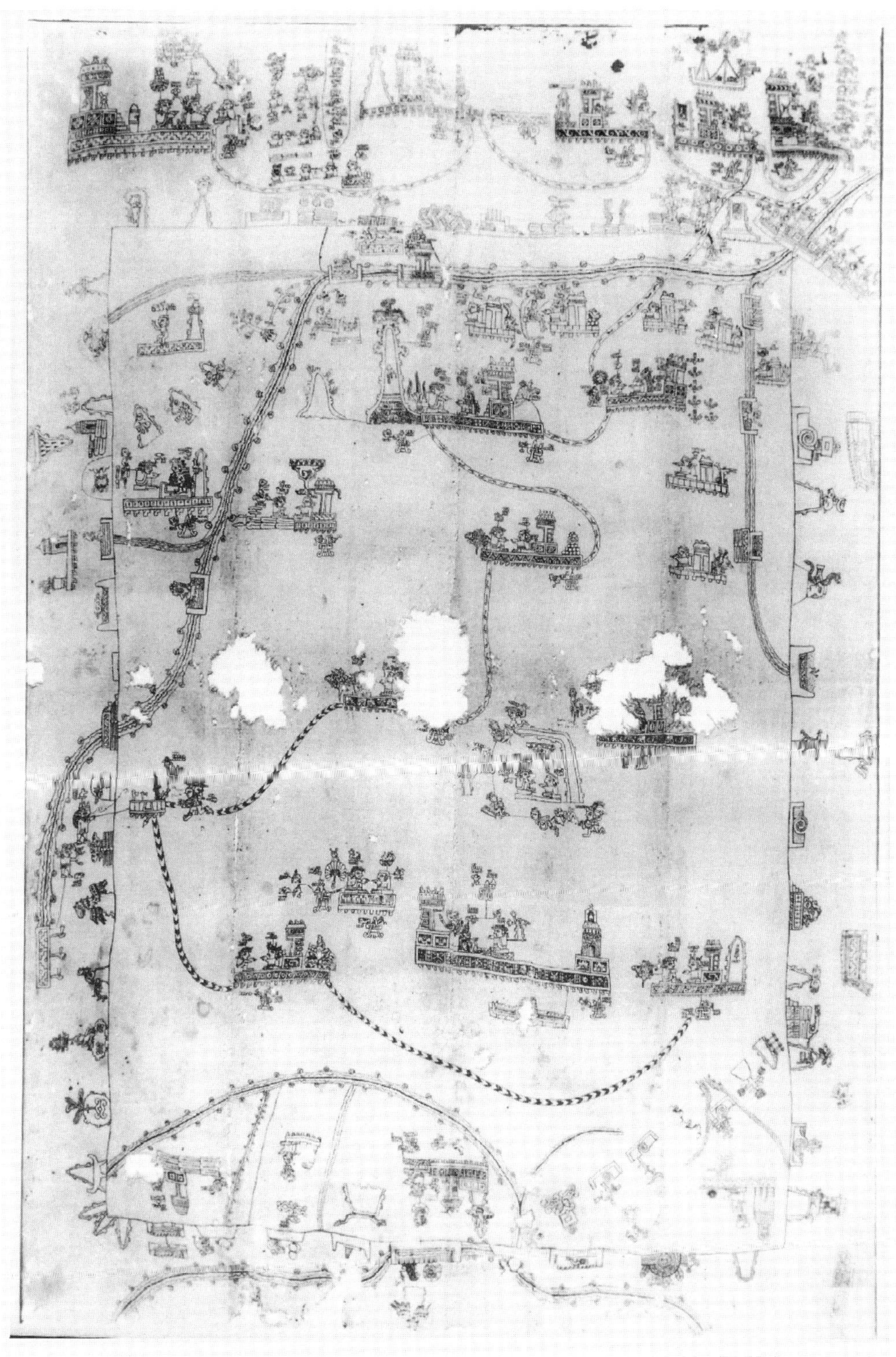

It is apparent that indigenous cartographic histories such as the Historia Tolteca-Chichimeca often combine different kinds of pictorial space onto one plane: the circle of boundaries makes us think we are looking at a flat, continuous geographic projection, but the historical scenes within often show us narratives set within temple interiors, or spaces we know to be three-dimensional. (Such a mixed pictorial space was hardly foreign to coeval Western maps, where narrative scenes or informative cartouches are often set within the map.) The spatial relationship of the rectangle of boundaries to a historical narrative is analogous to that of a frame to a picture—the frame defines and limits the space of the picture but is not seen by the observer as being the same pictorial space.

While the Historia Tolteca-Chichimeca is rather small (each leaf measures 30 x 22 cm), many cartographic histories are large panels of cloth the size of bedsheets; their size sug-

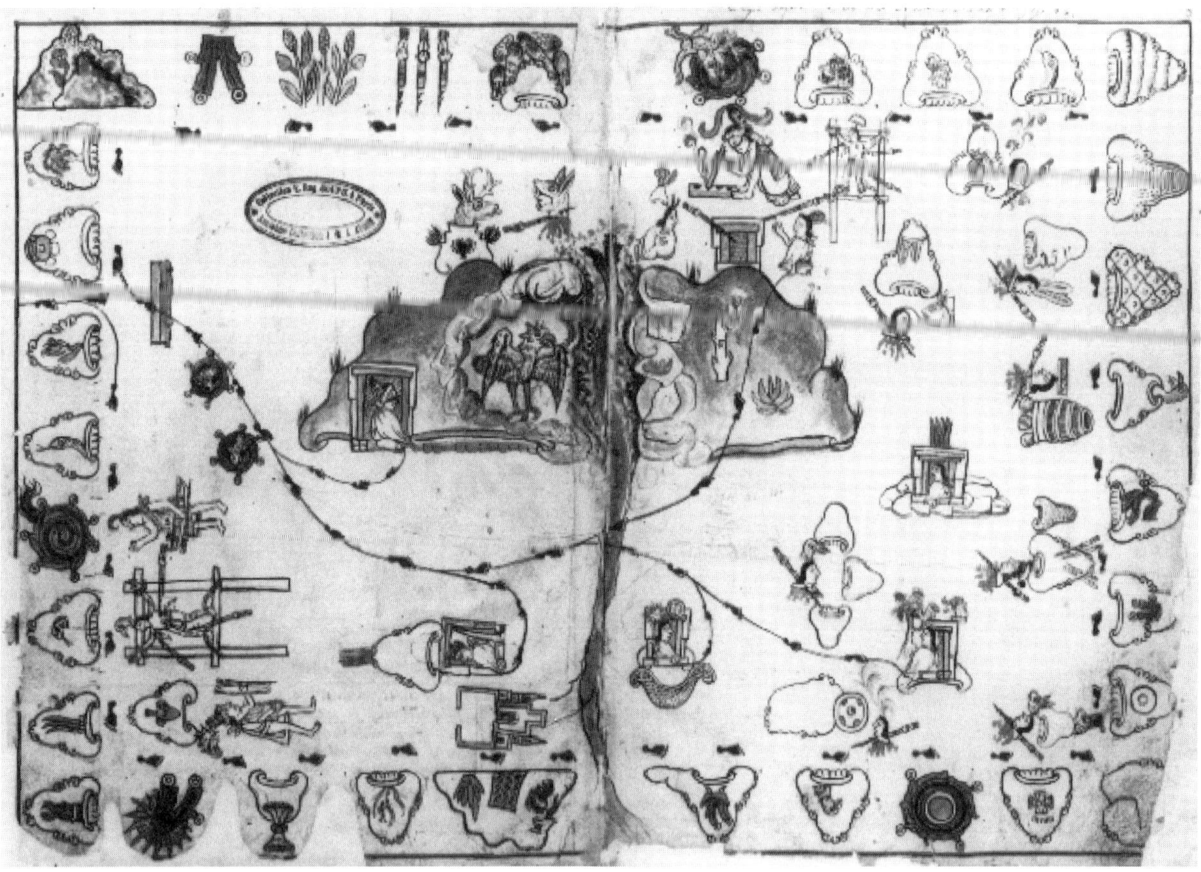

gests that they were meant to be seen publicly, hung out on the cabildo walls on public celebrations to visually remind community members of the shared history. Their iconography, too, suggests public viewing, for on all the cartographic histories, the pictorial symbols are clear and the historical narratives straightforward; there is little of the densely layered images we find in the ritual-calendric manuscripts that were meant to be read only by highly trained initiates. Any contemporary viewer could have easily read and understood a cartographic history such as the Lienzo of Zacatepec. From the colonial period up until modern times, indigenous communities are known to have presented cartographic histories to higher authorities to solve boundary disputes with neighbors, and it seems likely that even earlier they may have had the same purpose—to prove one community's claim to territory.[15]

As much as having a written record of boundaries may have helped each Amerindian community defend its lands from its acquisitive neighbors, the boundary map may have been equally effective in fostering a sense of collective identity within a population. In many native communities from the pre-Hispanic period to the present day, the walking of boundaries was an important ritual. As community members walked the earth to ensure the sanctity of boundary markers they were also, perhaps more important, suffusing themselves with a knowledge of this shared territory. Boundary maps, like *aide-mémoire,* encode the names of the boundaries visited during a perambulation; in addition, the boundary place-names are sometimes lined with a string of footprints, probably to emphasize that the map was a permanent record of an evanescent ritual of boundary walking. For instance, a string of footprints connects the logographic boundaries at the top of the Relación Geográfica map of the Minas de Zumpango (fig. 44). In the Historia Tolteca-Chichimeca, the accompanying text tells us that the footprints lining folios 32v–33r mark the leaders' path as they walked around their newly established boundaries (fig. 50). Thus, as much as the history that it contained, the boundary map itself was a record of a community-solidifying event, one that would be reenacted year in and year out, to rejuvenate the age-old communal ties.

In these maps, as their connection to perambulation suggests, it is the human presence that defines space, both through naming, and through movement. The mestizo writer

Figure 50 *(opposite).* Historia Tolteca-Chichimeca, fols. 32v–33r, c. 1547–1560. These folios, from a book narrating the history of the town of Cuauhtinchan, show the conquest and occupation of the Cuauhtinchan region by the Tolteca-Chichimeca, a Nahuatl-speaking group. A path defined by footprints shows the inroads made by the conquerors. The outer edges of the page are framed by pictographic toponyms, and the footprints at their bases seem to indicate a circumambulation. Size of the original: 30 x 44 cm. Photograph courtesy of Bibliothèque Nationale de France, Paris (Mss. Mexicain 46-50).

Diego Muñoz Camargo, who wrote Tlaxcala's Relación Geográfica, commented on how long distances were measured by, in Nahuatl, *cenecehuilli,* or the rests that the human body needed en route (RGS 4: 35). Thus, it is the pace of the human body that measured space and human language that defined it. These aspects of indigenous maps allow us to better understand the humanistic projections that underlay them.

THE CARTOGRAPHIC HISTORY AND THE RELACIONES GEOGRÁFICAS MAPS

After the conquest, in the decades before the Relaciones Geográficas were painted, indigenous cartographic histories remained a viable way for some communities to represent themselves, because in many cases communal self-perception and territories were not entirely disrupted by the colonial government. In the Relación Geográfica corpus, we can see the presence of the kinds of indigenous cartographic histories discussed above. In responding to the Relación questionnaire, two indigenous artists undoubtedly drew upon, perhaps even directly copied, the cartographic histories held within their cabildos. The refined maps of the neighboring towns of Teozacoalco and Amoltepec (Santiago Amoltepec, Oaxaca; figs. 10 [pl. 1] and 51 [pl. 6]) come from a region in Oaxaca that not only was removed from much of the colony's newer Spanish population, but also was the site of a fertile native tradition of manuscript painting (Caso 1977–1979; Smith 1973a; Spores 1984: 113–21). Both Relaciones Geográficas maps remain true to the main tenets of the indigenous cartographic history: they both define territory by an enclosure of logographic place-names representing boundaries and refer to common history, in this case in the form of the bonds of rulership. But while the Teozacoalco map provides an elaborate genealogy of the ruling lineage and its establishment, the Amoltepec map is more succinct, showing only the ruling couple.

Although the maps of Amoltepec and Teozacoalco were painted by different artists, both adopted a circular frame: each artist made an inscribed line with a compass and upon this rule placed the logographic place-names of the boundaries of each community. Teozacoalco represents its boundaries with forty-six logographic place-names arranged along this circle. Part of this boundary is defined by a wide river, the Cuananá, which marks the western boundary of Teozacoalco's lands. Within this fence of boundary markers, the map of Teozacoalco shows us its subject towns, set within a web of roads and rivers (fig. 10; pl. 1). Like the map of Teozacoalco, the map of its small neighbor Amoltepec (fig. 51; pl. 6) takes a circular form, its artist arranging nineteen beautifully drawn logographic place-names

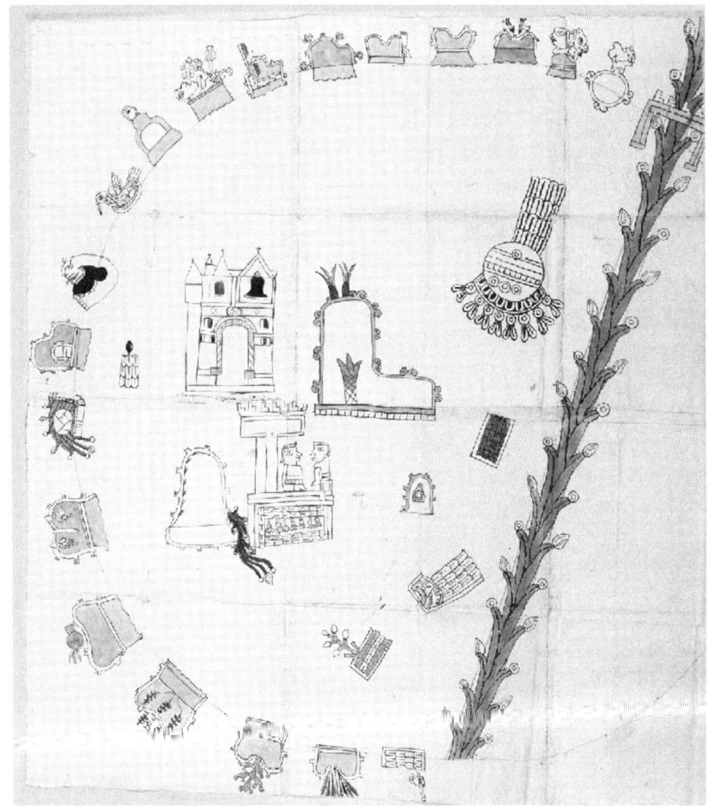

Figure 51. The Relación Geográfica map of Amoltepec, 1580. (See also pl. 6.) The indigenous artist of Amoltepec defined the town's boundaries with a two-thirds circle of toponyms, written with logographs. Inside this boundary are both the church of Amoltepec and the rulers' palace. The toponyms within the circle would seem to represent sites within Amoltepec's territory. Size of the original: 86 x 92 cm. Photograph: B. Mundy; reproduced courtesy of the Benson Latin American Collection, The General Libraries, The University of Texas at Austin (JGI xxv-3).

evenly around an inscribed circle. Cutting one arc off the circle is a river, which forms a natural boundary on one side of the town lands, just as in the case of Teozacoalco.[16]

The Teozacoalco map presents internal evidence that it was copied from an older, revised cartographic history. Not far from the town of Teozacoalco is Elotepec (San Juan Elotepec, Oaxaca), a town that once was its subject. Elotepec means "hill of the corn" *(elote)* in Nahuatl, while in Mixtec its name, which also translates as "hill of the corn," is *yucu ndedzi,* according to a sixteenth-century source (Alvarado 1962: 91; Arana and Swadesh 1965: 103; Smith 1973a: 176). On the Teozacoalco map, Elotepec is marked by a logograph that could represent either its Nahuatl or its Mixtec name, showing a squat hill beside which a stalk of corn grows. In the years before the map was painted, Elotepec was moved from under the aegis of Teozacoalco to that of Los Peñoles, the district to the east

(RGS 3: 45–53; Relación Geográfica map of Teozacoalco). The Teozacoalco map gives us a record of the boundary revision by using two sets of boundaries, one from before and one from after Elotepec was its subject. Appearing on the same inscribed circle as Teozacoalco's other boundaries are nine logographs naming its new boundaries with Elotepec. The old boundaries that embraced Elotepec's territory are added along an arc that rises like a crescent moon from the main circle.

Why would this artist have felt compelled to include two sets of boundaries, one of them old and inoperative, on this map? It seems likely that an earlier map, made while Elotepec was still under Teozacoalco's domain, initially showed one set of boundaries. After Elotepec was reassigned, Teozacoalco's cabildo revised its map by adding in the new boundary line. When the Relación Geográfica mapmaker made his map, he simply copied the two sets of boundaries that appeared on the earlier map.

Both within and outside of its encircling boundary frame, the Teozacoalco artist includes a history of the ruling lineages of the polity (figs. 10 [pl. 1] and 52). He (or she) pictures three vertical genealogies, each generation shown by a married couple who face each other, much as the two couples on page 36 of the Codex Zouche-Nuttall were portrayed (fig. 46). Occasionally a single male figure appears, marking the close of a dynasty. Only the third, and the shortest, of these genealogical lines appears within the inscribed boundary circle. As reconstructed by Alfonso Caso and other scholars, the first and longest of these genealogies, running along the left edge of the page, shows the four dynasties that ruled over the Mixtec town of Tilantongo up to the time the map was painted, Tilantongo being the source, via a migration of its rulers from Apoala, of many kingly lineages, including that of Teozacoalco. The second column begins to the right of the first, next to a temple that is the logographic place-name of Teozacoalco, with the last rulers of Teozacoalco's first dynasty. Their daughter marries the son of Tilantongo's rulers, who is shown, by means of a path of footprints, as he migrates out of its ruling house, to found the second dynasty of Teozacoalco. Above this Tilantongo-Teozacoalco couple, their dynastic line rises, as if it were a tree pushing heavenward, up through seven generations. The seventh rulers produce no living male heir, so their maternal grandson is grafted onto the ruling lineage, founding in 1321 the third dynastic house to rule Teozacoalco. The initiators of the third dynasty are pictured both outside the map's boundary and within it, next to Teozacoalco's church. Their successors appear above them, the ruling line progressing upwards for six generations, until the last of the line dies without an heir sometime in the latter half of the fifteenth century. Again a prince from Tilantongo picks up the dropped reins. His son, the final ruler shown on the map, was probably the ruler of Teozacoalco at the time the map was made in 1580.

While the Teozacoalco artist gives copious information about the royal lineages of his town, the Amoltepec artist includes only two figures sitting within a palace structure (fig. 51; pl. 6), who presumably were either the ruling couple who founded the Amoltepec lineage or else the rulers at the time the map was painted; the accompanying text provides no explanation, nor are these rulers named on the map.

Another element of the indigenous tradition that we can trace in these two Relaciones Geográficas maps is the idealized projection of the communities' boundaries. As we saw in the map from the Historia Tolteca-Chichimeca folios 32v–33r (fig. 50) and the Lienzo of Zacatepec (fig. 49), the principal means the mapmaker uses for making space visible is the

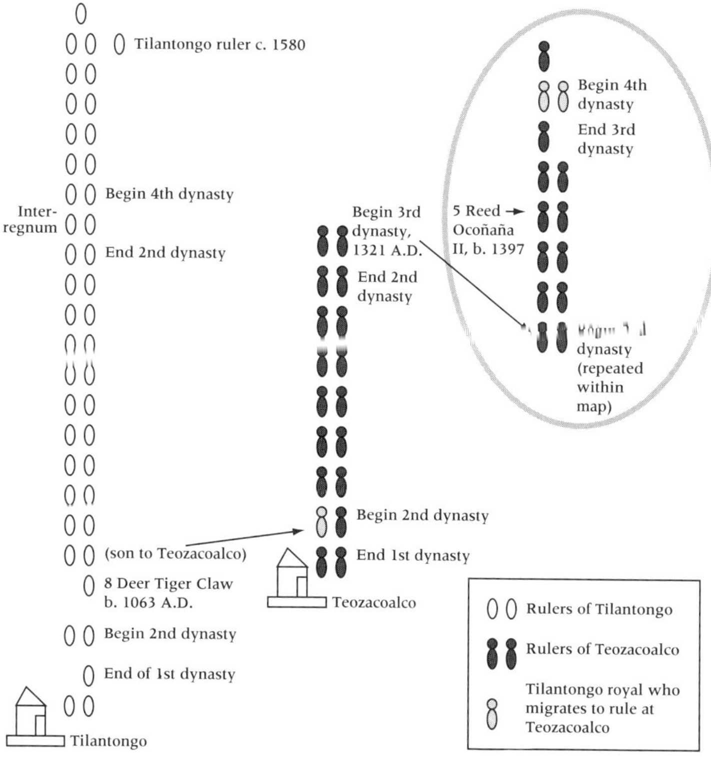

Figure 52. Diagram of genealogies of the Teozacoalco Relación Geográfica map. The Mixtec town of Tilantongo was the source of many ruling Mixtec lineages, so the ruling family of Teozacoalco used their Relación Geográfica map to document the connections. The Teozacoalco rulers at center and right are traced back to the twelfth century, and the Tilantongo rulers at left reach back a century earlier. The oval at right represents the territory of Teozacoalco, as it appears on the map.

frame of logographic place-names, which are arranged in an even pattern around the edge of the page. While in the Historia Tolteca-Chichimeca map the sequence of these place-names correlates to that of Cuauhtinchan's boundaries, their placement on the map does not correspond exactly to the planimetry of those boundaries on the ground (Robertson 1959a: 180). The area delimited by these boundaries has been mapped on a modern scale-model map by scholars, and it covers a region in central Puebla that is a rough polygon of about fifty kilometers across. Lacking adequate information about Teozacoalco's place-names, I have been able to only roughly plot the area shown in the Relación Geográfica map onto a modern topographical map in figure 53, based on the position of Teozacoalco's subject towns and the roads and rivers that appear in the map. Planimetrically, Teozacoalco's territory is an irregular oblong of enormous size: it is about thirty kilometers wide and seventy kilometers long.[17] The scale of the Relación Geográfica map—judging from the placement of towns—decreases as one moves farther away from Teozacoalco. I make such comparisons to planimetric maps not to impose an exterior criterion of planimetric accuracy upon these maps, but rather to suggest that the Teozacoalco map, with its boundary markers creating a geometrically perfect frame, uses a projection that I call "communicentric." Such a communicentric projection is centered on the heart of the community, often a palace or temple structure. On the map, spatial relations are manipulated to emphasize this center, often by increasing its size, and to yield an overall form for the community that is a geometrically perfect one. We see this in the Teozacoalco and Amotepec maps, where rulers and churches dominate the center and where the boundaries are drawn along a perfect circle. In the map of the Historia Tolteca-Chichimeca, the boundaries are arranged in an even rectangle around the amplified central place-name of Cuauhtinchan, or "place of the eagle."

Calling this a projection is, of course, a departure from the usual sense of the word, which implies a geometric base. There is none of that here. Instead, the rationale of this projection lay in its rhetoric, for the communicentric projection advanced the argument, visually, of a community's importance, unity, and perfection. Since so many of these boundary maps also contained narratives of communal foundations, it was as if the map proposed that the community, from the very moment of its inception, had been a perfect whole, an inviolate circle or rectangle. In addition, this communicentric projection was clearly rooted in each community's sense of itself as the center of things. This projection is one aspect of the humanism of Amerindian maps, which reflect a subjective human understanding of surroundings rather than the objective geometric overlay used in a survey map. Given this humanistic stance, it is no surprise that the Nahua preferred measuring in "cenecehuilli," or rests of the human body, rather than in leagues (RGS 4: 35).

Such a projection was widespread in Mesoamerica, where numerous idealized map frames, such as the circle of boundaries around the community center, are extant. We know, for instance, that the Maya also produced circular maps in the colonial period (Roys 1933: 125; Roys 1939: 87–9; Roys 1943: 175–94). The circular convention, in turn, may reflect an Amerindian concept of the local landscape, expressed by the Pueblo Indians, who saw their surroundings arranged in a great circle around the central town (Ortiz 1972).[18] The circular form was especially appropriate for a map of boundaries because circles are, as Arnheim points out, perfect forms, ones "hermetically closed off from [their] surroundings" (Arnheim 1983: 117). What better way to express boundaries than to show them lending a community a perfect form, creating around it a sealed enclosure?[19]

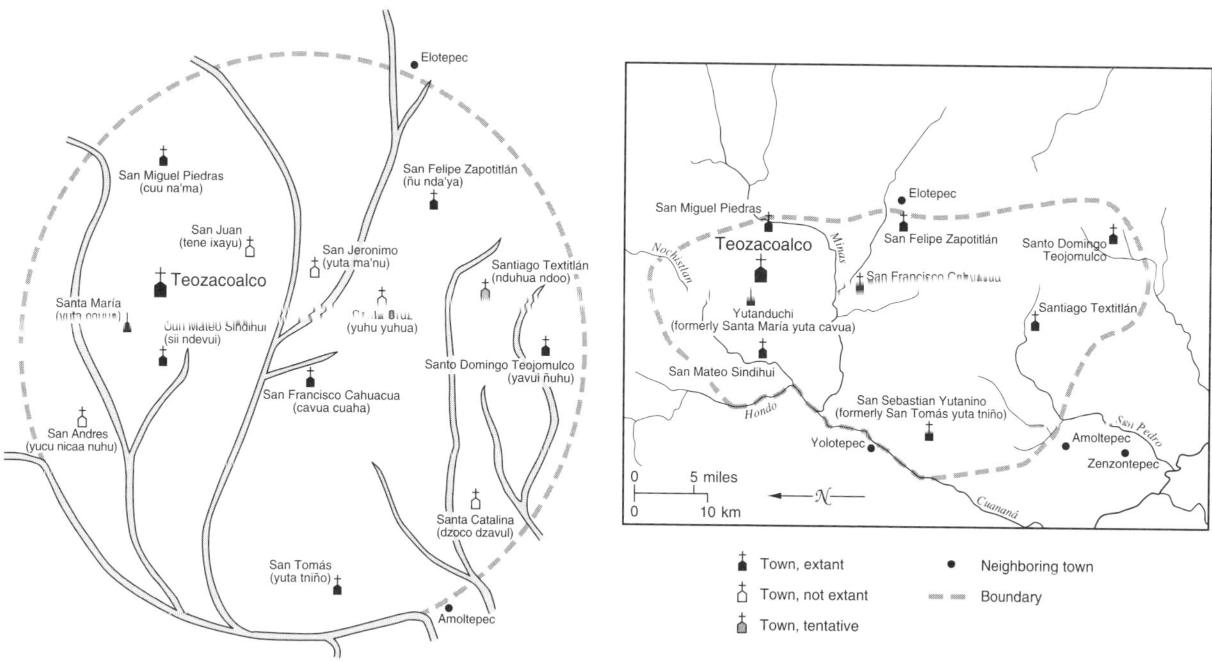

Figure 53. Comparison of the Teozacoalco region as shown on its Relación Geográfica map *(left)* with a modern topographical map. Maps reprinted from David Woodward and G. Malcolm Lewis, eds., *The History of Cartography,* vol. 2, bk. 3 (Chicago: University of Chicago Press, forthcoming). Prepared by the University of Wisconsin Cartographic Laboratory, Madison.

THE COMMUNITY MAP: SOCIAL SETTLEMENT MAPS

At the same time that a community could make itself visible through the conjunction of historical event and boundaries, another mode of self-presentation had found root in communities; these maps sought to show the territorial spaces along with the social hierarchies that ordered and defined them. Such maps were made primarily among Nahuatl-speaking peoples of Central Mexico. The Codex Mendoza fol. 2r is such a map, and the preface described how the physical space of Tenochtitlan was ordered by the city's social groupings, embodied in its ten founders (fig. 2).

Recent historical and anthropological studies have shown how precisely—and modularly—the Nahuas organized their societies. The historic Nahua groups in Central Mexico that have been closely studied were usually composed of multiple sets, with altepetl made of four or eight calpolli being the norm and with these subgroups arranged according to strict rules of preference and rotation. Sometimes these subunits comprised even smaller subunits themselves that were as rigorously structured and ordered. In the same fashion, altepetl also organized themselves into "complex altepetl," as the historian James Lockhart calls them. Lockhart describes them as "a set of altepetl, numerically and if possible symmetrically arranged, equal and separate, yet ranked in order of precedence and rotation, [that] constituted the larger state, which was considered and called an altepetl itself" (1992: 20–1).

Yet while an altepetl's political and social structure was rigidly ordered and the people that comprised it closely defined, its territory could be scattered, with scraps of territory sewn together, or the holdings of its subgroups could be intermingled like a crazy quilt. In other words, the orderly, modular blocks of society had no assured counterpart in orderly blocks of territory. For such scattered altepetl, the boundary maps that other communities made were of little use, since the community was not a neat block of territory and thus could not present itself as a unit fenced in with boundaries.[20]

How, then, were altepetl, or their subunits, to map the scattered fragments of their territory? It was as if two notions of self had to do battle—one, that of scattered and disorganized territory; the other, that of a rigidly ordered political structure. Mapmakers found a pictorial and cartographic solution to this dilemma in the social settlement map, which showed both the hierarchical and ordered internal structure of a community as well as the spatial relationship of those parts to each other. Social structure maps frequently are highly ordered compositions, as if the territory were filtered through the same grid that ordered society. Often pictures of palaces, or "lordly houses," represent the altepetl or the calpolli that made up society (in much the same way that in European maps, the church might represent the whole town), and this use was both metaphoric and literal, for the term "calpolli" meant both the social unit as well as the ceremonial building that was the center of the calpolli's festivities.[21]

Only a few social settlement maps like the one found in the Codex Mendoza survive today, certainly not enough to show that such community maps ever had the wide spread of the cartographic history. But such social settlement maps may have been better suited to represent larger, more complex communities, either altepetl comprising numerous calpolli or the complex altepetl of the Nahua, or communities in heavily populated regions whose territory was scattered across a landscape rather than compactly defined by a fence of boundaries. Not surprisingly, the densely peopled regions of Central Mexico, where social settlement maps were most likely to have been made, were also magnets for Spanish colonists, and thus the artists who made such maps may have been influenced early on by the different expectations of maps that their new Spanish audience held.

The Historia Tolteca-Chichimeca, which also contains a cartographic history, contains one of these social settlement maps on folios 26v–27r, showing us the ceremonial center of Cholula, defined by architecture and symbols (figs. 54 and 55).[22] Around it are arranged eight of the twelve calpolli of the Tolteca-Chichimeca that made up Cholula; the remaining four dwelt in the center. Each of these eight calpolli is shown as a small, frontally viewed palace with a logographic name (usually) on its roof, and each is flanked, with one exception, by a pair of rulers, or calpolleque.[23] Much like the Codex Mendoza folio 2r, this map is a historic map, that is, it purports to show the community around the time of the establishment of this order, and the text emphasizes that the map, like the book itself, is meant to preserve the memory of the past, "thus it is [shown], and thus we came to know our great-grandparents, our grandparents."[24]

The map of folios 26v–27r shows Cholula some centuries before the book was made, at such a remove that the historical perspective is like a camera lens—focused in the foreground, blurry in the distance. The principal calpolli, the first four in order of importance, are presented in the Historia in sharp-edged clarity; on this page they appear clearly named, arranged in order, counterclockwise around Cholula's center.[25] The next four calpolli were also distinctly remembered by the authors of the Historia Tolteca-Chichimeca, but as they seem to have coexisted in the center of Cholula, they are not portrayed separately on this map. The last four calpolli are of less importance in the text. One lacks a name, and the others appear twice, but each time in different sequence, as if the rigid order that once encased them had eroded. On the map's presentation of these four hapless calpolli in the Historia Tolteca-Chichimeca, two lack the clear logographic names that distinguished the first four. These last four calpolli are labeled on figure 55 as 9, 10, 11, and

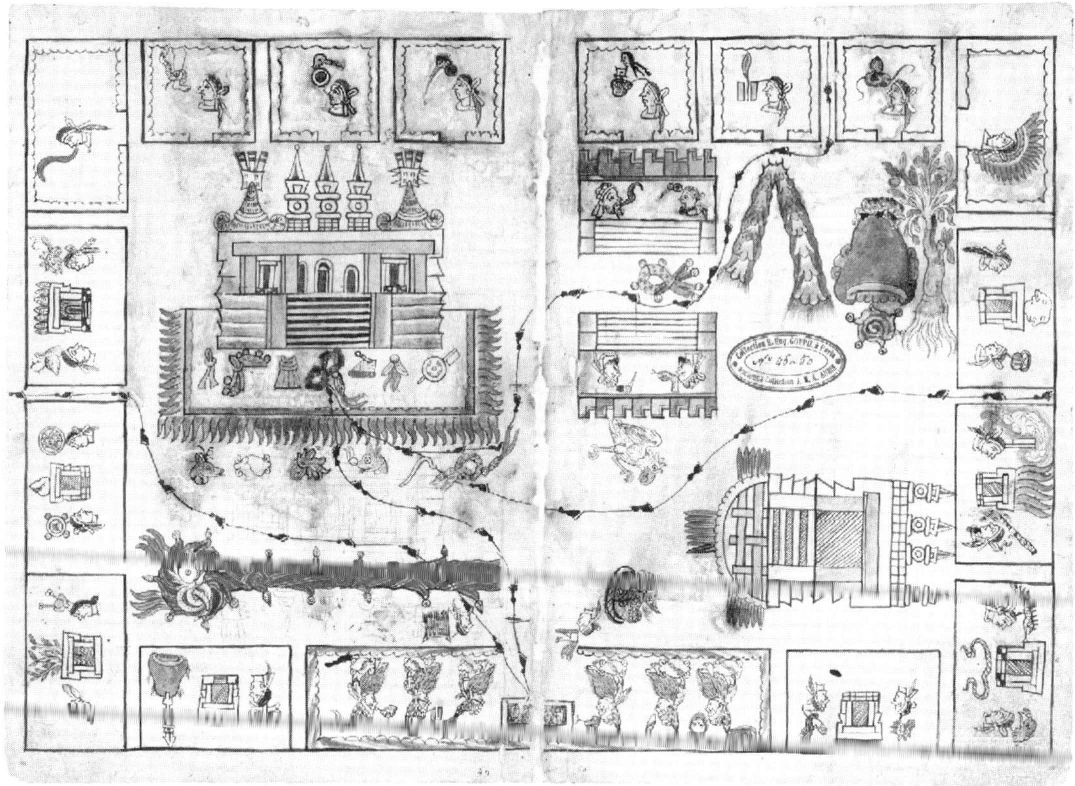

Figure 54. Historia Tolteca-Chichimeca, fols. 26v–27r, c. 1547–1560. These folios show the city of Cholula by means of emblematic architecture. The rectangular boxes lining the pages show leaders of the various parts of the city. Size of the original: 30 x 44 cm. Photograph courtesy of the Bibliothèque Nationale de France, Paris (Mss. Mexicain 46-50).

12. But the clarity of names, in these maps, is less important than the clarity of structure, and the clear divisions of the calpolli of Cholula and their sequence in a set rotation are clearly shown pictorially in the eight boxes that frame the center of the city.[26]

These pages in the Historia Tolteca-Chichimeca were not just a recounting of calpolli, because they also show the spatial layout of Cholula's center, with its temple, its ball-court, its academy, along with symbols further defining this central space. While we have no definitive evidence that the calpolli of Cholula shown on these pages were actually arranged around Cholula's center as they are in this representation, it seems likely that they, or their representative temple structures, were.

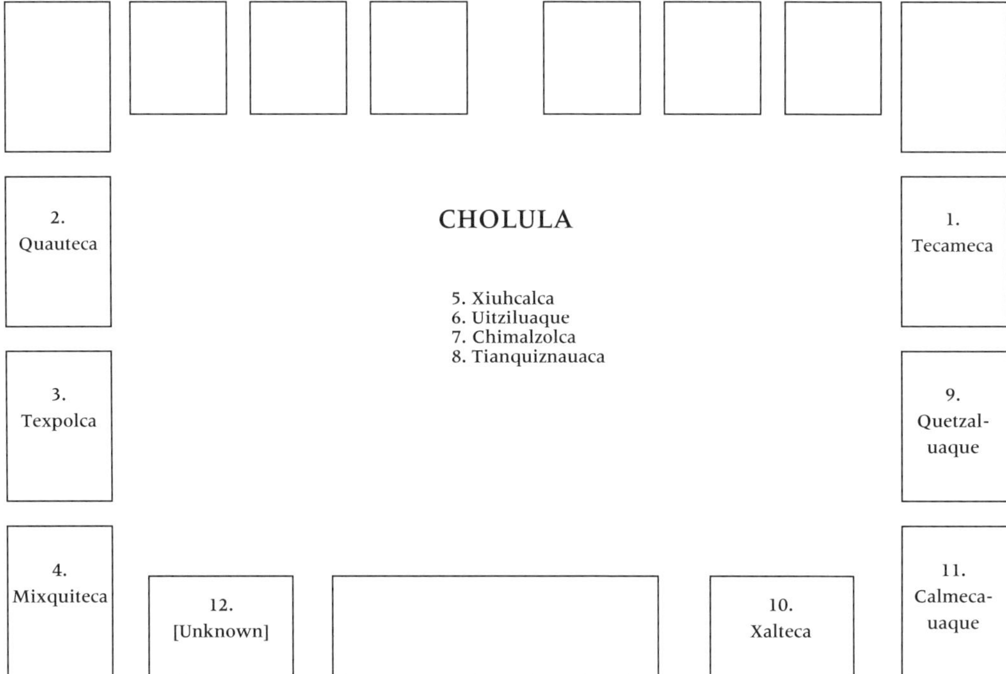

Figure 55. Diagram of fols. 26v–27r of the Historia Tolteca-Chichimeca. The text of the book designates a dozen calpolli populating Cholula, and this diagram locates where they are named on these pages of the manuscript. Four calpolli (5–8) are not named on these pages but may have occupied the center of the city.

What was the purpose of these maps? Their emphasis on the internal ordering of a community suggests that they were used within community cabildos; the textual associations and imagery of the maps also leads me to believe that they would have been made within cabildos. Here, members hammered out the social hierarchies and, more important, set up the positions of each of the calpolli within a strictly defined rotation that would keep the polity ordered and running. But showing the order of calpolli could be done much more simply, with a list or diagram; perhaps the most important function of these maps was to show the spatial relationships among the many parts of a single altepetl or the various subunits of a complex altepetl, because, as these maps show it, the territory of neighboring altepetl was often overlapping and interlaced.

The Map of Chichimec History offers another example of a social settlement map. We know it only through nineteenth-century copies, but the original came from the altepetl of Amaquemecan. Even in the known, and probably bowdlerized, copies, the logograph-

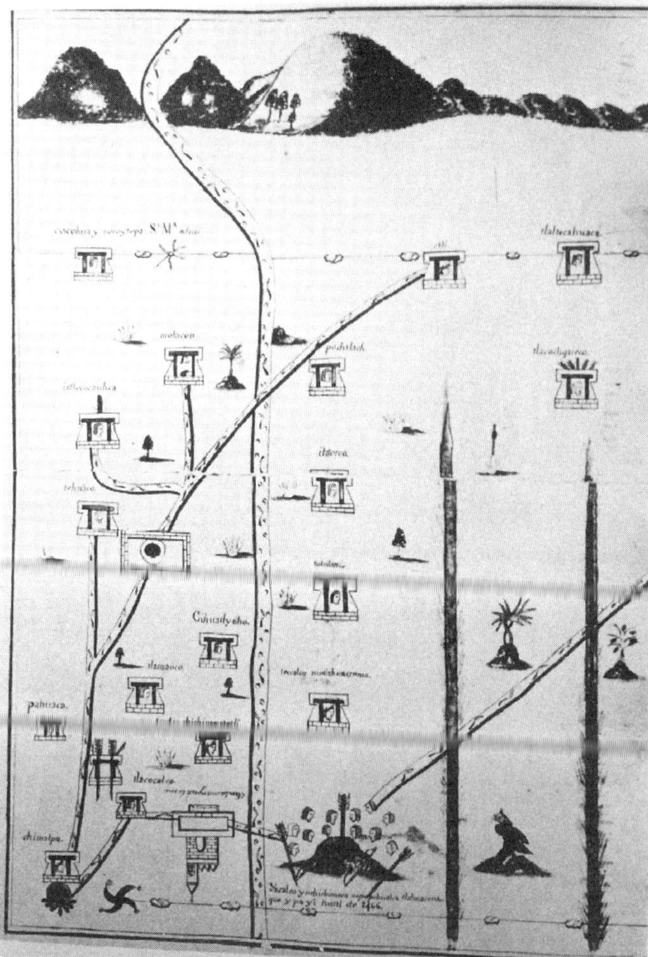

Figure 56. Map of Chichimec History. This map is perhaps a nineteenth-century copy of a lost original. The original came from the region of Amaquemecan, in the southern reaches of the Valley of Mexico, and was probably made in the sixteenth century. It shows how the various units of one polity arranged themselves spatially. Photograph reprinted from Handbook of Middle American Indians, vol. 14, figure 26 (University of Texas at Austin, 1975). Reproduced courtesy of the Department of Library Services, American Museum of Natural History.

ic place-name of Amaquemecan (a turkey on a hill, after Totolin, "Turkey," the altepetl's tutelary god) is clearly recognizable in the lower right (figs. 56, 57a, and 57b).[27] The map shows us a landscape creased by roads and rivers, and across it are scattered eighteen house structures. Amaquemecan, a part of Chalco, lies in the southern reaches of the Valley of Mexico, and while we know little of its territorial expanse in the sixteenth century, we know a good deal about its history and social structure from the work of the indigenous historian don Domingo de San Antón Muñón Chimalpahin Quauhtlehuanitzin (1579–1660). As Chimalpahin tells it, the altepetl of Amaquemecan comprised five units called *tlayacatl,* which in turn were made up of even smaller units called *tlaxilacalli* (fig. 58). We know the most about Chimalpahin's own tlayacatl, one called Tzaqualtitlan Tenanco, and its tlaxilacalli (Schroeder 1991).

Just as Chimalpahin writes his history from the vantage point of his own subunit, or tlayacatl, the Map of Chichimec History seems to be preoccupied with another tlayacatl of Amaquemecan called Tecuanipan (fig. 56), for it shows us many things, and it seems most concerned with showing in detail Tecuanipan, its subunits, the tlaxilacalli, and the spatial relationships among them. Since we know from historical sources that the calpolli and its subunits were not, strictly speaking, villages or hamlets, but rather groups of affiliated people, their spatial dimensions remain unclear. Did each of Tecuanipan's tlaxilacalli pos-

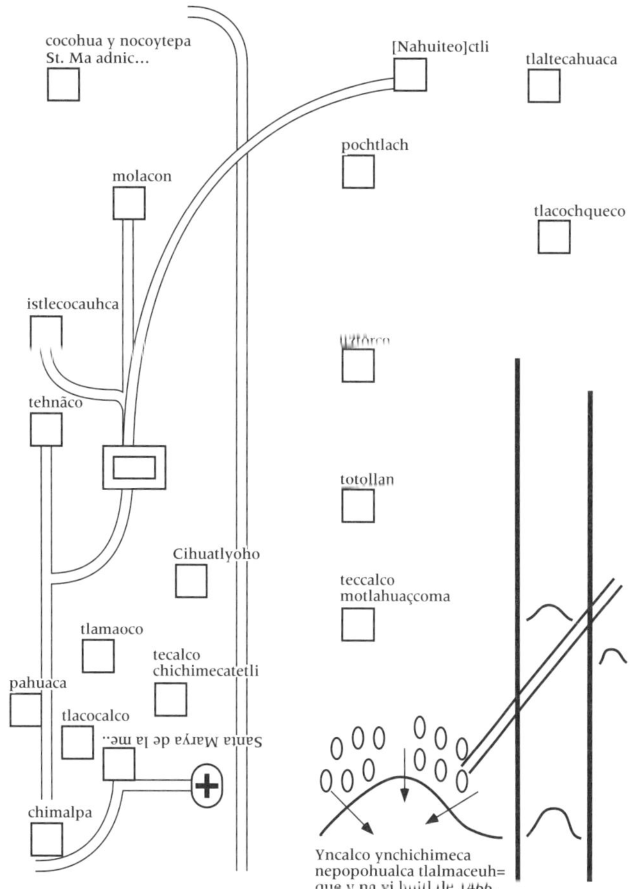

Figure 57a. Drawing after the Map of Chichimec History, with transcription of its inscriptions.

sess a clearly defined territory? Did Tecuanipan itself? Maps like this one do not definitively answer these questions, but they do give us a clearer idea of the spatial relationships of the parts of the altepetl. In the Map of Chichimec History, the tlaxilacalli of Tecuanipan shown are not clustered in some contiguous territory, but are scattered among the tlaxilacalli of other tlayacatl within the altepetl of Amaquemecan. Amaquemecan's territory, in turn, overlapped with that of neighboring altepetl within the complex altepetl of

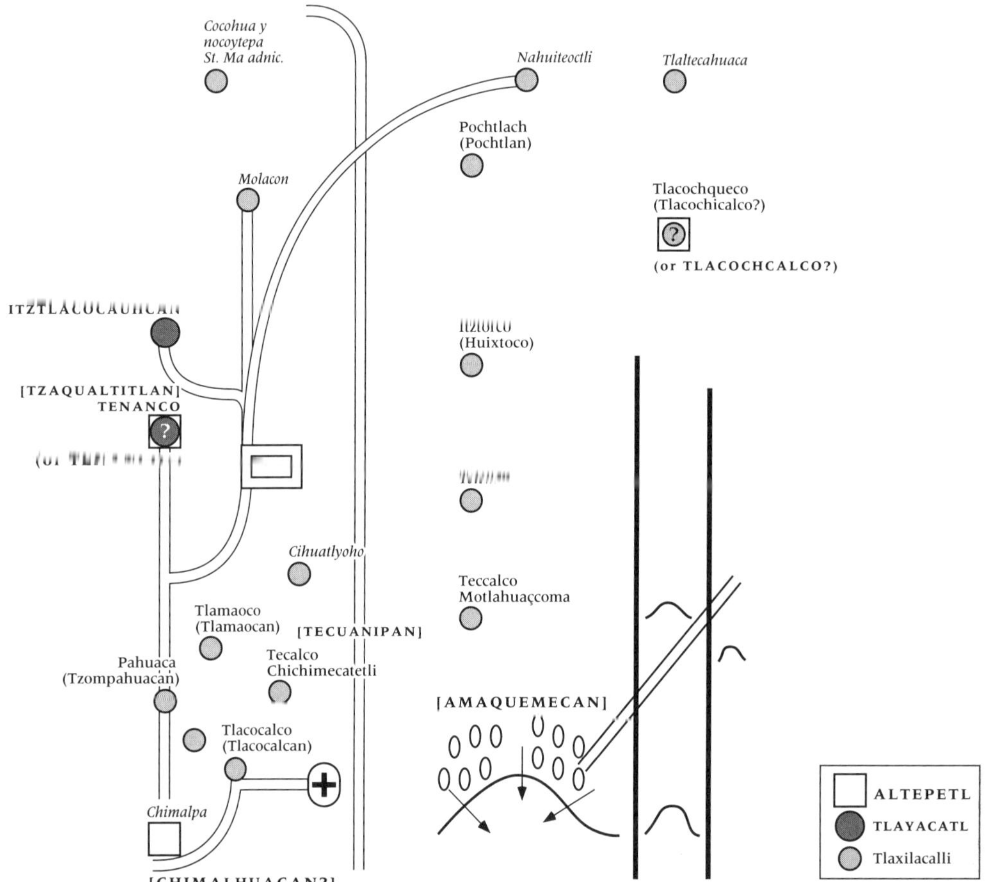

Figure 57b. Drawing after the Map of Chichimec History, identifying the sociopolitical units, be they altepetl, subunits (tlayacatl) or sub-subunits (tlaxilacalli), that appear on it. Italics show places whose identification is uncertain. Orthography of identified places, when different from that of the Map of Chichimec History, is shown in parentheses. Suggested identifications are shown in brackets.

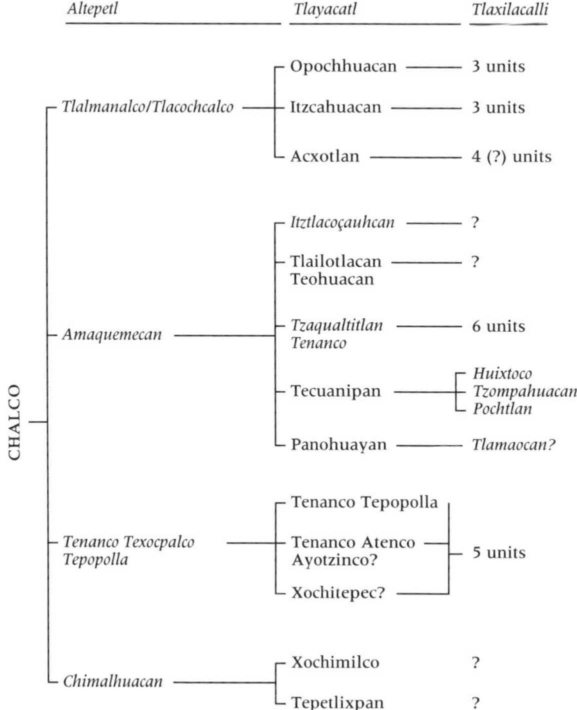

Figure 58. Chalco's sixteenth-century social structure. Chalco was a complicated polity, comprising four altepetl, one of which was Amaquemecan, and their composite subunits. This diagram is based on the reconstruction of sixteenth-century Chalco in Schroeder (1991). Names in italics are those represented on the Map of Chichimec History. Question marks signal uncertain identifications.

Chalco (figs. 57b and 58). Thus we find that, as the Map of Chichimec History shows it, in Chalco the units (altepetl), subunits (tlayacatl) and sub-subunits (tlaxilacalli), were jumbled together in a loosely defined common space.

Amaquemecan's various parts were spatially interlaced. The Map of Chichimec History shows us in addition how they were ordered by a political hierarchy. Two of the eighteen structures on the map are labeled *teccalco* ("place of the palace").[28] One is *te[c]calco chichimecatetli,* the other *teccalco motlahuaçcoma,* and these palaces seem to represent the dual leadership that once ruled Tecuanipan; we know for certain that one of Tecuanipan's lords bore the title *chichimeca tecuhtli* ("Chichimeca lord").[29] Inside these palaces, the human heads shown wear miter headdresses, which were worn by *tlatoani,* or rulers. Two of the other palaces pictured—named as Chimalpa ("Upon the Shield") and Nahuiteoctli ("Four Lord")—also contain men wearing miter headdresses and may show the rulers of other parts of Amaquemecan, although clear historical documentation is lacking.

In its original form, the Map of Chichimec History may have meant to record how such a complex net of settlements were reined in by a strict political structure. But because we know the Map of Chichimec History only through late colonial (and quite distant, it seems) copies, we can only glimpse how it may have functioned to align political structure with land tenure within the altepetl of Amaquemecan. Some pre-Hispanic vestiges remain: at the bottom of the map is a hill that is pierced by three arrows. This is most likely the hill of Chalchiuhmomoztli, now known as Sacromonte, that abuts the modern town of Amaquemecan. According to Chimalpahin, the original founding groups of Amaquemecan (who would in turn metamorphose into its component tlayacatl) settled on this hill before occupying the surrounding valley. Once on the hill, the settlers probably shot arrows towards the four directions in a ritual of foundation.[30] The Map of Chichimec History preserves echoes of Amaquemecan's foundation history, showing twelve men (their heads alone appear) gathered on the mountain before their descent, with the sets of bows and arrows on the hill, probably those used in the arrow shooting of foundation rituals; in this, it echoes the cartographic histories discussed in the previous section, which often included foundation histories.

The Map of Chichimec History in its present version gives enough information to speculate on what the original looked like. It showed not only the crazy quilt of the tlaxilacalli that made up Amaquemecan, but also the historical foundation which led to, and provided the rationale for, the strict order of Amaquemecan's component political units. Perhaps the original map also made the connections between political order and territorial layout more explicitly, but how it did so remains to be imagined.

THE SOCIAL SETTLEMENT MAP AND THE RELACIÓN MAP

Like the cartographic history, which would have been called to mind by the Relación Geográfica questionnaire's overall focus on communities in New Spain, the social settlement map may have been specifically evoked by item 9. This question asked for a list of the names of the cities and villages in a region, and item 10 asked for notable architecture. Correspondingly, these social settlement maps show both the parts of the polity, which the Spanish misunderstood to be separate villages or subject towns, and they often represent the parts of polity with *teccalli,* or lordly houses, corresponding to the architecture that item 10 desired.

The Relación Geográfica of Cholula (fig. 34; pl. 3) seems to have been one of the Relaciones Geográficas maps directly inspired by social settlement maps, the direct inheritor of the social settlement map of Cholula in the Historia Tolteca-Chichimeca, folios 26v–27r (fig. 54). On its surface, the Relación Geográfica map seems to show Cholula as a neatly organized grid, newly imposed by the mendicant friars, and arranged around their central and impressive monastery. While the map seems to show a small town of twenty-four urban blocks, these are, as Kubler has noted, "super-blocks," shorthand for the more numerous blocks needed to house the city's large population (Kubler 1985: 93). At the time of the conquest, Cholula held one of the larger populations outside the Valley of Mexico, holding about 100,000 within about eleven or twelve square kilometers; at the time the Relación Geográfica was made, Cholula had perhaps 9,000 people (Cortés 1986: 74; Gerhard 1986: 117; Peterson 1987: 71).

As the map shows it, six of the twenty-four urban blocks are dominated by churches, which are numbered and labeled as follows (fig. 34; pl. 3): (1) Sanct Miguel tecpan Cabezera; (2) Sanctiago Cabezera; (3) Sanct Joan Cabezera; (4) Sancta Maria Cabezera; (5) Sanct Pablo Cabezera; (6) Sanct Andres Cabezera. All of these churches are designated cabecera. Yet all are still found in modern Cholula and are quite closely grouped, within a kilometer or two of one another. Since cabeceras were head towns, centers for religious administration and religious proselytism, the presence of six cabeceras as well as a central monastery within a town of a few square kilometers is a remarkable event. A more common pattern would be a monastery with more widely scattered subject chapels, an arrangement shown on other Relaciones maps, such as Cuzcatlan (figs. 32 and 33) or Acapistla (fig. 31).

These six cabeceras were actually the colonial versions of Cholula's pre-Hispanic calpolli system. While all six cabeceras shown on the Cholula map today are known by Cholula's present residents as parishes, some preserve the Nahuatl names they bore in the pre-Hispanic period: there are Santiago Mixquictla (or Mixquitla); San Pablo Tecama; San Miguel Tianguisnahuetl; and Santa Maria Xixictla. Three of these Nahuatl names are place-names that correlate exactly to three of the twelve calpolli of Cholula discussed in the Historia Tolteca-Chichimeca and pictured on folios 26v–27r (Tecameca, Mixquiteca, and Tianquiznauaca; see figs. 54 and 55).[31] In fact, not only these three calpolli survived: scholars have found that all twelve calpolli survived, condensed and rearranged over time into the six cabeceras shown in the Relación Geográfica map. The regrouping is summarized in table 6. The numbers on the Relación Geográfica map are important because they preserve the rotational order, albeit in reverse and with a different starting point, that the calpolli were shown to have in the Historia Tolteca-Chichimeca folios 26v–27r (figs. 54 and 55).

Cholula seems to have had a tradition of mapping its social settlements, for another map contemporary with the Relación Geográfica map shows the component calpolli of Cholula. In this map, called the Codex of Cholula and dating to 1586, the calpolli are sym-

Table 6 Comparison of Cholula's Pre-Hispanic Calpolli and Colonial Cabeceras

Cholula Calpolli in Historia Tolteca-Chichimeca	Colonial Cabeceras shown on Relación map	Tecpan on 1586 Codex of Cholula
1. Tecameca →	5. San Pablo [Tecama]	Tecaman
2. Quauteca →	4. Santa María [Xixictla]	
3. Texpolca →	3. San Juan [Calvario]	
4. Mixquiteca →	2. Santiago [Mixquictla]	Mizquitla
5. Xiuhcalca 6. Uitziluaque 7. Chimalzolca 8. Tianquiznauaca →	1. San Miguel [Tianguisnahuetl]	Tianquiznahuac
9. Quetzaluaque 10. Xalteca 11. Calmecauahque 12. unknown →	6. San Andrés [Cholula]	

Notes: Numbers show rotational order; by the colonial period, the rotation has taken a new starting point, and reversed direction. Names in brackets are found in colonial and modern sources other than the Relación map.

bolized by *tecpan,* or "lordly places," the palaces of each leader, just as they are in the Historia Tolteca-Chichimeca and in maps of other regions, such as the Map of Chichimec History. Like the Historia Tolteca-Chichimeca folios 26–27r, the Codex of Cholula seems to show the town and its environs at an earlier point (or several earlier points) in its history (Simons 1967–1968). In the Codex, the center of Cholula is shown as the pyramid-hill, named as a *tlachihual tepetl,* just as it is on the Relación map. Around this central pyramid-hill are grouped various tecpan. The Nahuatl names appearing on the tecpan buildings of the 1586 Codex correlate to three of the six cabeceras of the Relación map (Tianquiznahuac = San Miguel Tianguisnahuetl; Mizquitla = Santiago Mixquictla; Tecaman = San Pablo Tecama). Most important, all three maps of Cholula show Cholula's polities, be they called calpolli or cabeceras or tecpan, following a similar pattern of rotation through time (refer to table 6).

Thus, at first glance this Relación map of Cholula shows a community ordered by a European grid and under the watchful eye of a European god, with no one living more than a block away from a church. Yet the artist's presentation of Cholula is firmly rooted in preconquest practice and reflects earlier social structure maps wherein the city was not an undivided whole but rather a league of various calpolli. In the 1586 Codex the calpolli are represented by their tecpan (palaces); in the Relación map, they are represented by parish churches. The rather abstract geometry of the Cholula Relación map is certainly more extreme than the grid of the city itself, but it allows the Relación Geográfica artist to show the highly structured political order of Cholula's internal parts at the time the map was painted.

Likewise, the Relación Geográfica of Cempoala seems also to have been inspired by a social settlement map, because it tries to reconcile the broad and irregular spread of Cempoala's territory with the political hierarchies that ruled the town (fig. 42; pl. 5). The Relación text reveals that Cempoala had four parts, which it lists as (1) Cempoala, (2) Tzaquala, (3) Tecpilpan, and (4) Tlaquilpa.[32] This quadripartite division with a highly specific ranking is typical of complex altepetl (Lockhart 1992: 15–8). Each of Cempoala's four parts, in turn, had either three or four subunits, or calpolli (fig. 59).

In contrast to the neat units into which we can parse Cempoala's social structure, the territories that were affiliated with each of these calpolli were scattered over a large region and were interlaced with the lands of Epazoyuca (Epazoyucan, Hidalgo), Cempoala's neighbor (fig. 60). The Relación Geográfica map is scored by red lines that divide the map into neat rectangles of differing proportions. The lines seem to mark the divisions between the lands held by Cempoala's fifteen calpolli and four altepetl as well as those of neighboring Epazoyuca.[33] Most of the places named on this map are now forgotten, for the region's population had declined and settlements were consolidated by the viceregal

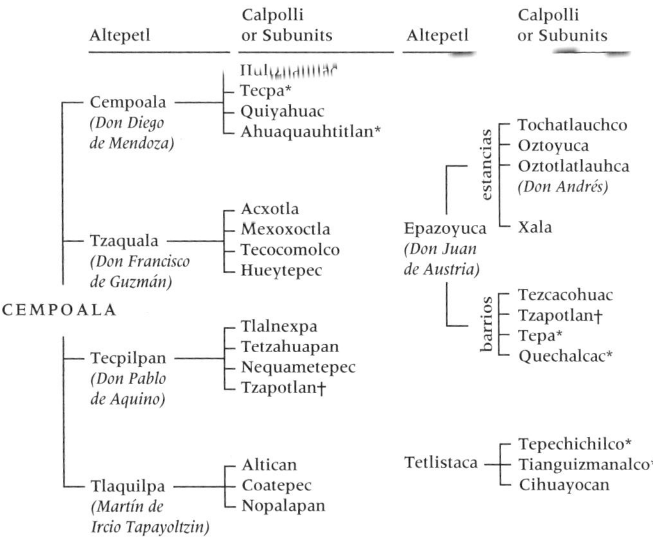

Figure 59. Diagram of social [illegible] in the Cempoala region. In their Relaciones and the accompanying maps, residents of Cempoala, Epazoyuca, and Tetlistaca provided an account of the units and subunits of their polities that is summarized here.

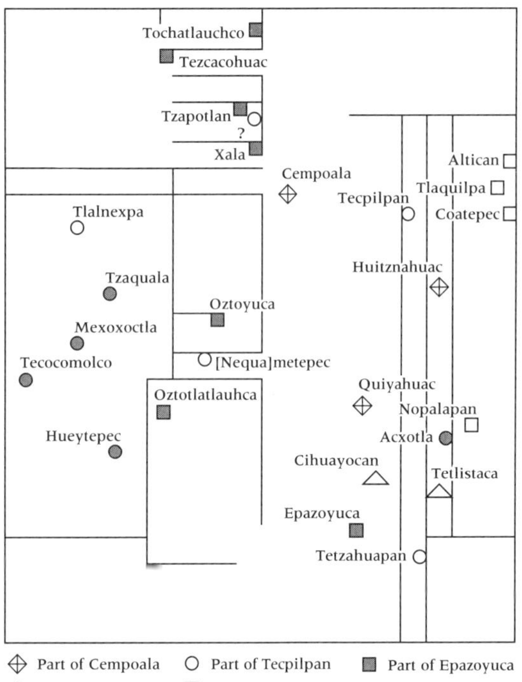

Figure 60. Diagram of the the Cempoala Relación Geográfica map. The Relación Geográfica map describes how land was parcelled out among Cempoala's various social units and those of its neighbors.

government in 1603 (AGN, Ramo de Congregaciónes, vol. 1, 59r–59v). However, the information on the map and in the Relación allows us to see land tenure in the region, with the parcels that belonged within the four altepetl that comprised Cempoala, as well as those of the adjacent communities, especially Epazoyuca, whose parts were described in the 1570s as so close that they were "nearly stuck together" to those of Cempoala (fig. 60; *Codice Franciscano* 1941: 14). While the altepetl may have thought of themselves as discrete entities, their lands, at least in Cempoala, were interlaced into a complex weave, and the complexity of land tenure in this region explains why a simple boundary map, like the one of the Historia Tolteca-Chichimeca with its fence of boundary markers, would be inappropriate, if not impossible, to create.

Like the map of Tenochtitlan found in the Codex Mendoza (fol. 2r; fig. 2), the Cempoala artist portrays regional rulers, and in doing so also depicts the complex social order of the region and its relationship to the territory shown (fig. 61). The rulers of the four parts of Cempoala are named in both the map of the Relación Geográfica and in its text; the text gives us their full names. Three of them, wearing *tilmahtli,* cloaks tied at the shoulder, and miter headdresses, a sign of nobility, face the plaza of Cempoala's church on the map. To the upper right of the church sits don Diego de Mendoza, gobernador of Cempoala. He faces a group of counters that stand for the number twenty-nine (the flag, *pantli,* stands for twenty); this count may refer to don Diego's age or to his years in office. Opposite him is don Francisco de Guzmán, gobernador of Tzaquala. To the right of the church is don Pablo de Aquino, gobernador of Tecpilpan. The fourth ruler, that of Tlaquilpa, is not seen in this group. Instead, he is pictured on the upper right edge of the map upon Tlaquilpa's land wearing a cloak of hide. The text names him as simply Martín de Ircio (not don Martín). The map identifies him, with both logographic name and gloss, as Tapayoltzin ("Honored Ball"), whose name is derived from *tapayoloa* ("ball") (Campbell 1985: 296), but not as Martín de Ircio.

The differences between the name and portrayal of Martín de Ircio Tapayoltzin and those of the other three rulers of Cempoala seem to represent ethnic differences within the community, because the Relación tells us that Tlaquilpa, where Tapayoltzin ruled, had many Otomi speakers, as opposed to the other altepetl, where Nahuatl was commonly spoken. Tlaquilpa, the last ranking altepetl, was probably the altepetl where the Otomi

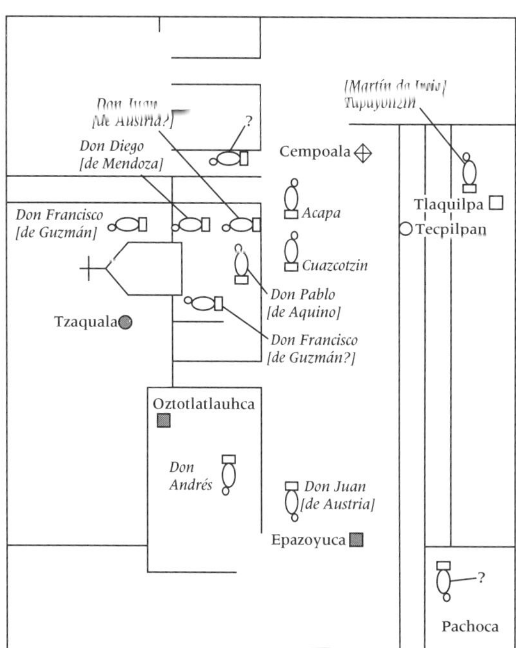

Figure 61. Diagram of the Cempoala Relación Geográfica map, showing indigenous rulers portrayed on it.

clustered, although the Relación text indicates there was a great deal of language mixing among the four parts of Cempoala.

Other rulers of the Cempoala region also appear on the map (fig. 61), indicating a political complexity that is not completely explained by the Relación text. Immediately below don Diego on the church plaza sits a figure named as don Juan. He is not identified in the text of the Cempoala Relación, but in all likelihood he is don Juan de Austria, named as the ruler of neighboring Epazoyuca in the text of its Relación Geográfica (RGS 6: 83). He appears on the Cempoala map a second time next to his town, Epazoyuca, at the bottom center of the map. Don Francisco, the ruler of Tzaquala, who faces the church plaza, is also shown again above the church, sitting within the lands marked as Tzaquala's. Yet another lord sits next to Pachuca, but he is not named. In addition, a don Andrés faces the church of Oztotlatlauhca, but he is not further identified in any Relación.[34]

Of a different order are the four men who wear cloaks of rough hide. One of them is Tapayoltzin; the others may also be Otomi leaders. Two of them appear in the center of the map, beneath the place-name of Cempoala, and may have been the leaders of Otomi calpolli within a larger altepetl. They have Nahuatl names rather than the Spanish names and titles of their counterparts, and they are named with logographs as well as alphabetic transliterations. The first is Acapa ("Upon the Reed"), from *acatl,* "reed," and *-pan,* a suffix meaning "upon," conveyed here by its homonym *pantli,* "banner." Below him is Cuazcotzin, the meaning of whose name I cannot identify. A third figure in a hide cloak sits behind don Juan near the church but his name is not transliterated.

With the nobles that cluster together around Cempoala's main Franciscan church, the map seems to show the circumstances of its own creation, as don Diego de Mendoza of Cempoala, don Francisco de Guzmán of Tzaquala, don Pablo de Aquino of Tecpilpan, and probably don Juan de Austria of neighboring Epazoyuca came together to answer the questions posed to them by the corregidor, Luis Obregón. His text echoes with their voices as they, and other "old and aged indians," tell of preconquest custom and life (RGS 6: 73). The artist of the map seems to have felt that their convocation was worthy of record and set them down on the map, like the official portrait of a summit meeting.

Not only does the Cempoala map tell us of the moment of its creation, it also bears the imprint of earlier versions of this map. At the center of the map is a picture of a palace and a Nahuatl gloss that reads *mexico tlatovani ytzcovatzi[n] ycha[n],* meaning "the house of the Mexica ruler, Itzcoatzin."[35] Itzcoatl (Itzcoatzin is the honorific form), who ruled the Aztec empire from 1428 to 1440, conquered most of the Cempoala region during his reign (RGS 6: 75); given the wealth of references to the pre-Hispanic past in both text and picture, this map may have been based upon an earlier one, one made in the wake of the establishment of Itzcoatl's domain over the region, when Cempoala's rulers, in the aftermath of the Aztec conquest, may have needed to reconfigure their territorial and social arrangements.

CONCLUSIONS

In answering the Relación Geográfica questionnaire, native mapmakers were not painting in a void. Rather, they were drawing on a rich tradition of mapping their communities that gave them cartographic histories and social settlement maps as models. Such maps allowed them to portray ideas of community that expressed themselves through the bonds of territory, history and social order. In a number of cases, the Relación Geográfica map is clearly traceable to these two strands of the native tradition. This chapter gave short shrift to writing in the maps, especially the logographic writing that indigenous peoples had used throughout the colonial period. Writing offers an arena in which we can examine at close range the complex cultural interchanges in the ways the people of colonial New Spain were coming to envision themselves, and will be the subject of the next chapter.

CHAPTER SIX

Language and Naming in the Relaciones Geográficas Maps

When the learned mestizo don Juan Bautista de Pomar of Tetzcoco (Mexico) penned his Relación Geográfica, he included a prescient lament to the king:

> And it is known that, if [indigenes] were to have alphabetic writing, they would come to understand many natural secrets, but, because the[ir] paintings are not capable of retaining the memory of that which is painted, [such knowledge] does not pass onward. When he who understands them best dies, his understanding dies with him (RGS 8: 86).

As Pomar wrote these lines, the war between traditional logographic writing and alphabetic writing, of which Pomar describes one battle, was at full tilt.

Looking at the Relación Geográfica map that came from the town of Misquiahuala, which lay about seventy-five kilometers from the Valley of Mexico, with its boundary of hill glyphs, its logographic toponyms and its depictions of native rulers, it looks as if logographic writing in 1580 still had the upper hand (fig. 62; pl 7). That the Misquiahuala artist would draw upon traditional models shows not only that he or she had such models at hand, but also that such models tenaciously offered valid means of representing the community; within Misquiahuala, this map was a true portrait. In making this map, the Misquiahuala artist had firm control of

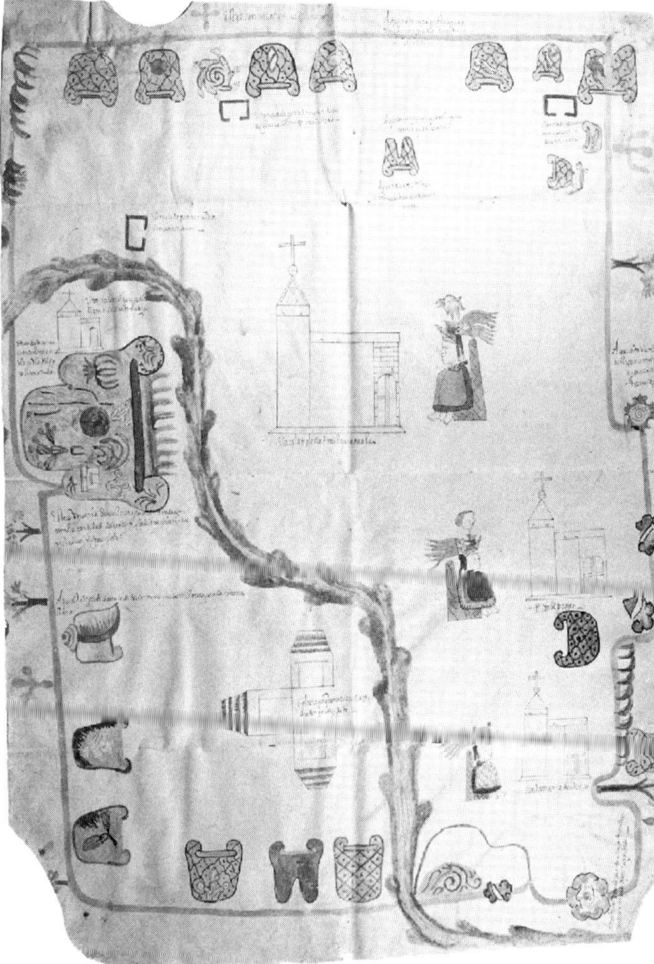

Figure 62. The Relación Geográfica map of Misquiahuala, c. 1579. (See also pl. 7.) This indigenous map is unusual in that it was painted on sized hide, unlike the other maps, which were painted on paper. Hide may have been easily available in this burgeoning ranching area. Indigenous toponyms line the sides of the hide, defining the local landscape. Size of the original: 78. 5 (top) x 55.5 (left) x 77 (bottom) x 47 cm (right). Photograph: B. Mundy; reproduced courtesy of the Benson Latin American Collection, The General Libraries, The University of Texas at Austin (JGI xxiii-12).

the pictorial representation as he or she moved within the image culture of New Spain, a culture that was both created for indigenes (by the friars) and by them.

But if we look at the Misquiahuala map remembering the somber tones of Pomar's lament, we see another aspect of the history of cartography of New Spain and the history of representations: the artist, while controlling the pictorial representations, did not write alphabetically: the inscriptions (used herein to mean alphabetic inscriptions) were written by Francisco Fernández de Córdova, a scribe who also wrote the text (fig. 63). In not being able to determine what the alphabetic text on the map should say, the Misquiahuala artist lost control of an important part of the written means of representation, and thereby the entire community of Misquiahuala, like others across New Spain, ceded the power to represent itself.

That the native artist did not write in Spanish alphabetically on this map, and the Spanish scribe did, was the outcome of a linguistic upset that had been set into play the moment the Spanish entered Mesoamerica. As the Spanish conquered the land, their language displaced Nahuatl, the tongue of the indigenous conquerors that we have come to call the Aztecs, to become the new language of the new Hispanic ruling elite. And just as surely as

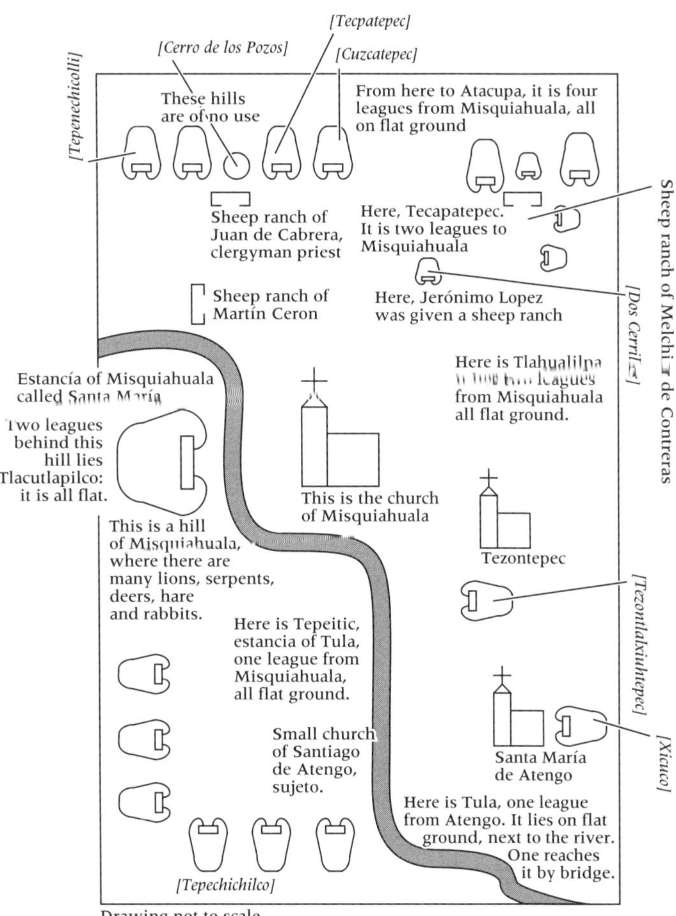

Figure 63. Drawing after the Relación Geográfica map of Misquiahuala, with translations of the Spanish inscriptions on the map.

Nahuatl was knocked off the top rung of the communication ladder, so was the largely logographic writing system used to express it. Now the alphabetic writing introduced by the Spaniards held sway.[1]

Looking at language in the maps as it manifests itself in writing gives us a way of seeing beneath the calm, unruffled surface of pictorial representation, which we examined in the previous chapter, to the swirling currents beneath. The Relaciones Geográficas show us how the shifts in writing and language that followed in the wake of the conquest could lead directly to the dispossession of New Spain's native communities. In the Relaciones Geográficas, we find both Nahuatl and Spanish languages and both alphabetic and logographic writing used as different authors, indigenous and Spanish, added to the same map. With each addition, each author added a new veneer of meaning over a previous one. Because we can often distinguish the work of different authors and see the different layerings of images and texts, we can trace how the information that maps carried and the meanings they had shifted as they traveled out of the sphere of the indigenous cabildo and into the world that the Spanish empire dominated.

On the maps, much of the written language of whatever type is nominative, an understandable bias, since naming is the principal means humans use to filter the raw material of space into the sphere of their cognition: once given a name, an otherwise undistinguished space becomes a place. Naming is at the heart of mapping, since with a name, a place can be singled out and then represented on and with the map. Most of the words on the Relaciones Geográficas maps, be they written alphabetically or logographically, are toponyms. Well before this group of maps was created to represent different polities across Mesoamerica, naming was as necessary to mapping as the vibrating of strings is to a Beethoven symphony. We saw how crucial naming is in our examination of the landscape created out of logographic names that appears on page 36 of the Codex Zouche-Nuttall (fig. 46). Through an examination of language and writing, most specifically as they were manifest in naming in the Relación Geográfica corpus, we shall see how indigenous communities had language and power slip from their control.[2]

NATIVE LANGUAGE AND LOGOGRAPHIC WRITING

At the time of the Spanish conquest, Mesoamericans had at their disposal a well-developed writing system, one that was largely logographic, with its signs standing for entire words or component morphemes. Logographic writing is often called "picture writing" because its signs are representational images, that is, the picture of a fish stands for the word "fish." In maps, we glimpse this written language mainly in the logograms used to write place-names, as in the place-name of Xilotepec discussed in the chapter 5 (fig. 45).[3] Because Mesoamerican writing was composed of images, with pictures of things used to symbolize a particular word or part of a word on maps, this picture-writing was closely knitted into its pictorial, or nontextual, substrate. There is no better way to realize the intertwining of word and image than to try to separate the two. We clearly distinguish between the word and the image on the maps that we use, but to do so on a work like page 36 of the Codex Zouche-Nuttall is a greater challenge (fig. 46). Is the U-shape that defines the Apoala Valley, for instance, the image of that valley (as it seems to be), or is it a name of a place written logographically, that is to say, a toponym? Are the two rivers images of the valley's two rivers or (as they seem to be) are they toponyms of rivers? The distinction between text and image is rarely sharp: in Mesoamerican written works, text is interspersed with images, not set off in a separate cartouche or box, and in addition, the same images that function as pictures (or nonlinguistic symbols) can also, in another context, be meant to be words (or linguistic signs). Even manuscript scholars are divided on the question of where to draw the line between the written word and the pictorial image: one camp favors reading almost all images as written words; the other, in which I include myself, chooses to interpret only a small portion of images as written words.[4]

The words expressed logographically by the people of Central Mexico were frequently specific dates, personal names, and toponyms; their relationship to one another and to their pictorial framework conveyed a narrative. Such a writing system is, while specific about individual words, quite unprescriptive of longer narratives. It contains a plethora of nouns but only a few verbs, and no pronouns, articles, adverbs, or adjectives. That written language in Central Mexico should have followed the path it did is no reflection on the cultural adequacy of its Mesoamerican creators. Rather, it is the writing system of a society that held oral expression to be the supreme means of communication. In Central Mexico, the reader of the text fleshed out the bare script in much the same way as a jazz musician improvises on a simple score, using it to trigger his or her memory and incite his or her imagination, to give each text

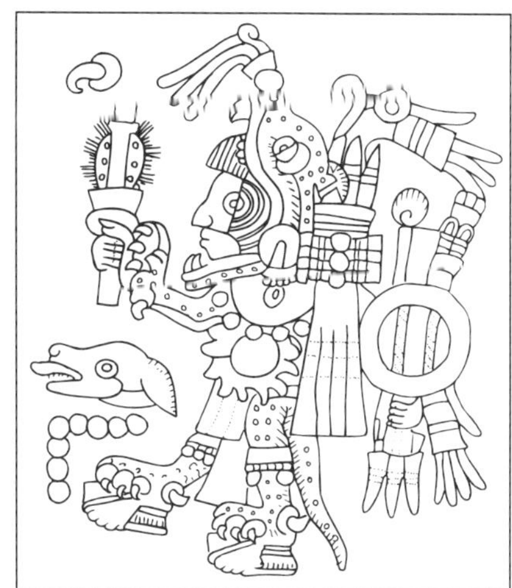

Figure 64. The Mixtec warlord 8 Deer Jaguar Claw. His calendrical name appears next to his foot at left. Drawing after the Codex Zouche-Nuttall, page 43.

full meaning in performance. It is no accident that Nahuatl speakers called their leaders *tlatoani,* meaning "speaker," and that *Nahuatl* is thought to mean "clear speech," or that *pohua* means both "to read" and "to recite" (Lockhart 1992: 327).

In addition to allowing full play to the powers of the reader/reciter, Mesoamerican logographic writing could to some degree slip through the linguistic borders that balkanized Central Mexico, and smuggle meaning from one language group to another.[5] Take the powerful Mixtec warlord 8 Deer Jaguar Claw, named in part after the calendrical name of his birthday, and written with eight dots connected to the picture of a deer's head (fig. 64). Upon seeing his image, a Mixtec speaker would call him *na cuaa,* a Nahuatl speaker *chicuei mazatl,* but both would identify him as the same person. No such interlegibility exists between different language speakers when it comes to alphabetic script. The date 1/5, written alphabetically in Spanish as "cinco de enero," bears little relation to its English form of "January fifth."

NAMING AND POWER

The fact that Central Mexico shared a common writing system never meant that indigenous languages were all on equal footing. Rather, in the late fifteenth century, Nahuatl reigned supreme: it was the language of the Aztec conquerors and as a result, the second language of non-Nahua elites throughout Central Mexico who had learned the imperial tongue. Doña Marina, a slave whom Hernán Cortés acquired on the Gulf Coast in 1519, was a Putun Maya woman who was able to speak Nahuatl as well as her mother tongue (Cortés 1986: 73; Díaz del Castillo 1956: 66–8; López de Gómara 1966: 56–7). Using her as an interpreter, Cortés was able to negotiate with elites and rulers throughout Central Mexico, eventually banding them against the Culhua-Mexica.

For the Culhua-Mexica in Tenochtitlan, the Nahuatl language was the currency of empire: not only did they encourage non-Nahuatl-speaking elites to use their tongue, they also imposed Nahuatl names upon their tributary states, giving each a Nahuatl name if it lacked one. We see these connections between naming, renaming, and possession clearly made in the Codex Mendoza, a book from about 1542 that contains the map of Tenochtitlan (fig. 2) that was discussed in the preface. Within this book, running from folios 17 to 55, is a hand-drawn copy of a preconquest tribute list that showed the goods that vassal states had once owed periodically to the pre-Hispanic Aztec state (fig. 65). This part of the Codex Mendoza was the Aztec tax collector's bible. These pages of the Mendoza are one of the two surviving native records of the extent of the Aztec domain, the other being a close relative, the Matrícula de Tributos.[6]

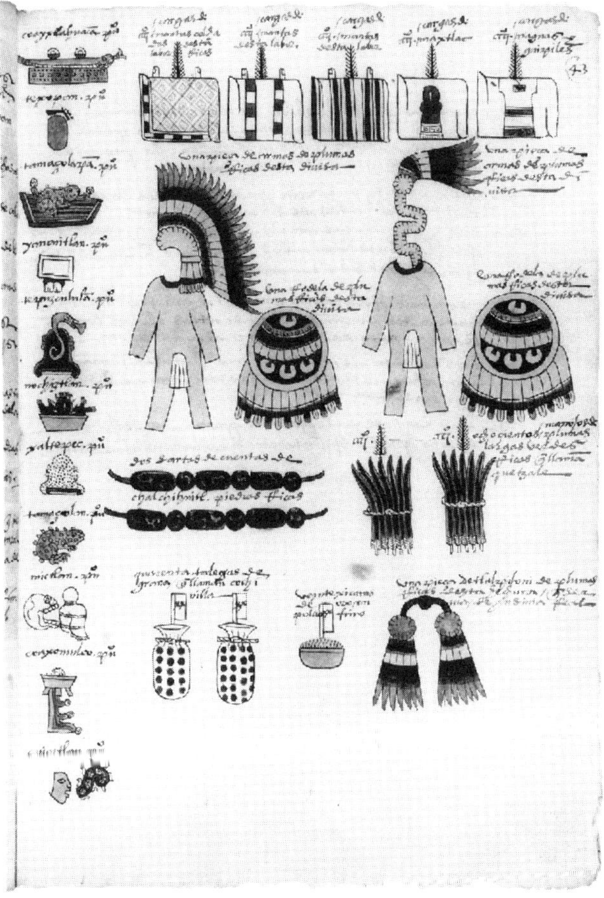

Figure 65. The Codex Mendoza, fol. 43r, c. 1542. Size of the original: 32.7 x 22.9 cm. Photograph courtesy of the Bodleian Library, Oxford (MS. Arch. Selden. A. 1 fol. 43r).

This being so, how did the Aztec represent their domain? In the Mendoza, the Aztec empire is shown as listings of place-names. Each page shows one group, sometimes more, of place-names, listed by region, which run along the side and at times the bottom of the page. For instance, the names of tributary community-kingdoms of a region in Oaxaca appear along the left edge of folio 43r (fig. 65). At center are the luxury goods demanded of this region as tribute. The tribute list, the only representation of their entire empire that the Culhua-Mexica were known to have made, shows us that the Culhua-Mexica in effect collected place-names from conquered communities, using these signs to represent their empire. In many respects, the pages of the tribute list are a written parallel to the temple within Tenochtitlan's main temple precinct, wherein deity images of conquered commu

Table 7 Comparison of Logographic Place-Names from the Relaciones Geográficas Maps and Their Counterparts in the Codex Mendoza

Place-Name and Nahuatl Etymology	Relación Logograph	Mendoza Logograph	Mendoza placement and explanation
Acapistla "place of points" *yacatl* = nose *yacapitztli* = point *-tlan* = place of			Appears on fols. 8r and 24v, as *"Yacapichtlan."* The ant, or *azcatl,* of the Mendoza version seems to be a phonetic marker of the initial "aca" sound.
Cempoala "place of twenty" *cempohualli* = twenty			Appears on fol. 21v as *"Çenpoalan."* In both the Relación and Mendoza, the head is that of a Totonac, with whom Central Mexicans associated Cempoala.
Culhuacan "place of the Culhua" *colli* = curved *-hua* = possesive *-can* = locative			Appears on fol. [illegible] as *"Colhuacan."* The Culhua were an ethnic group; this logograph is expressed with a homonym, here portrayed as a hill with a curved top.
Guaxtepec "on the hill of the guaje tree" *huaxin* = guaje tree *tepetl* = hill *-c* = locative			Appears on fols. 7v and 24v as *"Huaxtepec."* The tree shown in both Relación and Mendoza is heavy with red seed pods.
Macuilsuchil "Five Flower" *macuilli* = five *xochitl* = flower			Appears on fol. 44r as *"Macuilxochic."* Macuilxochitl was a central Mexican deity, but is not directly portrayed here. Instead, flowers and counters express the name.

Table 7 *(continued)*

Place-Name and Nahuatl Etymology	Relación Logograph	Mendoza Logograph	Mendoza placement and explanation
Misquiahuala "place of mesquite circles" *mizquitl* = mesquite *yahualtic* = circular		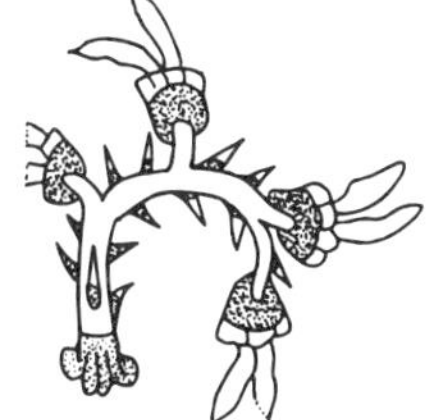	Appears on fol. 27r as *"Myzquiyahuala."* The Mendoza version shows a bent mesquite bush, while the Relación shows a hill symbol enclosing a circle and various cactus and agave plants.
Papantla "place of the pepe birds" *papanes* = type of birds *papatli* = matted hair of priests *pantli* = banner *-tlan* = place of	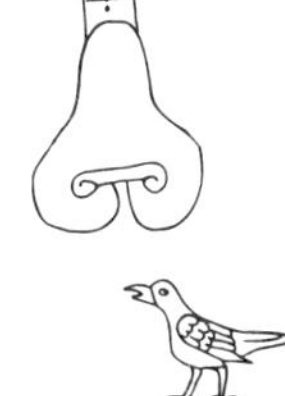		Appears on fol. 52r. The Mendoza logograph employs both a banner and a hank of hair, while the Relación uses the bird.
Tehuantepec "jaguar hill" *tecuani* = wild animal, jaguar *tepetl* = hill *-c* = locative			Appears on fol. 13v as *"Tequantepec."* Logograph is quite similar on Mendoza and Relación.
Teutenango "place of divine walls" *teotl* = deity, divine *tenantli* = wall *-c* = locative	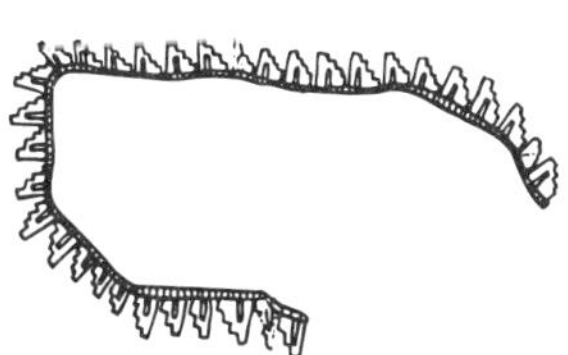		Appears on fols. 10r and 33r as *"Teotenanco."* Both logographs use the same stepped motif to signify "wall."
Texupa "upon the blue" *texutli* = blue *-pan* = upon	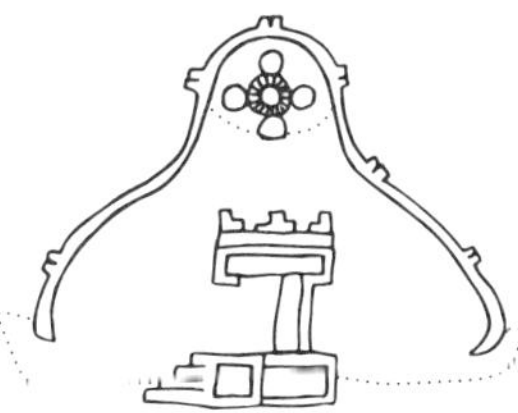		Appears on fol. 43r as *"Texopan."* The Mendoza version uses a foot over a blue circle to convey the idea of "upon."

nity-kingdoms were held, as if hostages; on the tribute list it is the place-name that stands captive, a representation of the vassal state.

In another light, the place-names in this Aztec tribute list also are parallel to the pictures of cities with which Philip II lined the halls of the Alcázar, for both Aztec place-names and Spanish cityscapes served as emblems of regions under state control (fig. 5). While the presumed artist of the Alcázar cityscapes, Anton van den Wyngaerde, was somewhat oblique about each city's contribution to the maintenance of the Spanish state, the Aztec tribute list is direct; next to each regional grouping of place-names are pictures of the tribute—the necklaces of jade beads, the war costumes of ocelot skin, the bundles of iridescent quetzal feathers, the sumptuously woven mantles—owed to the conquering Culhua-Mexica lords.

While the conquered kingdoms roped together into the Aztec tribute state may have spoken Mixtec, Zapotec, Otomi, or another language, their central Mexican overlords took it upon themselves to give these foreign states names in Nahuatl before entering them onto the master list of empire, such as the one copied into the Codex Mendoza.[7] Typical is the case of Texupa: after the Culhua-Mexica conquered the Mixtec town known to its residents as *ñuu ndaa,* meaning "place of blue," they rechristened it Texupan, a Nahuatl name meaning "upon the blue" (from *texutli,* "blue," and *pan,* "upon"; Alvarado 1962: 96; Molina 1977: fol. 18r; RGS 3: 220), and registered it in the Mendoza (in the top left of fol. 43r) as a blue oval topped by a footprint to convey the idea of "upon" (fig. 65; table 7). While the Nahuatl *Texupan* is a rough translation of the indigenous *ñuu ndaa,* other cases show us that the Culhua-Mexica also transliterated names in other languages into Nahuatl, and sometimes gave places Nahuatl names that bore little relation to the indigenous name, as far as we know.[8] Whether through translation or transliteration, the Culhua-Mexica were intent on the symbolic domination of subject communities through the imposition of a Nahuatl name, which they wrote down logographically in tribute lists like the one that inspired the Codex Mendoza pages.

The Mendoza pages are the view of the Aztec empire from the central vantage of Tenochtitlan, and they show the hubris that marks all empires as places were conquered and then renamed, as local leaders were cowed and tribute exacted. Conquered altepetl and kingdoms that previously were autonomous were reduced to a logographic place-name and a list of goods on the pages of the Mendoza, and this presentation mirrors to some degree how Moteuczoma and other Culhua-Mexica rulers envisioned their tribute states. In some respects, the native maps of the Relaciones Geográficas are the antithesis of the Mendoza pages, in that in the native maps, as in the community maps that preceded them, we abandon the distorted perspective of the imperial hub to take up the

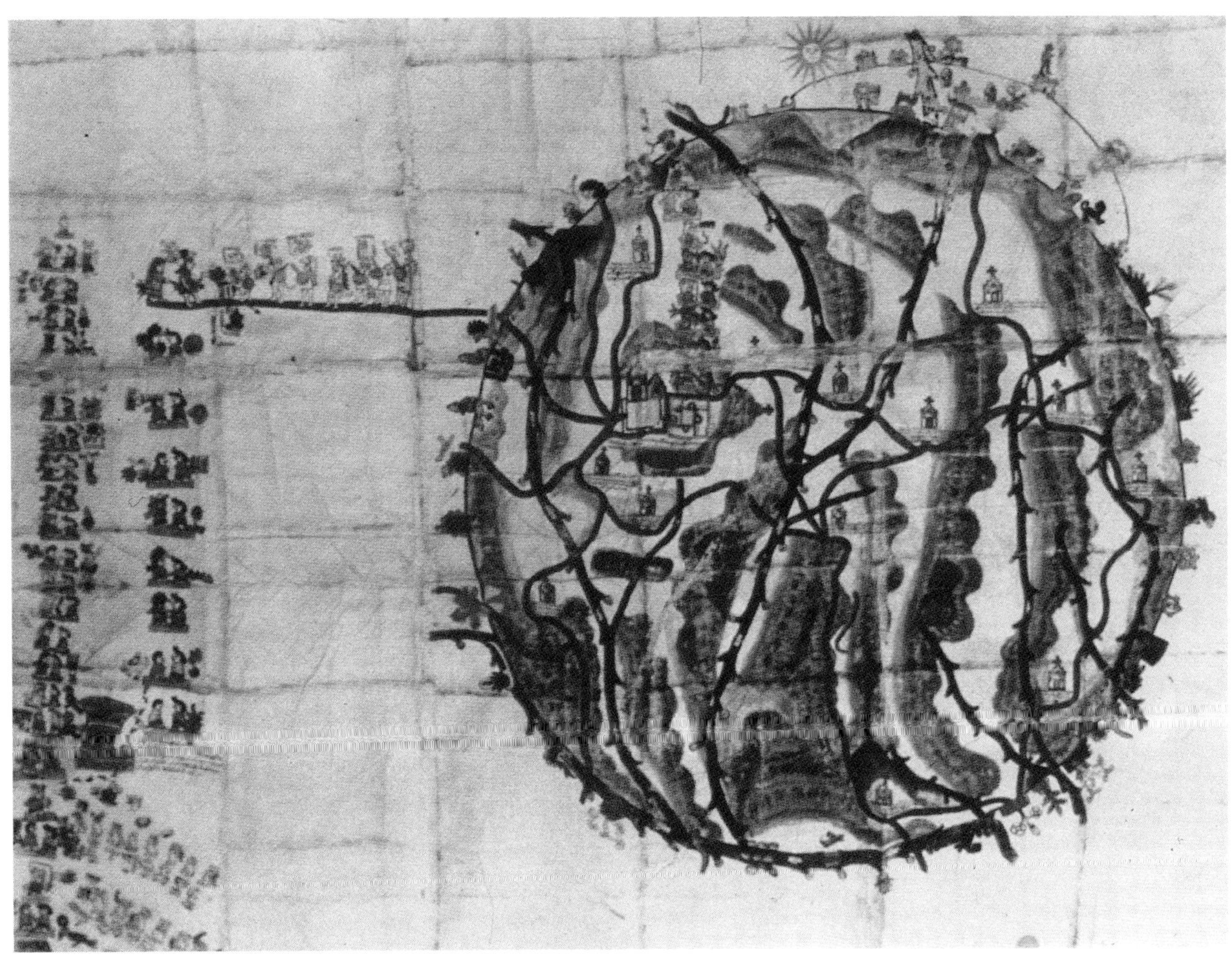

Plate 1. The Relación Geográfica map of Teozacoalco, 1580. (See fig. 10, p. 26.) Photograph courtesy of the Benson Latin American Collection, The General Libraries, The University of Texas at Austin (JGI xxv-3).

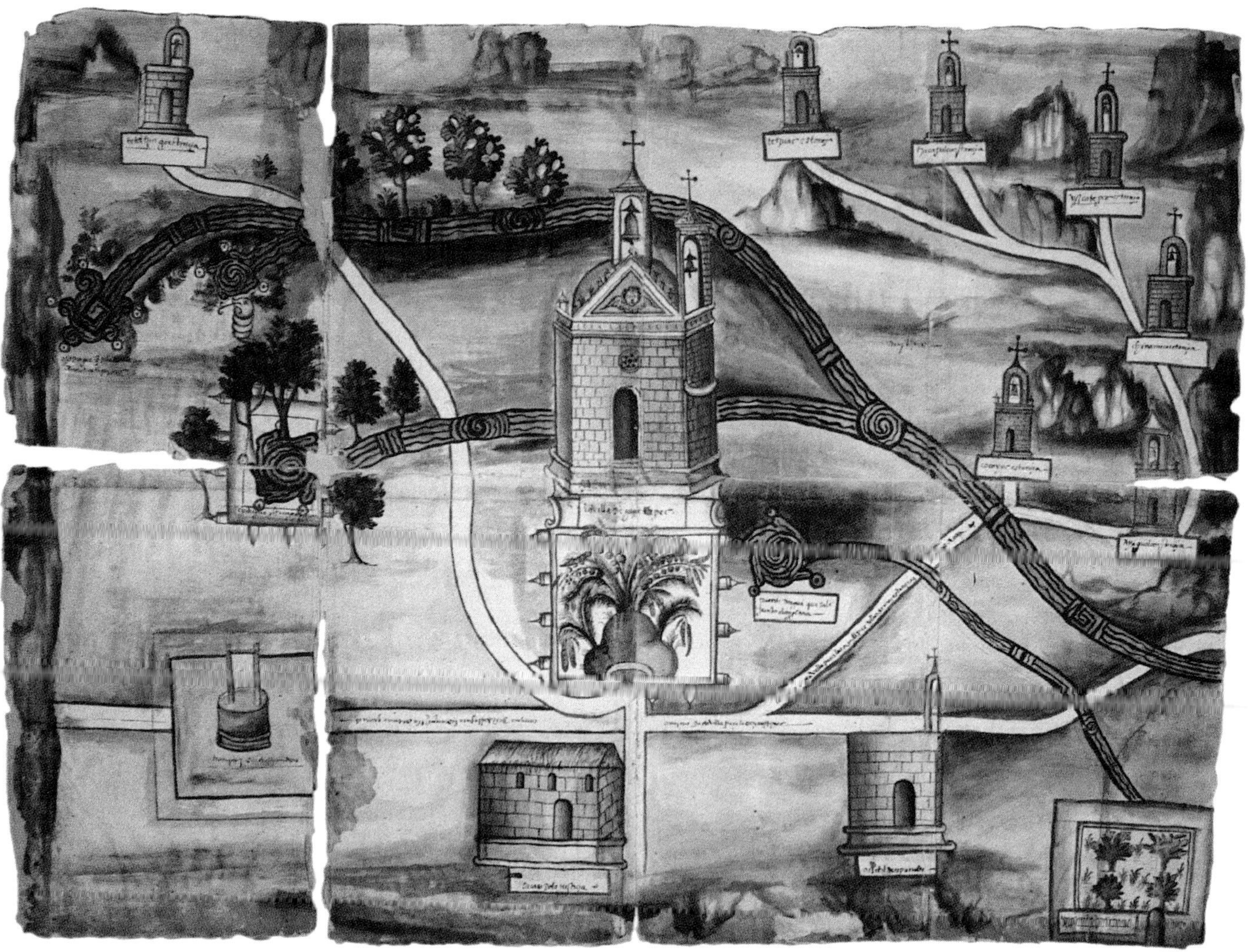

Plate 2. The Relación Geográfica map of Guaxtepec, 1580. (See fig. 30, p. 68.) Photograph: B. Mundy; reproduced courtesy of the Benson Latin American Collection, the General Libraries, The University of Texas at Austin.

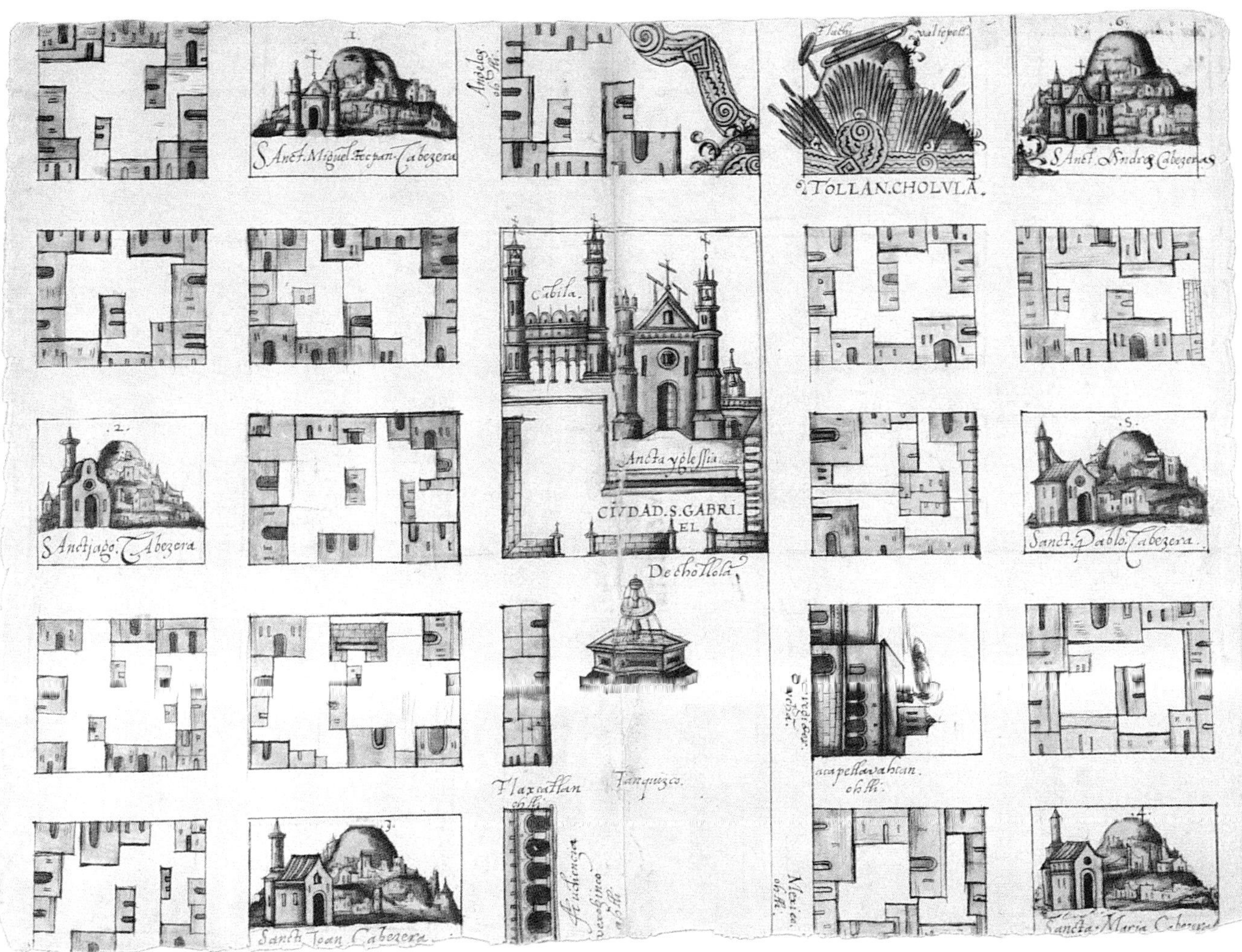

Plate 3. The Relación Geográfica map of Cholula, 1581. (See fig. 34, p. 72.) Photograph: B. Mundy; reproduced courtesy of the Benson Latin American Collection, The General Libraries, The University of Texas at Austin.

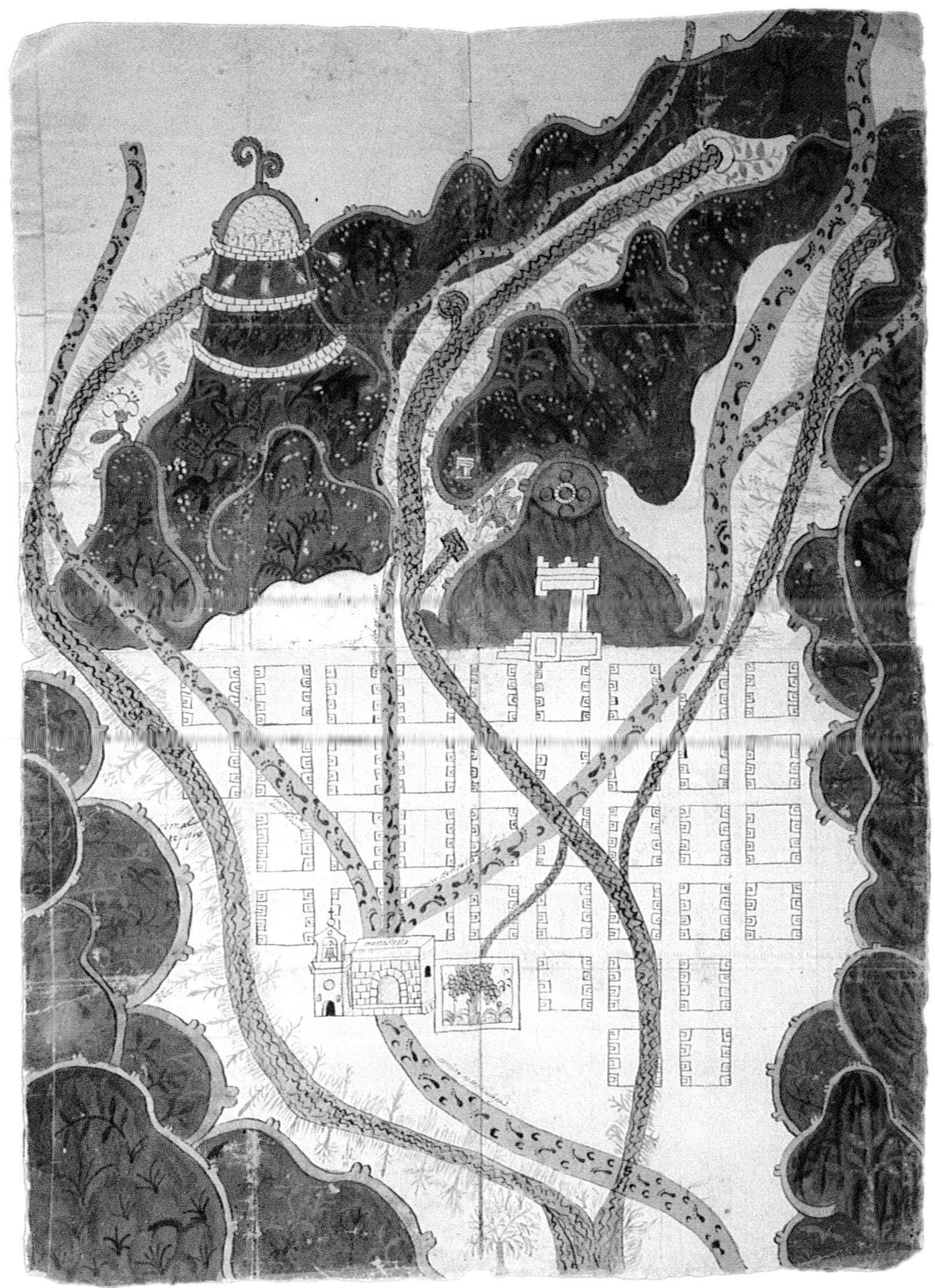

Plate 4. The Relación Geográfica map of Texupa, 1579. (See fig. 38, p. 80.) Photograph: Juan Jiménez Salmerón; reproduced courtesy of the Real Academia de la Historia, Madrid (9254/ 4663xvii).

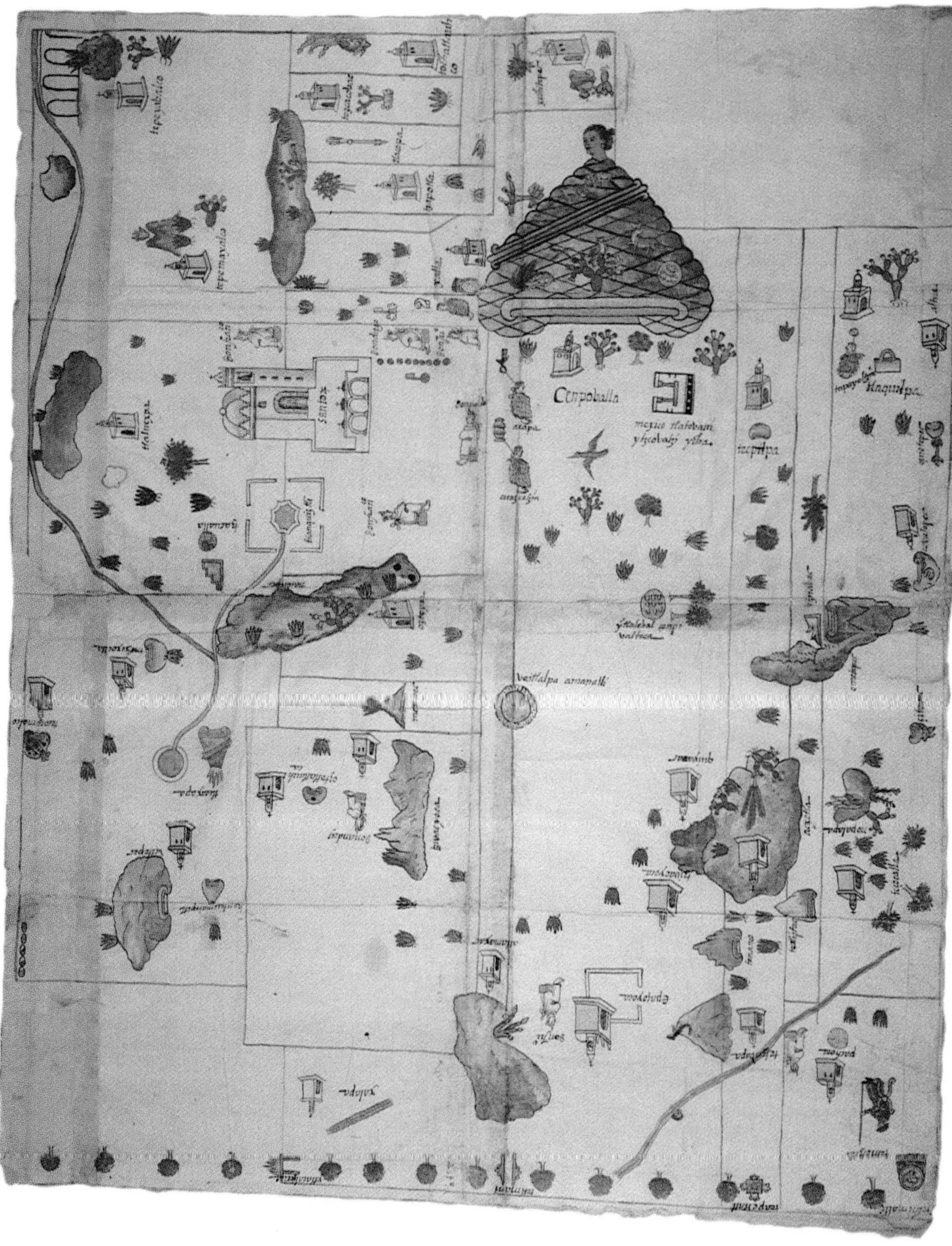

Plate 5. The Relación Geográfica map of Cempoala, 1580. (See fig. 42, p. 95.) Photograph: B. Mundy; reproduced courtesy of the Benson Latin American Collections, The General Libraries, The University of Texas at Austin.

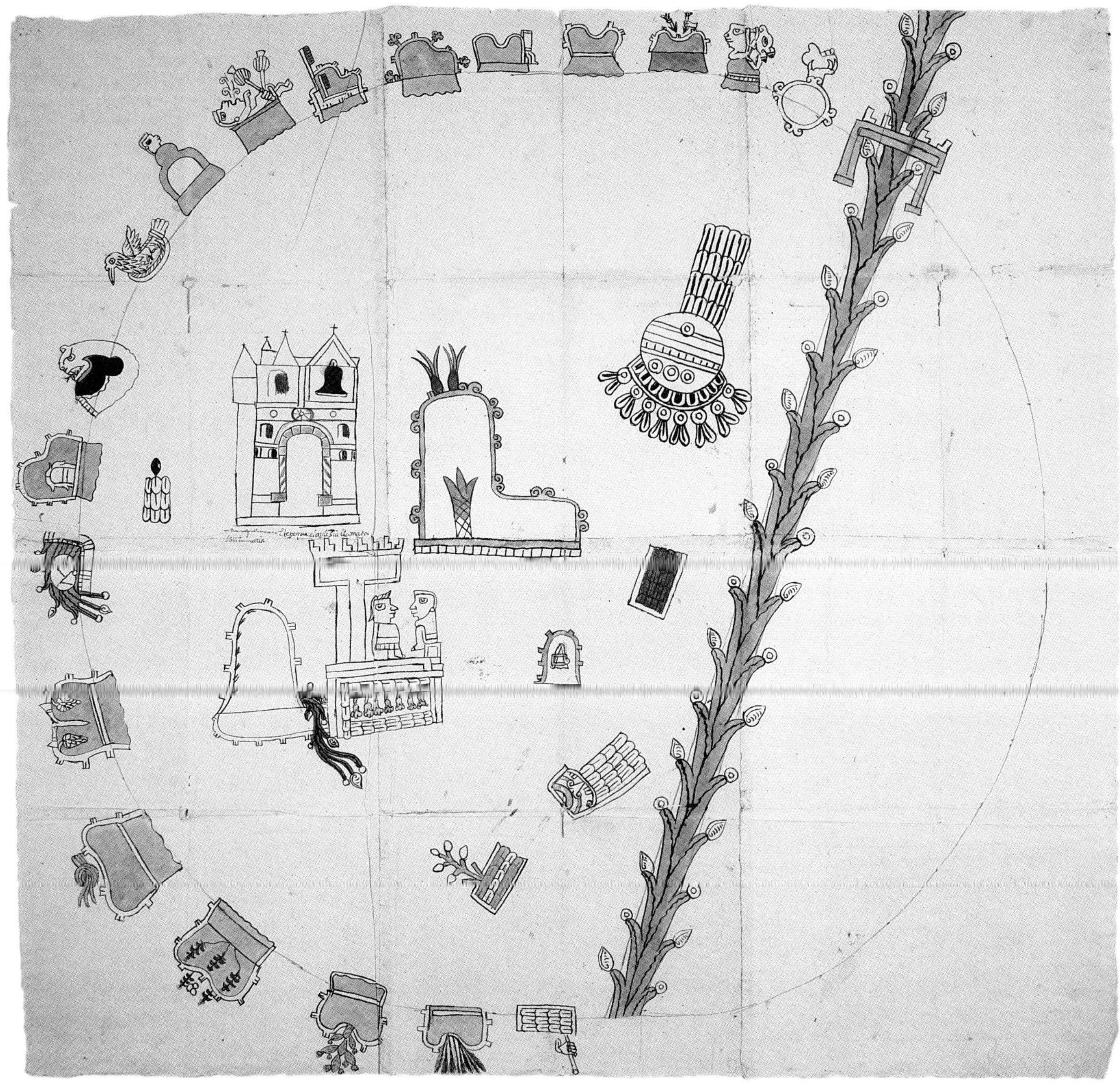

Plate 6. The Relación Geográfica map of Amoltepec, 1580. (See fig. 51, p. 113.) Photograph courtesy of the Benson Latin American Collection, The General Libraries, The University of Texas at Austin.

Plate 7 *(opposite)*. The Relación Geográfica map of Misquiahuala, c. 1579. (See fig. 62, p. 136.) Photograph courtesy of the Benson Latin American Collection, The General Libraries, The University of Texas at Austin.

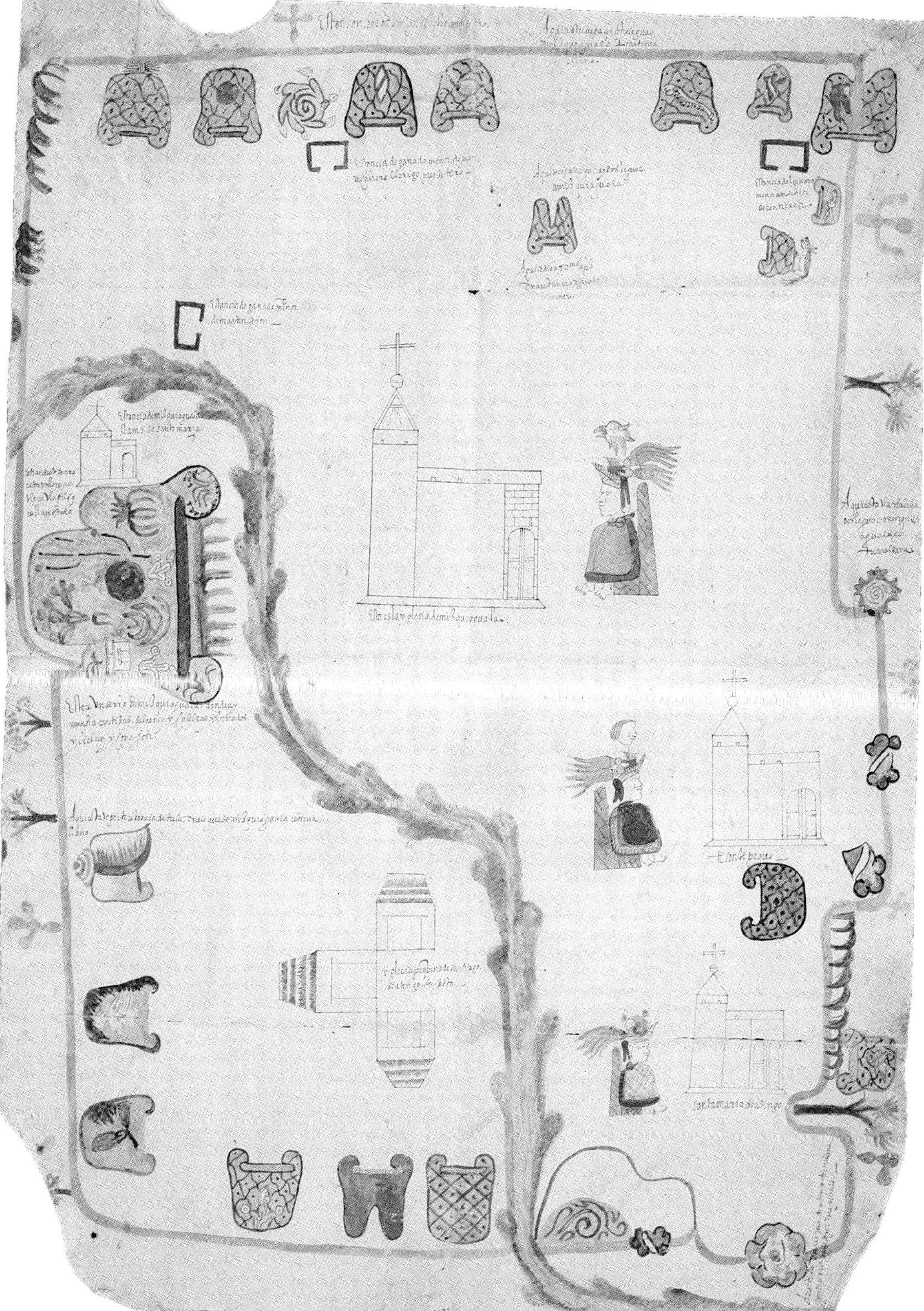
Esta es la yglesia demiẞquiaguala.

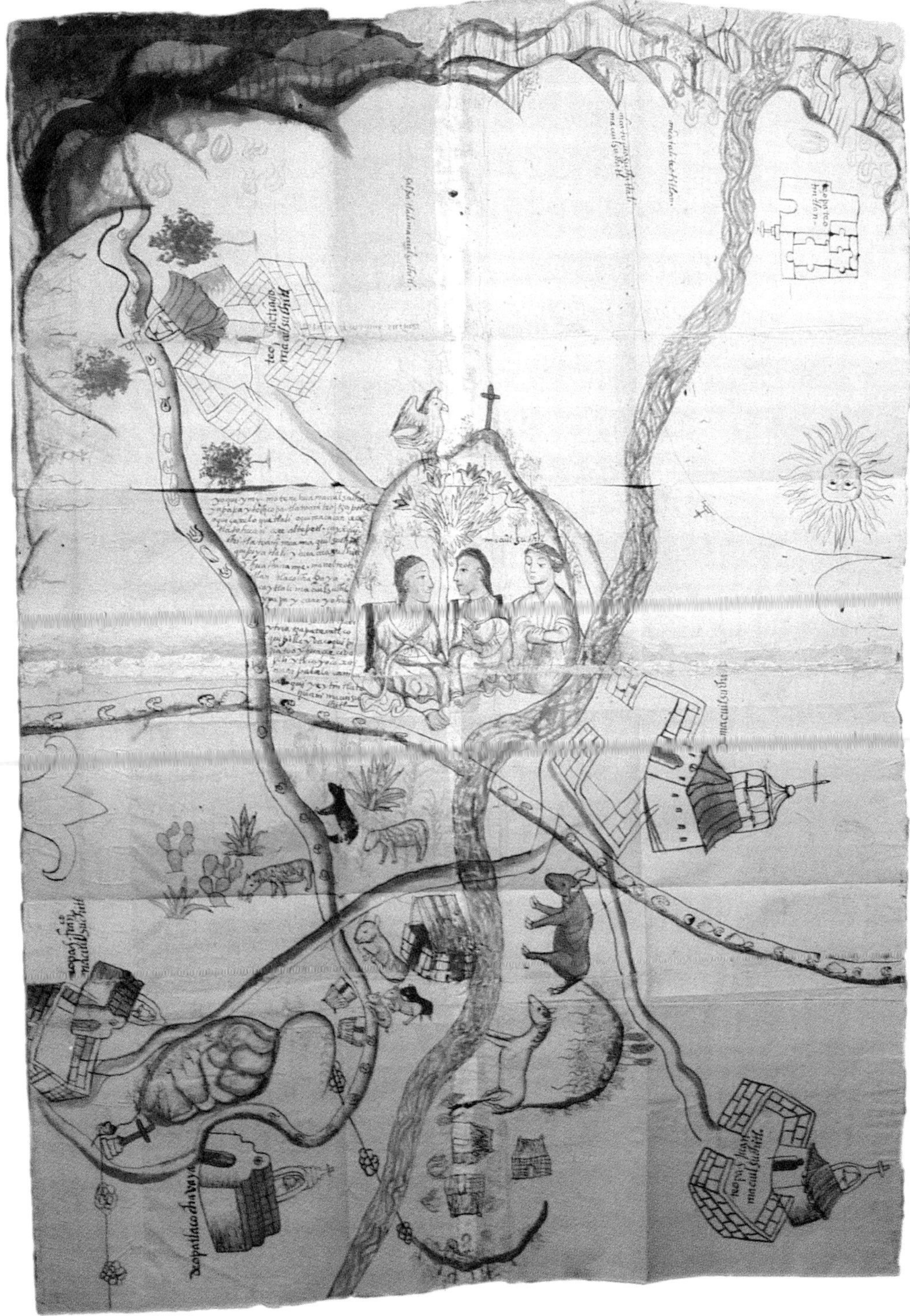

Plate 8. The Relación Geográfica map of Macuilsuchil, 1580. (See fig. 79, p. 162.) Photograph: Juan Jiménez Salmerón; reproduced courtesy of the Real Academia de la Historia, Madrid (9254/ 4663xix).

distinctly local and parochial vantages offered by scores of communities in New Spain, although, of course, some forty years after the creation of the Codex Mendoza. In the Relaciones Geográficas, the provinces counter the empire.

PLACE-NAMES IN THE RELACIONES GEOGRÁFICAS MAPS

On one central issue, however, the imperial Codex Mendoza and the parochial Relación map concur, for both depend on logographic place-names as the primary means of symbolizing communities. Logographic place-names—that is, logographic writing used to create indigenous toponyms—abound in the Relaciones Geográficas maps. Native artists favored logographic writing to represent in particular the name of the altepetl or community-kingdom they were depicting. On many of the native maps, a logographic place-name representing the community's name occupies the center, the most important point, on the maps. Logographic place-names retain their distinctive forms and set of referents even on maps that otherwise embraced European conventions. The Relación Geográfica map of Guaxtepec (fig. 30; pl. 2) was painted by a native artist who was probably trained at Guaxtepec's Dominican monastery and exposed to European prints therein, which doubtless inspired the background of the map, a painted landscape, done in delicately shaded colors. Set against this illusionistic landscape, the logographic place-name of Guaxtepec vies with the Dominican monastery as being the map's central feature. The Nahuatl place-name, meaning "hill of the huaxin tree," is composed of an evenly rounded tepetl symbol on whose top sits a luxuriant *huaxin* tree, a canopy of feathery leaves over heavy seed pods.[9] This symbol was exactly how the Codex Mendoza artist represented Guaxtepec on both folio 7v and folio 24v, where he or she shows a hill topped by a tree whose three branches are heavy with red seed pods (table 7). On another Relación Geográfica map, from Cempoala, the large logographic place-name of Cempoala is central, a tepetl topped by the head of a man wearing regional dress (fig. 42; pl. 5). A similar figure, with hair bound with a red cord and ears and nose adorned with turquoise jewelry, is used along with a tepetl symbol to denote Cempoala in the Codex Mendoza (fol. 21v; table 7).

It is the map from Muchitlan (Mochitlan, Guerrero), of all the maps in the corpus, wherein logographically written place-names are nearest to the roles they played in preconquest manuscripts, where they were the primary conveyors of the map's information (fig. 66). Each of the thirteen settlements has a logographic place-name which flanks a conventional frontal building, a symbol of "town" or "community." The Muchitlan artist was well versed in the preconquest symbol system; many of the town names refer to different species of trees, and the artist's pictographs capture their differing attributes, allowing the reader to distinguish one kind of tree from another (fig. 67). The map conveys its information not only by logographic writing, but by pictorial devices. For instance, the logographic place-name of Muchitlan occupies the central position on the map, indicating its importance, and its dependencies cluster like satellites.

As is the case with other maps, the Muchitlan map's logographically written place-names often incorporate the symbol for tepetl, or hill, even though the morpheme *tepe* might be absent from the place-name itself. In fact, only one place-name of the thirteen portrayed, that of San Lucas Tepechocotlan, has the *tepe* morpheme in its name.[10] In the

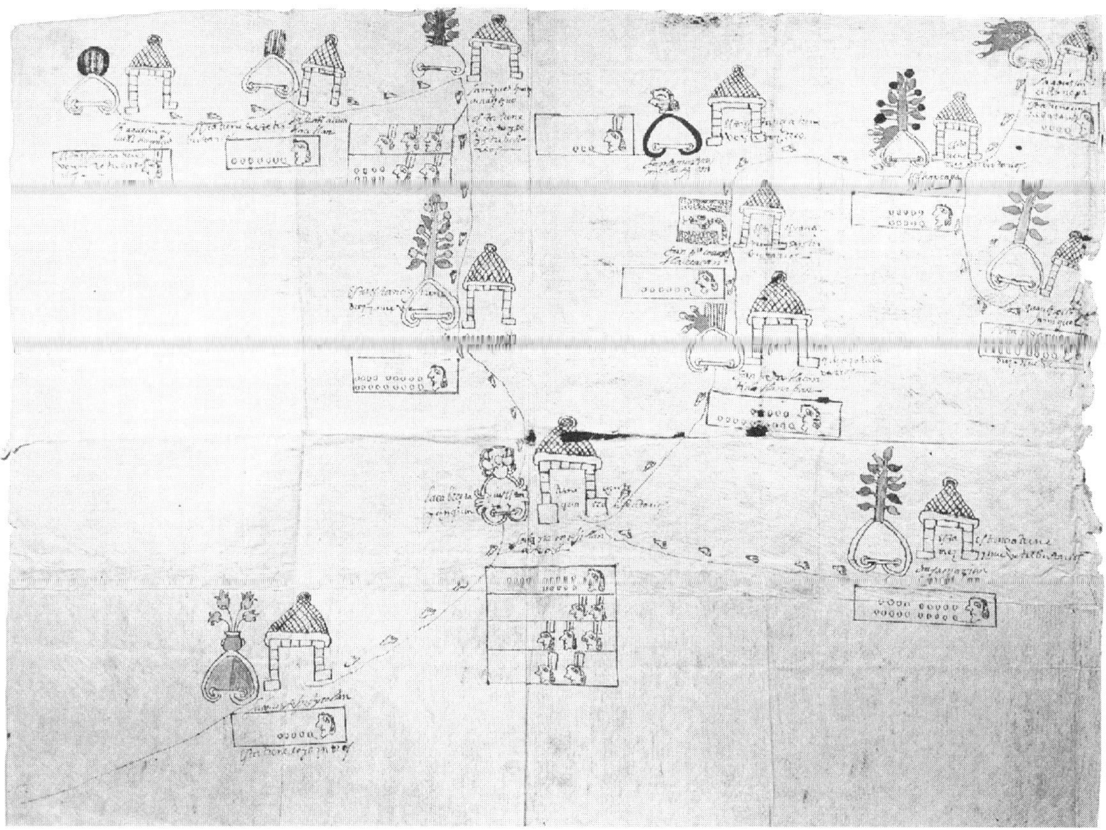

Figure 66. The Relación Geográfica map of Muchitlan, 1582. Size of the original: 56.5 x 77 cm. Photograph: B. Mundy; reproduced courtesy of the Benson Latin American Collection, The General Libraries, The University of Texas at Austin (JGI xxv-13).

Figure 67 *(opposite)*. Toponyms on the Relación Geográfica map of Muchitlan.

Place name and Nahuatl etymology	Logograph	Explanation
S. Ana Muchitlan "Place of the 'Mochil'" *Cuamochil* = a medicinal plant *-tlan* = place of		A four petalled flower sits on top of a tepetl symbol, perhaps the flower of the cuamochil.
S. Lucas Tepechocotlan "Place of the Fruit Hill" *tepetl* = hill *xocotl* = fruit *xocotli* = pot *-tlan* = place of		Logograph shows a pot holding four fruits on top of the tepetl symbol. The pot phonetically reiterates the "xoco" reading. A similar fruit appears in a picto-graph for Xocotitlan in the Codex Mendoza, fol. 10v.
S. María Chichilan "Red Place" *Chichiltic* = something red *chichic* = bitter *-tlan* = place of	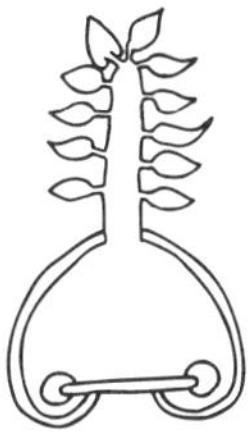	A tree rising out of a tepetl symbol, bears leaves of blue, yellow, and red. This may be a chichic-cuahuitl, a tree from whose bark comes quinine.
S. Agustín Citlanapan "Cornfield River" *cintlan* = cornfield [illegible] *apantli* = river		A flow of water, rising out of a tepetl symbol, encloses two stalks of corn, which bear dried ears. The town's name was also writ-ten Cintlanapa.
S. Andrés Quauhtamaltitlan "Among the Tamale Tress" *cuahuitl* = tree *tamalli* = corncake, tamale *-ti-tlan* = among	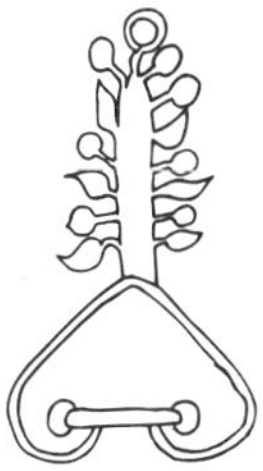	A tree bears fruits looking like round corn cakes. It grows out of a tepetl symbol.
S. Pedro Tlacontintlanapan "River beneath the Sticks" *tlacomeh* = sticks *-tzintlan* = beneath *apantli* = river	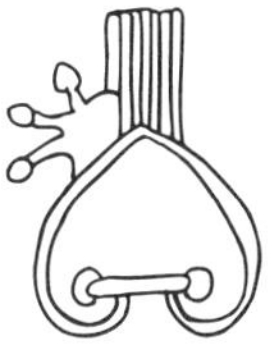	At the top of a tepetl symbol, water spurts out from among a row of sticks.
San Agustín Yohualtianquizco "Night Marketplace" *yohualli* = night *tianquiztli* = market *-co* = locative		A dark disc with speckled tripar-tite interior sits at the top of a tepetl symbol. It could be the cir-cular symbol that stands for mar-ket or the eye-dotted black disc that symbolizes the night sky.

Place name and Nahuatl etymology	Logograph	Explanation
S. Francisco Ahuatlacotlan "Place of the Thorny Sticks" *ahhuatl* = long thorn *tlacotl* = stick *-tlan* = place of		At the top of a tepetl symbol, three staffs bristle with thorns.
S. Miguel Huitzquauhtzinco "Place of Thorn Trees" *huitztli* = thorn *cuahuitl* = tree *huitzcuahuitl* = medicinal tree *-tzin* = honorific *-co* = locative	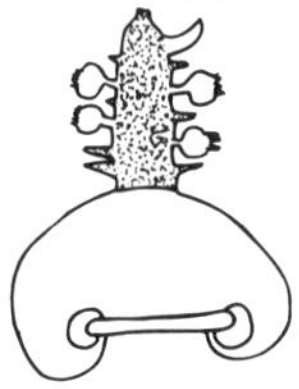	A tree, growing out of the top of the tepetl symbol, bears fruits and thorns.
S. María Mictlantzinco "Honored Place of the Dead" *micqui* = a dead person *mictlan* = the underworld, place of the dead *-tzin* = honorific *-co* = locative		The head of a person, set at the top of a tepetl symbol has eyes closed in death and face painted with stripes, characteristic of the dead and victims of sacrifice.
S. Miguel Quauhxilotlan "Place of the Cuauhxilotl Tree" *xilotl* = green ear of corn *cuahuitl* = tree *cuauhxilotl* = a tropical tree *-tlan* = place of	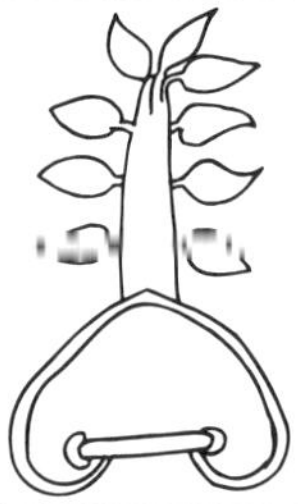	A tree, rising from the top of a tepetl symbol, bears pale green leaves that look like ears of corn.
S. Pedro Coaixtlahuacan "At the Snake River" *coatl* = snake *ixtlahuatl* = field *-can* = locative		A snake crawls upon ground marked with dots, the stippling designating cultivated land
S. Juan Tliltzapoapan "River of the Black Sapodilla" *tlilzapotl* = black sapodilla tree *tlilli* = black *tzapotl* = sapodilla *apantli* = river	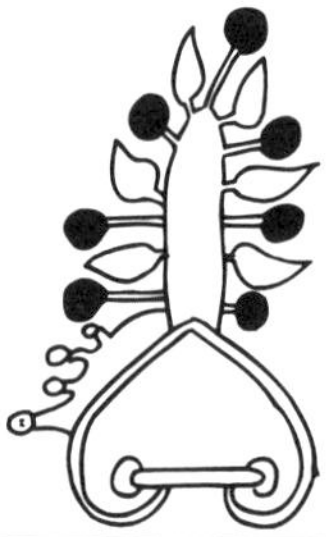	A tree with round black fruits, set at the top of a tepetl symbol, has a spring of water at its base.

Figure 67 *(continued)*

Muchitlan map, as in other maps, the tepetl symbol has what H. J. Prem has tagged a "determinative" or classificatory function (1992). Here the tepetl symbol is neither a logographic nor a phonetic transcription of the written word, but is a classifier, symbolizing the class of nouns (toponyms) to which the word belongs. In Mixtec maps and manuscripts, it is not the hill *(yucu)* symbol that often acts as a determinative, but a stepped fret sign, standing for *ñuu,* or "place" (Smith 1973a: 39).

Not only could the tepetl symbol be a determinative, when combined with a water symbol, it also could be the sign for altepetl, a word used to designate a community-kingdom in Nahuatl and that literally meant "water, hill," as these two elements were perhaps the definitive criteria—agricultural and defensive—for settlement. In the Relación Geográfica map from Misquiahuala (fig. 62; pl. 7) the place-name of this town incorporates a tepetl symbol from which water *(atl)* flows; neither atl or tepetl appear within the name "Misquiahuala." Instead, they seem to designate *this* place-name, out of all those appearing on the map, as that of the altepetl. In the map of Cholula (fig. 34; pl. 3), the mapmaker does the same, drawing at the top center of the map a river of running water (atl) looping under an elaborately drawn tepetl to create a sign for altepetl. The symbol flanking the central palace on the Mixtec Relación Geográfica map of Amoltepec (fig. 51; pl. 6) is also a linguistic sign of altepetl, as it shows a hill glyph from which water flows. Yet altepetl was not a Mixtec term, and perhaps on this map was a concept borrowed from Amoltepec's pre-Hispanic overlords, the Nahuatl-speaking Culhua-Mexica.

Linguistic signs on maps were elastic; they could mark the existence of place, its name, as well as the status of altepetl. Their varied attributes made them more than ways of writing names, and certainly contributed to their longevity in the postconquest period. So did their primacy in symbolizing the community itself on painted manuscripts. The extent of the use of logographic place-names in the Relación Geográfica corpus leaves little doubt that the native artists at work held them to be an essential means of representing their communities. In the Relación Geográfica map of Acapistla, a town that lies only a few kilometers from Guaxtepec (fig. 31), the landscape is littered with logographs to name places. Most incorporate the tepetl symbol, which shifts back and forth between determinative and logographic functions. The logographic place-name of Acapistla, also spelled Yecapixtla, lacks the *tepe* morpheme, but incorporates a tepetl symbol, perhaps as a marker of its status as altepetl. This Nahuatl name means "place of the point," from *yacapitztli,* "point," and it is logographically rendered in the bottom left corner of the Relación Geográfica map. The last of a range of hills is depicted with a nose protruding from its side; the word for "nose," *yacatl,* is a partial homophone of yacapitztli.

Here again, we see that an artist of the Relación Geográfica corpus represented the com-

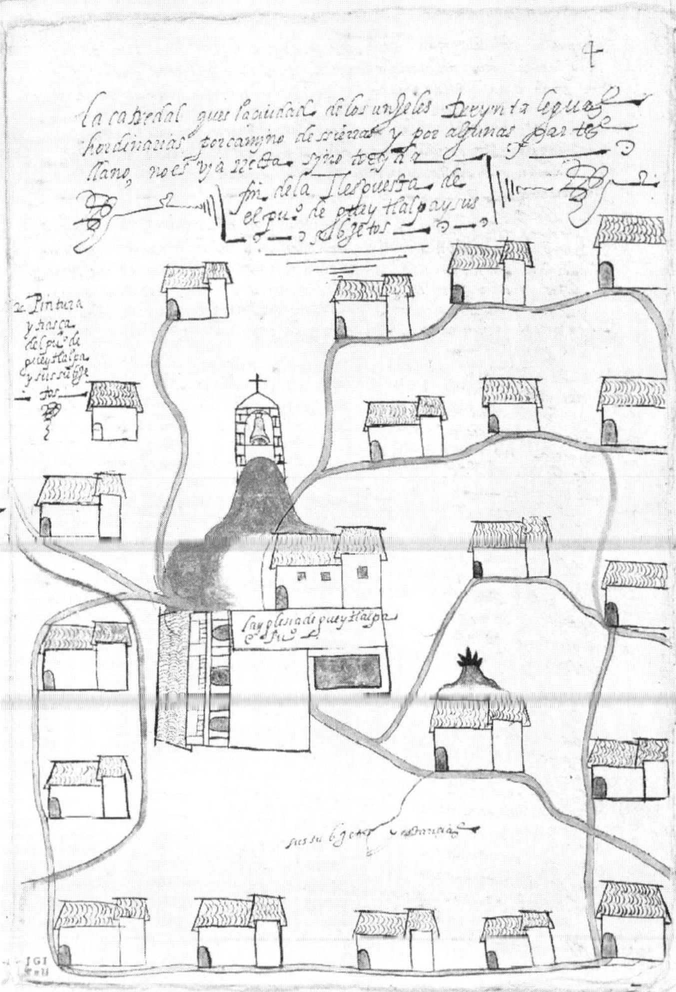

Figure 68. The Relación Geográfica map of Gueytlalpa, 1581. Size of the original: 24 x 21.5 cm. Photograph: B. Mundy; reproduced courtesy of the Benson Latin American Collection, The General Libraries, The University of Texas at Austin (JGI xxiv-5).

munity in a fashion similar to the way that it was registered in the Codex Mendoza (table 7), not because he or she could have known this source, but because both artists drew on long-standing ways of community representation through written place-names. In the Codex Mendoza, Yecapixtla is symbolized by a tepetl out of whose side protrudes a nose, yacatl, below which crawls a black ant. (Readers of the Codex Mendoza could have confused the place-name of Yecapixtla with that of Tepeyacac, "hill nose" [cf. Codex Mendoza, fol. 10v], so the Mendoza artist added the ant, *azcatl,* to Yecapixtla's place-name as a phonetic marker, a nearly homophonic reiteration of the two first syllables of Yecapixtla so as to distinguish it from Tepeyacac.)[11]

In addition to the rendering of "Yecapixtla," the Relación Geográfica artist includes a number of other place-names written logographically. In the upper left, the town of Suchitlan, also spelled Xochitlan, "flower place," is represented by a chapel next to a flowering plant, *xochitl.* In the center left edge, Achichipico is shown by a *chichitl,* "owl," who spits out water, *atl,* on top of a hill symbol. Slightly above Achichipico, the town of Aya-

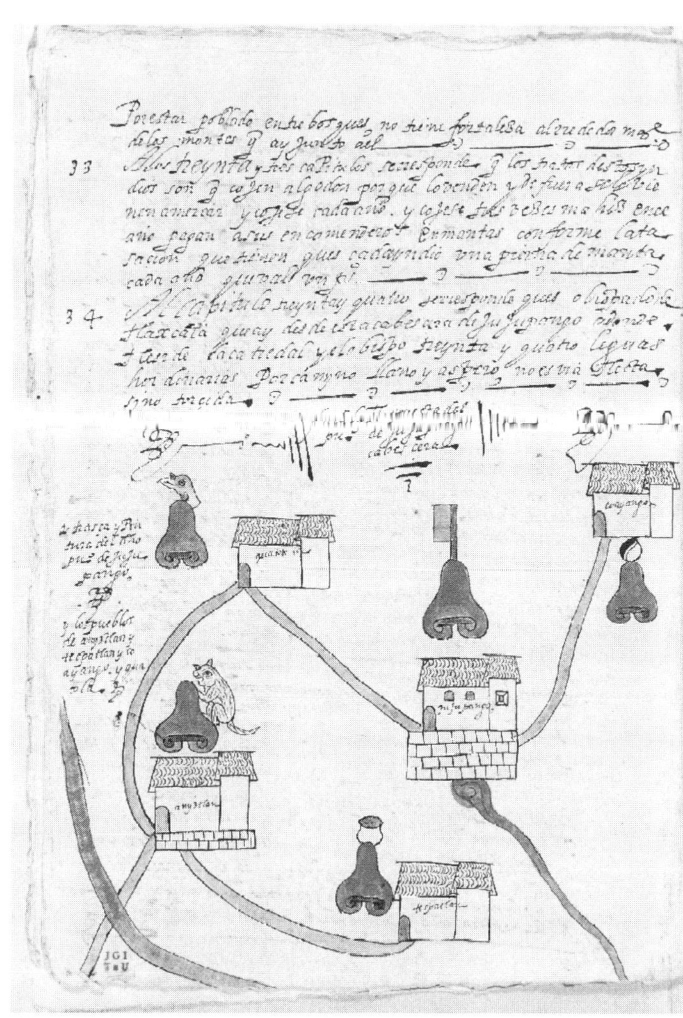

Figure 69. The Relación Geográfica map of Jujupango, 1581. Size of the original: 18 x 21.5 cm. Photograph: B. Mundy; reproduced courtesy of the Benson Latin American Collection, The General Libraries, The University of Texas at Austin (JGI xxiv-5).

panco is shown by a tepetl containing a bird that I have been unable to identify. Pazulco, in the bottom right corner of the map, is shown by a whorl of thread or twigs on top of a hill symbol to represent *pahzoltic,* "something twisted or tangled," or *pahzolli,* "briar patch."

To drive home the point that indigenous toponyms, written logographically, were a primary means of communal representation, both before the Spanish conquest as well as after, we should consider a group of seven spare maps, all drawn by one artist, that were

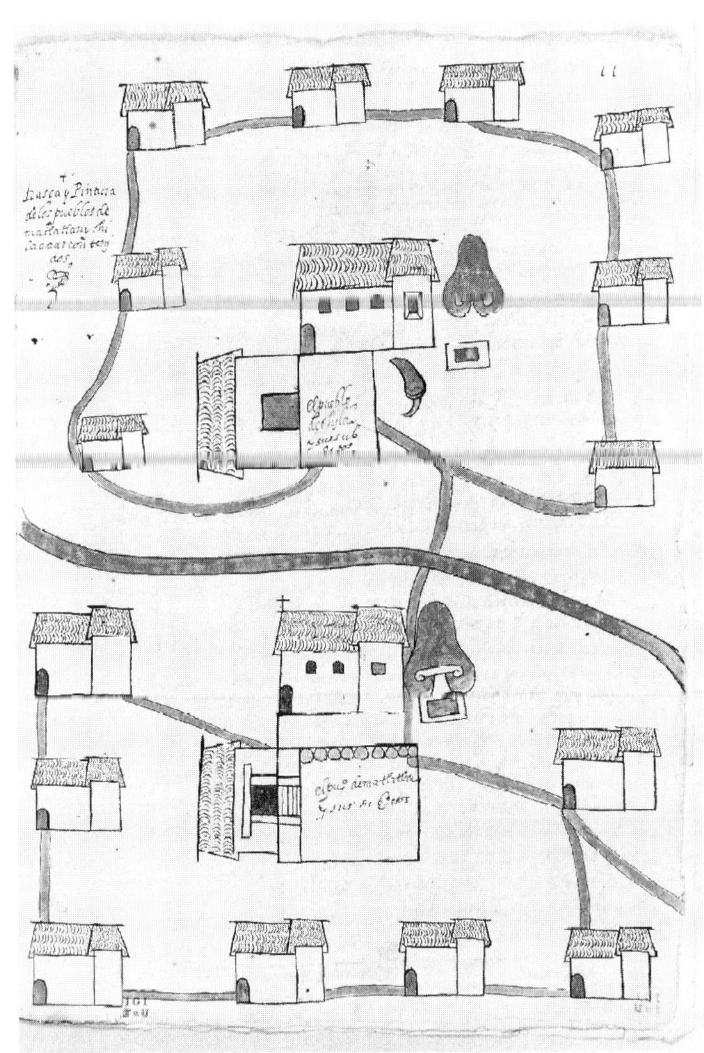

Figure 70. The Relación Geográfica map of Matlatlan and Chila, 1581. The chili pepper at top and the ten *(mahtlactli)* dots at lower center stand for toponyms. Size of the original: 30.5 x 21.5 cm. Photograph: B. Mundy; reproduced courtesy of the Benson Latin American Collection, The General Libraries, The University of Texas at Austin (JGI xxiv-5).

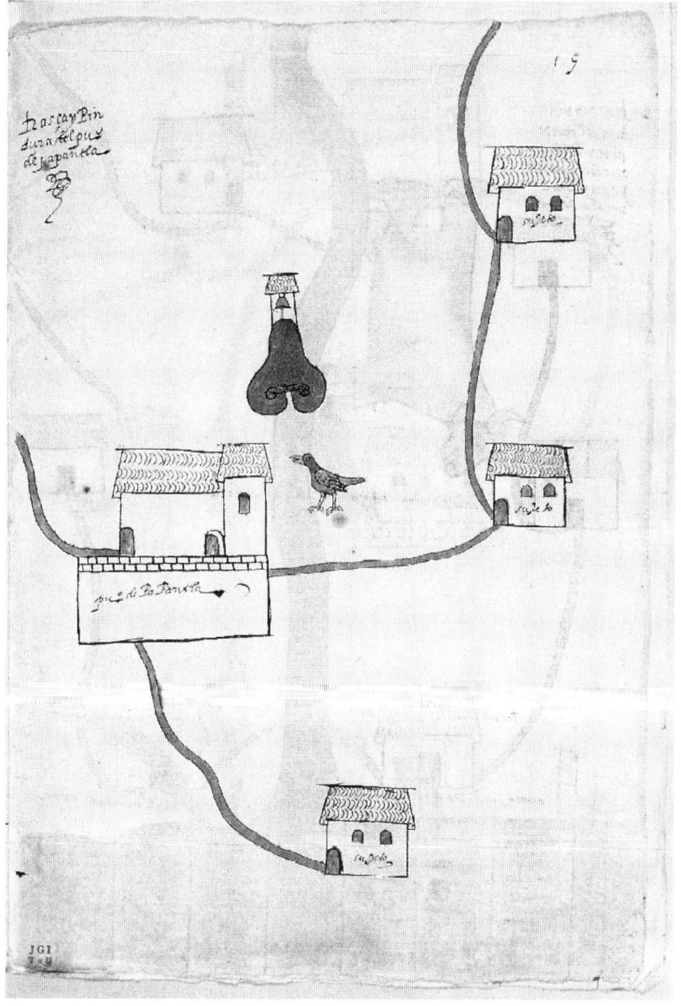

Figure 71. The Relación Geográfica map of Papantla, 1581. The black bird, a *papane*, is a logograph for the town's name. Size of the original: 30.5 x 21.5 cm. Photograph: B. Mundy; reproduced courtesy of the Benson Latin American Collection, The General Libraries, The University of Texas at Austin (JGI xxiv-5).

interspersed with the text of the Relación Geográfica from the region of Gueytlalpa (Hueytlalpan, Puebla; figs. 68–74). Although one map represents Gueytlalpa, another Jujupango (Jojupango, Puebla), and others Matlatlan and Chila (Chila, Puebla), Papantla (Papantla de Olarte, Veracruz), Tecolutla (Veracruz), Tenanpulco-Matlactonatico (Tenampulco, Puebla), and Zacatlan (Puebla), five of these maps are cast in the same mold, except that each features one or more distinctive logographic place-names; the artist uses these as his or her primary means of differentiating and identifying. The remaining

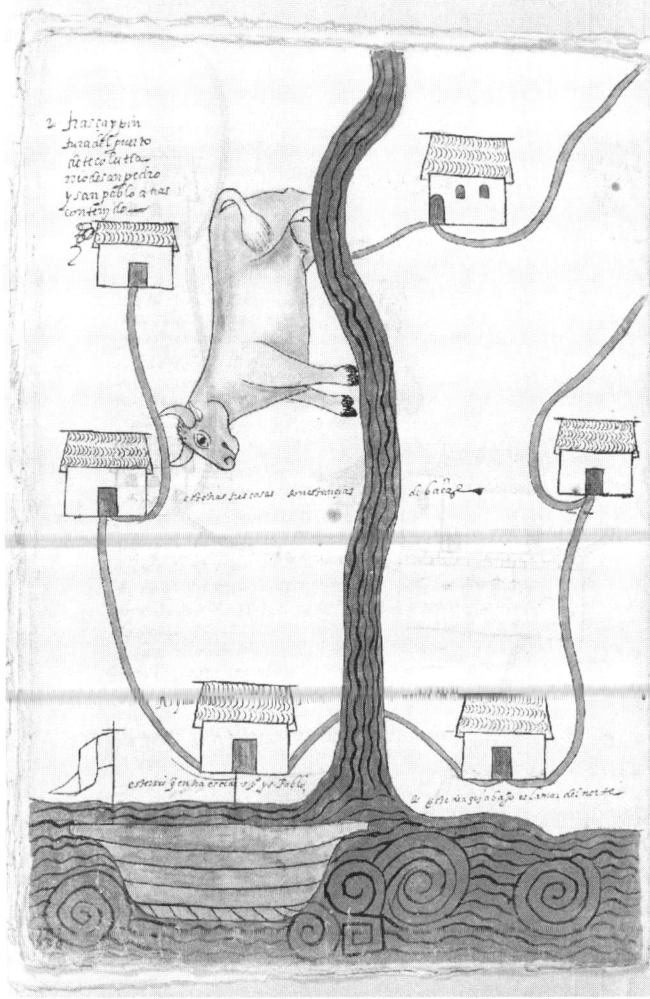

Figure 72. The Relación Geográfica map of Tecolutla, 1581. Size of the original: 30.5 x 21.5 cm. Photograph: B. Mundy; reproduced courtesy of the Benson Latin American Collection, The General Libraries, The University of Texas at Austin (JGI xxiv-5).

two, Tecolutla and Tenanpulco-Matlactonatico, are more distinctive because in the former, the artist pictures the sea, while in both, he or she draws steer, using them as a way of designating cattle ranches (figs. 72 and 73). The Nahuatl logographs on all these maps are simple and direct. The map of Jujupango (fig. 69) shows five towns, each with a logograph representing the town's name. Jujupango, or Xoxopango, at center, is marked by a banner *(pantli)* whose green *(xoxotic)* color has greatly faded since 1580. Quatototla, or Cuatotola, at top left, is marked with the head *(cuaitl)* of a bird *(totolin)*; Amiztlan, "place of the water lions," (RGS 5: 167) is marked at bottom left with a feline *(miztli)*.[12] Many of the maps of Gueytlalpa use the tepetl symbol in its determinative form. In the map of Tenanpulco and Matlactonatico, "place of 10 days," a determinative tepetl is topped by a sun *(tonalli)* to stand for "day" and flanked by ten *(mahtlactli)* dots (fig. 73). In the map of Zacatlan, the grassy plants *(zacalli)* used to name the town lie apart from the tepetl symbol (fig. 74); the former is logographic, the latter, determinative.

In a map from the Valley of Mexico, from a town called Chicoalapa, the artist's use of a

Figure 73. The Relación Geográfica map of Tenanpulco and Matlactonatico, 1581. Size of the original: 30.5 x 21.5 cm. Photograph: B. Mundy; reproduced courtesy of the Benson Latin American Collection, The General Libraries, The University of Texas at Austin (JGI xxiv-5).

logographic place-name is one of its most distinctive features (fig. 75). The artist draws a grid to show the town and below it places a local bird, the *chicuahtototl* (RGS 6: 169), drinking from a spring *(apantli)* to stand for Chicoalapa.[13] Likewise, in the map of Tetlistaca, the exquisitely drawn place-names are the most distinctive features of the map (fig. 43). Each of the four settlements are marked by the union of a conventionalized church with a logographic place-name. Tetlistaca translates as "place of white stone" and is shown by a craggy hill-tepetl symbol with a white peak; Tepechichilco, "hill of the chili pepper," is shown

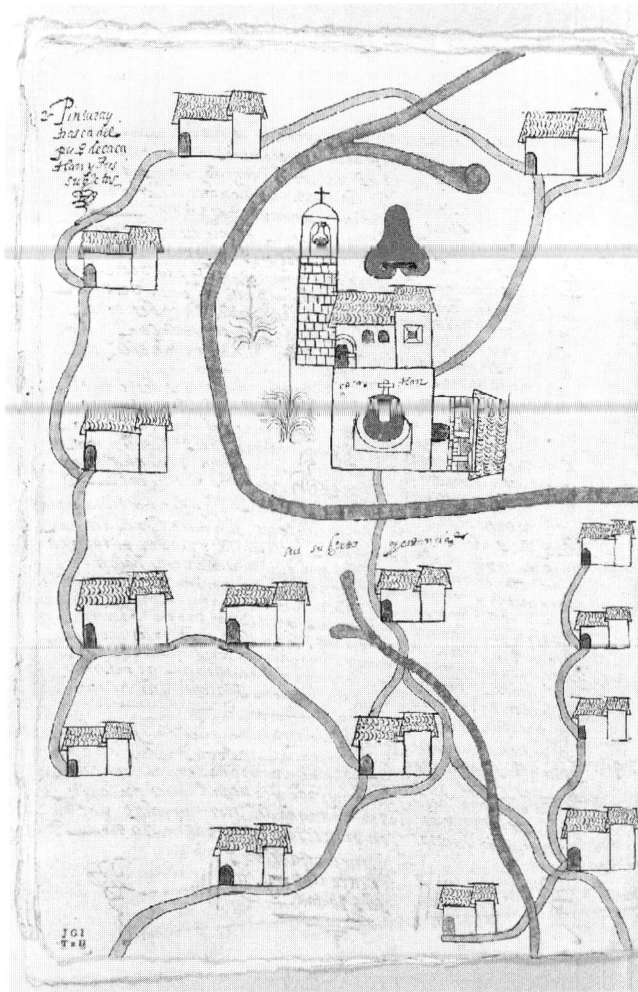

Figure 74. The Relación Geográfica map of Zacatlan, 1581. Grassy *zacalli* plants that lend the town its name grow to the left of the church plaza. Size of the original: 30.5 x 21.5 cm. Photograph: B. Mundy; reproduced courtesy of the Benson Latin American Collection, The General Libraries, The University of Texas at Austin (JGI xxiv-5).

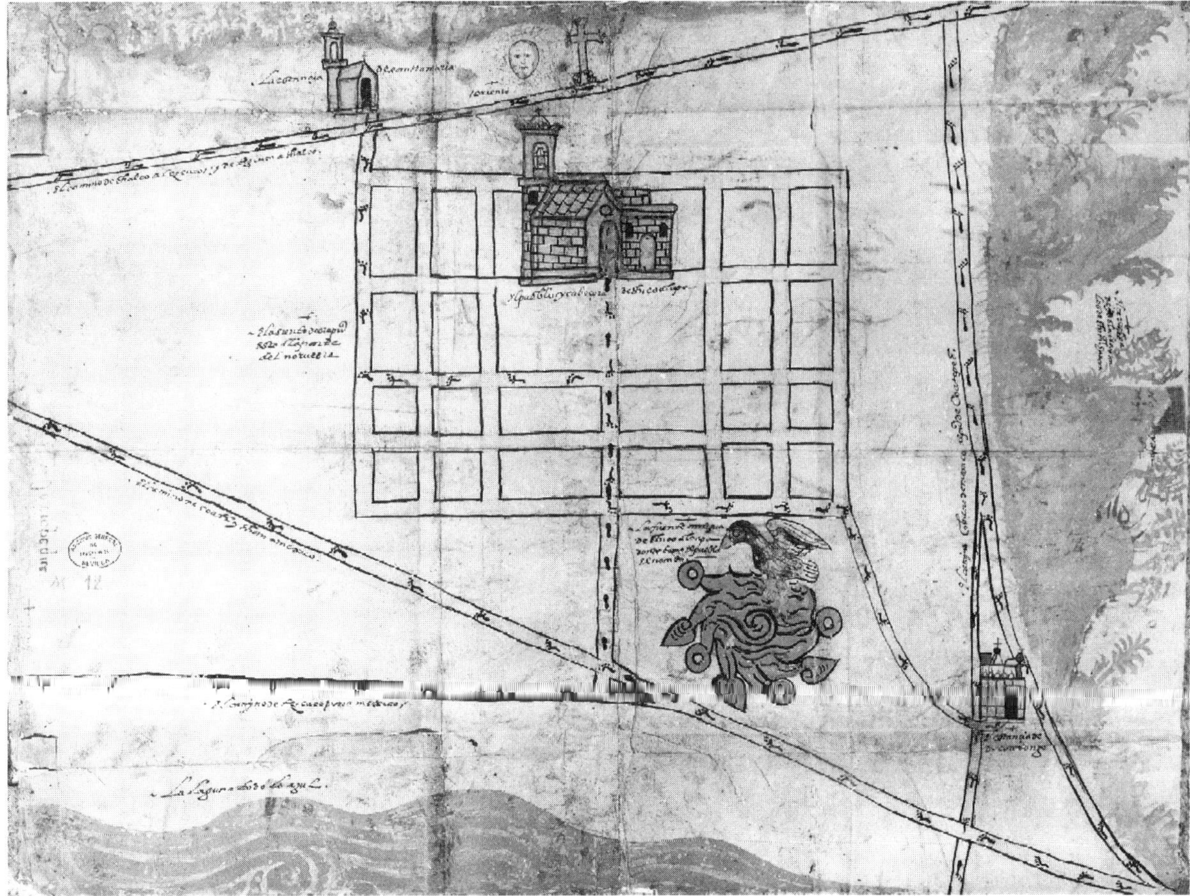

Figure 75. The Relación Geográfica map of Chicoalapa, 1579. Size of the original: 43.5 x 60 cm. Photograph courtesy of the Archivo General de Indias, Seville (Mapas y Planos, 12).

by a tepetl sign topped by a red chili; Tianquismanalco, "place of the market," is represented by the round symbol for marketplace; Cihuayocan, "place of womanhood," is shown by an alternate name, "place of the shoots of a palm tree," probably "Zoyaquiyoca," which was a near homophone in Nahuatl.[14] In his or her drawing of logographic place-names, the Tetlistaca artist accommodates both the native symbol of a hill—an even bell shape—and the European illusion of one—a craggy protuberance—with great success. The Tetlistaca painter tried to convey a sense of the vegetative abundance of the region by decorating his or her map with pictures of cacti and budding trees, as one would find in a European drawing, but the plants themselves conform to native style, shown with leaves above and roots below (Robertson 1959a: 22).

MIXTEC PLACE-NAMES

For the maps from Guaxtepec, Acapistla, and Chicoalapa, the Nahuatl names they used to represent the altepetl and its subunits were all born of local soil, since these were all primarily Nahuatl-speaking towns. But for non-Nahuatl speaking communities, Nahuatl names were not ones they had chosen for themselves, but ones that had been imposed on them by their Aztec conquerors; not only did the Aztecs represent their subject towns with names translated into Nahuatl, as in the Codex Mendoza, they also compelled indigenous communities to use their Nahuatl names, at least officially. Indigenous communities continue to use the names imposed on them by the Aztec conquerors to the present day. In 1580, in its Relación Geográfica, the Mixtec town of *ñuu ndaa* called itself Texupa, its Nahuatl name, and today the town is called Tejúpan (fig. 38; pl. 4). Even specific hills shown on the Texupa map bear the imprint of Aztec conquest and renaming with Nahuatl monikers. One of the hills outside the town's grid is named in the text as Comaltepec (a Nahuatl name meaning "place of the griddle") as well as by an alphabetic inscription on the map. The map also shows a prominent hill ringed with three layers of brick fortifications, and within the last layer is a row of human heads with arrows pointing at each of the heads, as if shot from below. This hill fortress was likely to have been a defensive site used by the town during the preconquest battles with Chocho-speaking peoples, mentioned in the text (RGS 3: 221). The Relación reveals that the hill carried a Nahuatl, not Mixtec, name in calling it Miahualtepec, meaning "hill of the corn tassel." The map's artist writes the same name logographically with the double-curlicued tassel at the top of the hill.[15]

While Mixtec towns like Texupa may have accepted a Nahuatl name for their dealings with the world outside their community, it is clear that Aztec renaming often failed to penetrate the logographic writing used within the community. On the Texupa Relación Geográfica map (fig. 38; pl. 4), the central pictogram best represents the town's indigenous Mixtec name, *ñuu ndaa,* "place of blue," rather than its Nahuatl counterpart, "upon the blue," that is entered in the Codex Mendoza (see table 7). The logographic place-name on Texupa's Relación map is a hill symbol edged with blunt nubs, Mixtec style, and is topped with a quincunx motif—representing a turquoise jewel—set in an area of lighter green. Within the hill symbol is a native-style temple drawn in profile to represent *ñuu,* or "place of" or "town," and here "blue," *ndaa,* is represented by the turquoise jewel (Smith 1973a: 60).

The Teozacoalco map, likewise, logographically writes its Mixtec name, not its Nahuatl one. The town's Nahuatl name is a corruption of Hueitzacualco, which means "place of the large temple or pyramid," from *huei,* "great or large," and *tzacualli,* "small hill, temple, pyramid" (RGS 3: 142–3). Yet the logographic place-name for Teozacoalco on the map (figs. 10 [pl. 1] and 76) represents its Mixtec name, *chiyo ca'nu,* which means "large or great altar or foundation" (Alvarado 1962: 95; Caso 1949; Smith 1973a: 57). The symbol shows a stepped fret sign, to represent *chiyo,* "altar or foundation," and a small figure bending or breaking the frieze, to represent *ca'nu.* In the toponym, *ca'nu* means "large or great," but is represented by its homophone, which means "to break or cut" (Smith 1973a: 57).

In the case of the Amoltepec map, the other elaborate native map from a Mixtec-speaking town, both Mixtec *(yucu nama)* and Nahuatl (Amoltepec) names mean "hill of the soap plant," and are reflected in the central symbol, which shows a hill out of which grows an agave-like plant (figs. 51 [pl. 6] and 77).[16] But *yucu nama,* the name the Aztec seized for transliteration into Nahuatl, seems to have been only a secondary name; its appearance in the Relación map is the only known example of it, for it does not appear in any of the Mixtec codices (Smith 1973a: 67).[17]

This Mixtec community held back a primary name, *yodzo yuhu,* or "plain of the flowers," from its Nahuatl-speaking conquerors. However, the Mixtec informants who helped in compiling the Relación Geográfica text divulged that "plain of the flowers" was the name of the site of the town (RGS 3: 148). It is this name that the mapmaker uses to identify the town on its Relación Geográfica map, and this name, not *yucu nama,* that other indigenous documents use to refer to Amoltepec.[18] Traditionally, in Mixtec manuscripts, the rulers' palace is flanked by or sits upon the logographic place-name of the town. In the Amoltepec Relación Geográfica map, the rulers' palace lies well below the "hill of the soap plant" symbol. But the palace does rest upon a plain of flowers, the plain *(yodzo)* shown by its homonym *yodzo,* meaning "large feather" (represented on the map by a cartouche outlined with large, overlapping feathers) within which dangles a row of flower buds, or *yuhu* (figs. 51 [pl. 6] and 77).[19]

Figure 76. The logographic place-name of Teozacoalco. Drawing after the Relación Geográfica map of Teozacoalco, this logograph represents the town's Mixtec name.

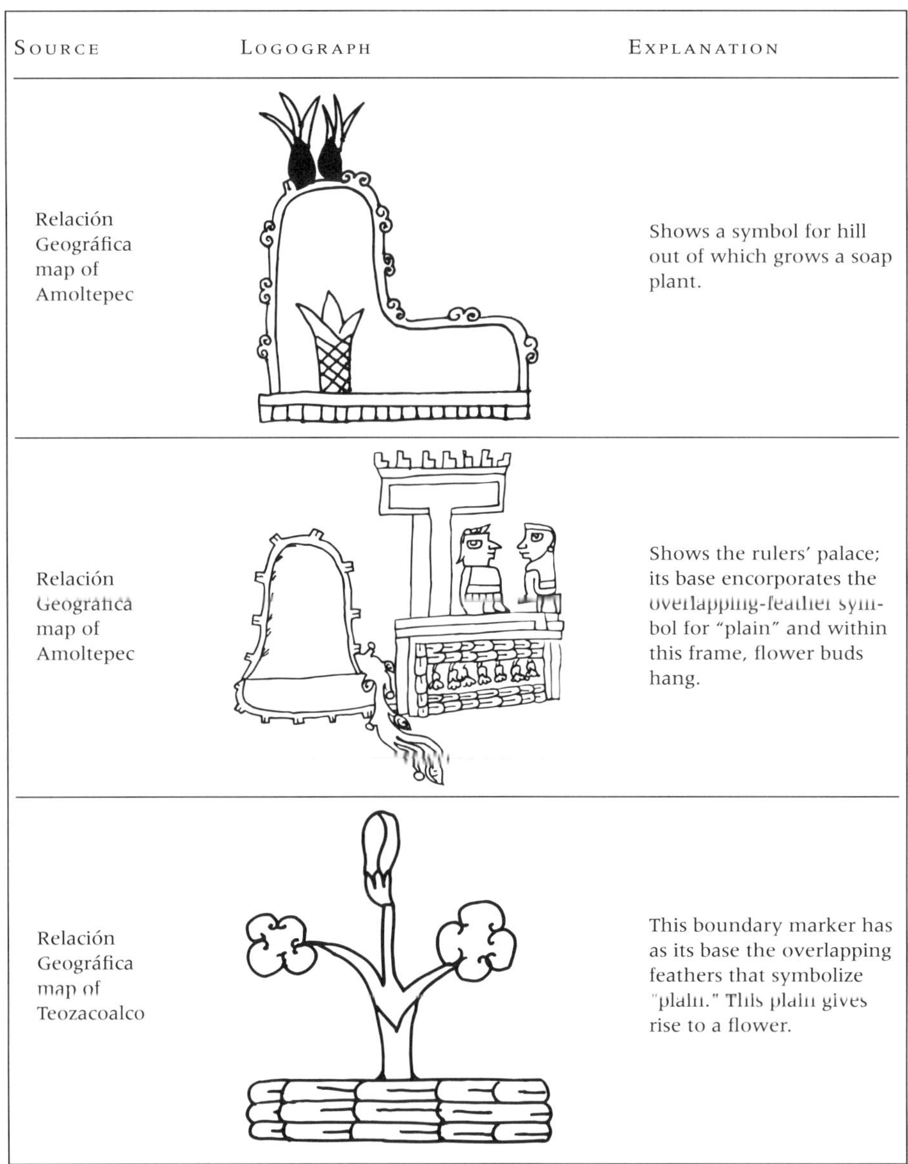

Source	Logograph	Explanation
Relación Geográfica map of Amoltepec		Shows a symbol for hill out of which grows a soap plant.
Relación Geográfica map of Amoltepec		Shows the rulers' palace; its base encorporates the overlapping-feather symbol for "plain" and within this frame, flower buds hang.
Relación Geográfica map of Teozacoalco		This boundary marker has as its base the overlapping feathers that symbolize "plain." This plain gives rise to a flower.

Figure 77. Logographs of Amoltepec. The upper two are logographs drawn from the Relación Geográfica map of Amoltepec. The top one represents *amoltepec,* a Nahuatl name meaning "hill of the soap plant," while the lower one, at the base of the palace, represents *yodzo yuhu,* a Mixtec name meaning "plain of flowers." The logograph used in Relación Geográfica map of Teozacoalco is seen at bottom, and it also represents *yodzo yuhu.*

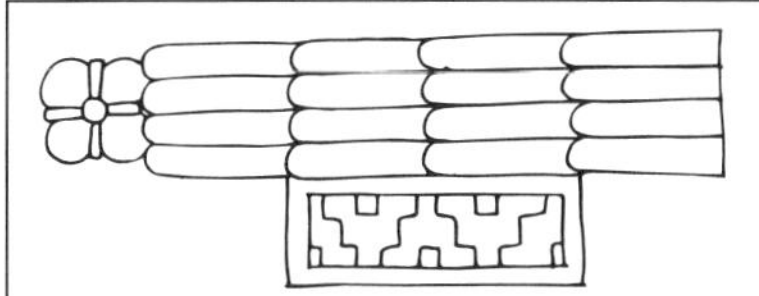

Figure 78. Logograph from the Lienzo of Yolotepec. This symbol shows a flower, *yuhu,* attached to a plain, *yodzo,* symbolized by overlapping feathers. The rectangular box below the plain is filled with stepped frets and is a symbol of "place." Thus this composite symbol is a place-name, read as *yodzo yuhu.*

It is this logographic place-name—*yodzo yuhu,* or "plain of flowers"—that the Relación Geográfica map from neighboring Teozacoalco uses to refer to Amoltepec and its lands (fig. 10; pl.1). Among the boundary symbols at the bottom of the Teozacoalco map is a plain (again shown with a mat of overlapping feathers) out of which grows a flowering plant (fig. 77). This symbol lies exactly where Amoltepec would fall on Teozacoalco's borders. Unlike most of the other pictographs inscribed on the circle of the Teozacoalco map, which we assume to name boundary markers, two main roads lead directly to *yodzo yuhu,* as they would to a settlement, but not to a boundary marker.[20]

The "plain of flowers" symbol also turns up in the Lienzo de Yolotepec, a sixteenth-century pictorial manuscript that was taken out of Amoltepec in 1889 (Caso 1958: 41).[21] Caso identified the central logographic place-name of the cotton lienzo as that of Yolotepec, the town bordering Amoltepec to the north (Caso 1958: 43; RGS 3: 147). While the arrangement of places on the lienzo is not planimetric, the content of the lienzo does deal with migrations to Yolotepec (Caso 1958) and it is likely to mention nearby Amoltepec, where the manuscript was found. Below the sign for Yolotepec on the lienzo, a place sign shows a feather-mat plain attached to a flower and a stepped fret; this symbol again probably refers to Amoltepec (fig. 78).[22] Thus, in these Mixtec maps, the logographic place-names all reflected Mixtec names; in the case of Texupa and Teozacoalco, these were the names that were roughly transliterated into Nahuatl, but in the case of Amoltepec, the primary name of the town, "plain of flowers," seems to have been used only within the Mixtec-speaking world.

ZAPOTEC, OTOMI, AND TOTONAC PLACE-NAMES

While Mixtec communities logographically wrote their Mixtec names on the Relaciones Geográficas maps, resisting the Nahuatl names imposed upon them, the same is not true of the Zapotec-speaking communities in Oaxaca or the Otomi communities in Hidalgo, or the Totonac-speaking communities in Veracruz and Puebla, who made Relaciones Geográficas maps. For example, the map of Macuilsuchil (San Mateo Macuilxochitl, Oaxaca), from a town near Oaxaca in the Tlacolula Valley (fig. 79; pl. 8), has its place-name at

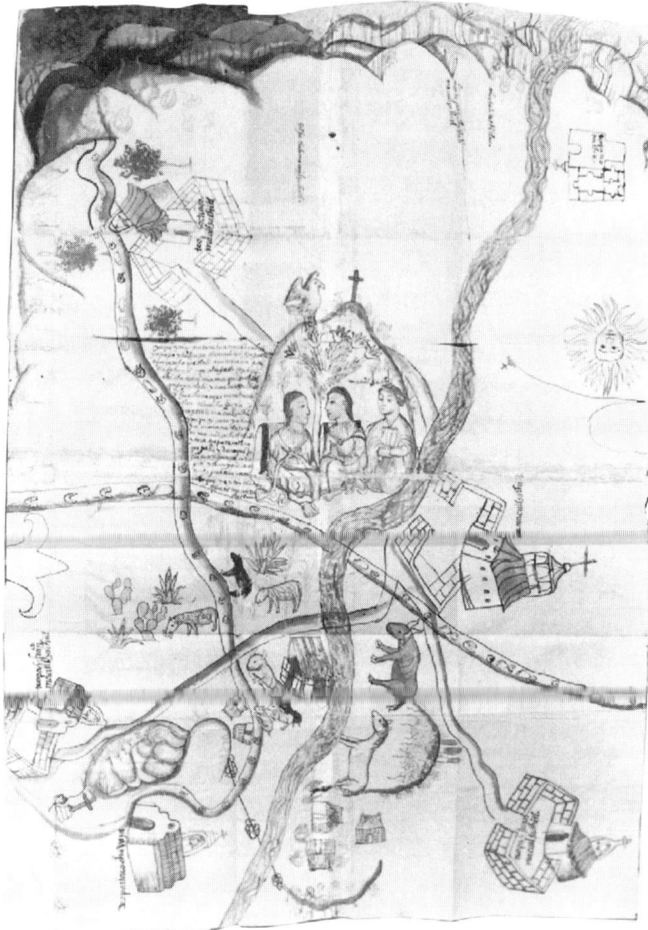

Figure 79. The Relación Geográfica map of Macuilsuchil, 1580. (See also pl. 8.) This map, from a Zapotec-speaking town in Oaxaca, includes a long inscription written in rough Nahuatl. At its center, three rulers are sheltered under the branches of a flowering tree. In Nahuatl, Macuilsuchil means "five flower," and the five flowers of the tree are this artist's rendering of that name. Size of the original: 85 x 61.5 cm. Photograph: Juan Jiménez Salmerón; reproduced courtesy of the Real Academia de la Historia, Madrid (9-25-4/ 4663-xix)

the map's center. Here, the artist draws a hill within which sit three rulers of the town; his concern with rulership is consistent with the native tradition. Above the rulers is a bunch of five *(macuilli)* flowers *(xochime)*. This logograph is an imperfect rendering of the calendrical day-name "5 Flower," which is more correctly shown in the Codex Mendoza (fol. 44r). The artist of the Mendoza version draws a flower and below it five circular counters (table 7), "5 Flower" being the name of a deity worshipped in Oaxaca (Paddock 1982: 346). But to represent "5 Flower," or Macuilsuchil, the artist of the Relación Geográfica map chooses to represent, albeit incorrectly, the Nahuatl name "5 Flower" over the indigenous Zapotec name of the town, *Quiabelagayo*, that informants in the town supplied to the writer of the Relación text (RGS 2: 330). Quiabelagayo, a name that apparently combines "rock," "serpent," and "five" (Paddock 1982: 346), was not directly transliterated into the Nahuatl calendrical name "Macuilsuchil," because "5 Flower" in Zapotec is *Pélloo* or *Yolao* (Whitecotton 1982). There is no doubt that the meaning of the Zapotec name was known when the map was painted, because informants to the Relación Geográfica writer said that the Zapotec name related to *"cinco piedras grandes,"* "five large rocks," near the town (RGS 2: 330).

Figure 80. The Relación Geográfica map of Suchitepec, 1579. At the upper left of this map, two plants flower on top of a hill, yielding the name Suchitepec, "flowery hill place." This map's artist also painted four other maps of communities in the Suchitepec region to be included in the Relación corpus. Size of the original: 84.5 x 59.5 cm. Photograph courtesy of the Archivo General de Indias, Seville (Mapas y Planos, 29).

The large map of Tehuantepec A (fig. 27), from a Zapotec town on the Isthmus of Tehuantepec, also uses the logographic place-name representing its Nahuatl name, Tehuantepec (see also table 7). Behind the Tehuantepec church a man-eating beast *(tecuani),* represented by a jaguar, scales the peak of a hill (tepetl). This name is shown rather than its Zapotec name, *Guixegui,* which means "hill of fire" (Barlow 1943: 157).[23] Likewise, the name of the Zapotec and Chontal-speaking town of Suchitepec (Santa María Xadan, Oaxaca; fig. 80) identifies itself on its map with a pictograph of its Nahuatl name, *Suchitepec,* meaning "flowery hill place." According to residents, the town's indigenous Zapotec name had been forgotten (RGS 3: 59). From the hill in the upper left of the map, two elaborate flowers, *xochime,* grow, transforming this quadrant of the landscape into a symbol of the town's Nahuatl name.

In the towns of Otomi speakers, often the indigenous town names were not even registered in the Relación Geográfica text. Tetlistaca, for example, supplies only Nahuatl names in its Relación Geográfica text, and depicts only Nahuatl names on its map (fig. 43). And the mainly Totonac towns of the Gueytlalpa group are all identified on the maps by Nahuatl place-names; no Totonac names are registered by the writer of the Relación Geográfica (figs. 68–74).

This discrepancy between the Mixtec maps and the Zapotec, Otomi, and Totonac maps in the corpus may have been the legacy of the pre-Hispanic period. While we know that Mixtecs created a wealth of written manuscripts, we have no such record of Zapotec- or Otomi- or Totonac-speakers doing the same in the immediate pre-Hispanic period.[24] While the Mixtec regions clearly used logographic writing as a way of recording their names, we do not know to what degree other peoples used writing in this period. Perhaps the Nahuatl speakers, who came, conquered, and imposed new names, may have also promoted a reliance on written forms of communication among their subjects, kindling the use of writing in their provinces. As this might have been the case, provincial artists may have adapted the forms—including place-names—that their conquerors introduced.

NAMES AND THE ARRIVAL OF THE SPANISH

By the time the Relación Geográfica corpus was assembled, New Spain housed a hierarchy of social groups, and language and writing also became hierarchical (table 8). Fluent, alphabetically written Spanish was, for the most part, the province of educated Spaniards and Creoles, including the clergy and the corregidores who created the Relaciones Geográficas. Some members of the native elite and professional scribes also could speak and write Spanish.

Table 8 Linguistic/Literate Interplay in Sixteenth-Century New Spain

Group	Languages Spoken	Writing	Interlegibility of Writing Systems
Spaniards	• Spanish	Alphabetic	
Nahua elite	• Nahuatl • Some Spanish • Other indigenous language	Alphabetic Logographic	
Non-Nahua Native Elite	• Non-nahuatl indigenous language • Spanish • Nahuatl	Alphabetic Logographic	
Commoners	• Indigenous language (Nahuatl and others)	Logographic	

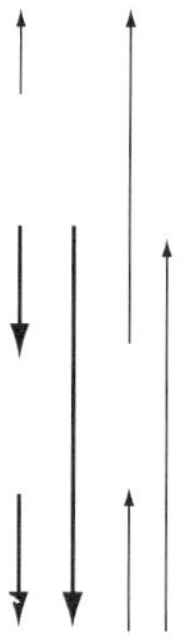

On the whole, though, native elites who could write alphabetically would be most likely to write in Nahuatl. The friars had successfully inducted the elites into the ranks of the alphabetically literate but had taught them a native language. Of all native languages, the friars preferred Nahuatl, not only because it was the language of the Valley of Mexico, where the largest Spanish settlement was clustered, but also because the friars found that the native elite across Central Mexico were mostly Nahuatl speakers (Klor de Alva 1989). Nine Relaciones Geográficas maps—Cholula, Gueguetlan, Macuilsuchil, Cempoala, and the five from Suchitepec—contain indigenous language texts written alphabetically, and all are in Nahuatl, despite the fact that Macuilsuchil and Suchitepec were Zapotec-speaking towns, and Suchitepec's four subject towns spoke Chontal. The maps bespeak a bilingualism of the elite that was a subject of comment in the Relaciones Geográficas texts as well (RGS 5: 153, 166).

For native commoners, who were not, on the whole, taught to write alphabetically, pictorial literacy was their means of written communication. Thus levels and types of literacy that largely followed class lines were a defining factor in the creation of audiences in New Spain (table 8).

For many of the Relaciones Geográficas maps that were painted by indigenous artists, it is clear that while a native painter was the principal author, other hands added the written texts, adding another layer of meaning to the map. In their finished state, as we know them today, some parts of all of the maps would be easily accessible to different audiences, each defined by their differing levels of literacy. For instance, the Teozacoalco resident who saw the map of his or her town could probably read many, if not all, of the logographic place-names representing Mixtec names forming its boundary circle (fig. 10; pl. 1); the Spanish corregidor could read the brief Spanish texts written on the map. The layering of writing systems had another side effect: a map, such as the Teozacoalco map, was accessible in its entirety to almost no one: the native commoner would have had little success deciphering the Spanish text on the map, while the Spanish corregidor would find little specific meaning in the linguistic signs used for boundaries. The audience who could read and understand fully a map like the one from Teozacoalco was limited to a handful of highly educated members of the native elite who were alphabetically literate in Spanish and both logographically and alphabetically literate in their native tongue. Even though they possessed the fullest understanding, these elites were never intended to be the maps' only audience; they were the accidental beneficiaries of many-handed authorship.

Not only were many of the Relaciones Geográficas maps not fully legible to most, but their illegibility took on a different coloring for different audiences, given New Spain's hierarchy of language and writing. To the native commoner, the alphabetic script in Spanish would have been a meaningful cipher, that is, he or she would have understood the importance of such script to Spaniards and the native elite, and perhaps would even have been aware that such a script was an insurmountable barrier to knowledge. The Spaniard, in contrast, would have felt little interest in logographic writing. Such glyphs were the ciphers of the "indios" and of little consequence to him. In fact, even such a pro-Indianist as fray Bartolomé de Las Casas denied that logographs were true writing (Las Casas 1974: 42). Notably, none of the Spanish or Creole writers of the Relación Geográfica corpus ever mentions that the accompanying map contains a writing system or code that is beyond his purview; he understood that it was through his Spanish text that he could communicate the information he felt important to the audience of Spanish-speakers, the only audience that, to him, really mattered.

Although hierarchies of language had existed before in Mesoamerica, their effect was mitigated by a logographic writing system that allowed all readers a kind of passage toward understanding written works, no matter what language they knew. The Spanish conquest changed all this, not only by introducing a new language, but also by establishing a new way of writing that allowed no access to nonspeakers. The barrier imposed by alphabetic writing would matter particularly to native mapmakers. When the Spanish and the friars spread across New Spain, they, like the Aztecs before them, initiated a program of renaming. Spaniards rechristened communities, linking names of saints or the Virgin Mary to indigenous, usually Nahuatl, settlement names. Texupa became Santiago Tejúpan, after Saint James, who, as Santiago Matamoros, was a popular saint in Spain of the Reconquista. Culhuacan became San Juan Evangelista Culhuacan; Muchitlan, Santa Ana Muchitlan.

Names in other indigenous languages, such as *ñuu ndaa* for Texupa, were passed over and relegated to local usage, not only as a habit left over from pre-Hispanic times, but also because the new colonizers chose Nahuatl as the lingua franca of the indigenous world.

Other scholars have dwelt on the violence implicit in the Spanish program of renaming, as the new Spanish and Christian names erased the landscape created by indigenous toponyms (Anders et al. 1992: 36–7). But renaming was not foreign to indigenous practice, and in the earlier Aztec campaigns, Mixtec communities such as Texupa had found ways of resisting the imposition of Nahuatl names, at least in part, by enshrining Mixtec names in writing. The real violence of the Spanish conquest of the landscape came not in imposing names, but in imposing names that were written alphabetically. Although names like "Santa Ana" or "San Juan Evangelista" could be written logographically, and occasionally were, they were most often rendered alphabetically.[25] In the map of Muchitlan (figs. 66 and 67), for example, the logographs represent only the Nahuatl parts of the place-names, but not the Spanish saints' names. Part of the reason, I think, has to do with the weight of tradition. Indigenous artists who drew upon existing pre-Hispanic or early colonial manuscripts would not have found names written alphabetically therein. Also, many native painters may not have been alphabetically literate.

The imposition of alphabetically written names on maps radically changed the act of viewing for a member of the indigenous audience. Before the conquest, he or she could probably write and certainly read logographic place-names; when place-names were used as the most important way of representing a community on a map, that representation was available to all community members. But as alphabetic literacy came, and reached the hands of only the very few, this new way of representing a name was a cipher to most indigenous viewers. In looking at their Relación Geográfica map (fig. 29), for instance, Culhuacan residents could read the *colhua* (after *colli,* meaning curved) sign from which their name derived, but the letters "S. Juan Evangelista" would have put them at a loss. Mapmakers who were not alphabetically literate could no longer perfectly represent the communal name. In changing names, the Spaniards played the same game as the Aztecs, but in changing the writing, they rewrote the rules and effectively removed a primary means of representing a community—through a written place-name—from the reach of New Spain's indigenous community.

If we look again at the maps with this idea of their being differently legible, we see that, depending on the audience, they were read in very different ways. Thus, when we look at the map from Teutenango (Tenango de Arista, Mexico), "place of divine walls," we see that the native artist has used the repeating stepped motifs standing for "wall" *(tenamitl)* to show the enclosure of pyramidal ruins that covers a hill above the town (fig. 81).

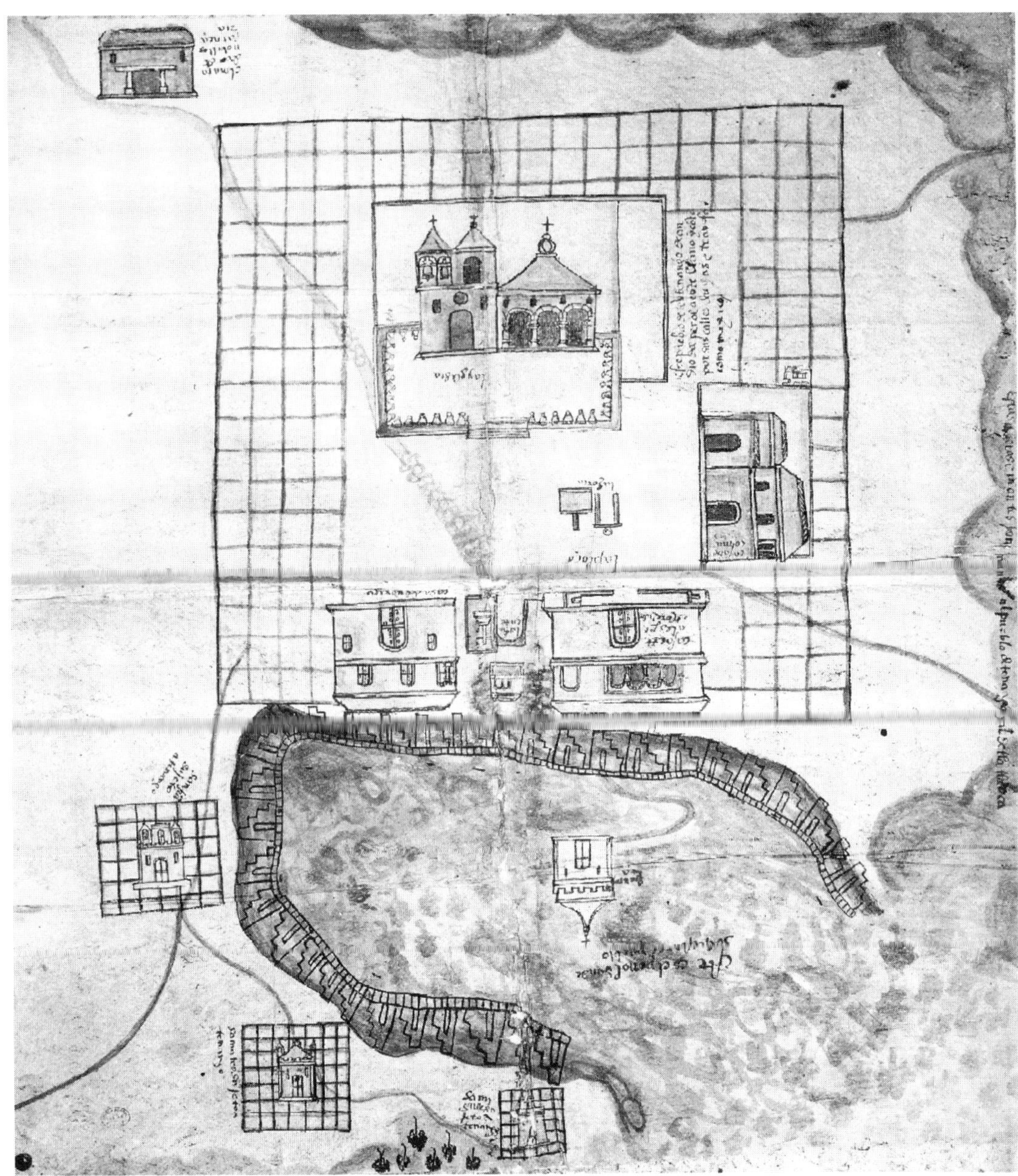

Figure 81. Relación Geográfica map of Teutenango, 1582. The town of Teutenango was laid out on an even grid in the wake of the Spanish conquest. The original pre-Hispanic settlement of the town lay on the adjacent hilltop, seen below the town on the map. Size of the original: 75 x 68.5 cm. Photograph courtesy of the Archivo General de Indias, Seville (Mapas y Planos, 33).

Perhaps the artist wanted to suggest logographically the town's name, somewhat as it appears in the Codex Mendoza folio 33r, where the stepped wall symbol (for tenamitl) rises above a sun-disc (symbolizing *teotl,* "divinity"; see table 7). In this map, as in others, the pictorial symbols emphasize the landscape of the past: the great walls that crowned the hill above Teutenango had been abandoned some time before the map was painted.[26] All who viewed the map could see, in the neat gridiron of the town, how ordered Teutenango was, with each of its dependencies dominated by a church. But the indigenous viewer was denied access to the names of the three outlying dependencies seen at the bottom left of the map, which are expressed alphabetically by their saints' names alone. The native viewer of the map may not have been able to distinguish among San Francisco, San Mateo, or San Miguel; his or her only clues to their identity lay in their placement in relation to Teutenango. The same is true of the map of Chimalhuacan (fig. 35). Although its artist preserves the altepetl's logographically written name at the center of the map (a hill symbol containing a striated circular shield—*chimalli,* in Nahuatl), all of its outlying settlements carry only saints' names, written alphabetically. Their meaning would have been impenetrable to any native viewer who was untrained in reading alphabetically.

Looking specifically at the relationship between alphabetic and logographic writing on the maps makes us realize the yawning gulf between the alphabetically and the logographically literate. Consider the two maps of Cuzcatlan (figs. 32 and 33), close copies, both done by the same native artist who framed much of the map with a border of logographic toponyms meant to represent names of boundaries. One of the two maps (called Cuzcatlan A) has twenty such toponyms, while the other (Cuzcatlan B) has only eleven. Although this artist could write logographically, and no doubt most of Cuzcatlan's residents could read the boundary names, he or she almost certainly could not write alphabetically, because after the artist finished the maps, he or she turned them over to the Spanish-speaking scribe, who filled in the names of the towns. This scribe, in turn, probably could not read the logographic boundary names and he made no attempt to transliterate them, leaving the task to befuddled modern scholars.[27] The Cuzcatlan maps are typical in that the scribe who wrote the Relación also added their alphabetic texts. In other cases, the responding corregidor or alcalde mayor wrote on the maps. Because these men controlled alphabetic writing, they could directly shape, by adding texts of their own, how the most powerful audience, Spanish officials, would understand these maps.

In the map of Guaxtepec, the native artist painted a lush landscape, with its imposing monastery, fruit-filled gardens, rushing streams; below each settlement the artist drew a rectangular cartouche (fig. 30; pl. 2). It was these boxes that were the artist's means of communicating with a Spanish-speaking audience, and it was these boxes that the artist left blank, perhaps being unable to write in the alphabetic letters that Spaniards would understand. With these blank boxes, the Guaxtepec artist ceded his or her role as the primary arbiter of the map's meaning. Into this void, the Guaxtepec corregidor stepped. It is his words, written in these boxes, that fix one meaning of the map, the meaning that would show to Spanish viewers what the map meant.

THE NAHUATL INSCRIPTIONS

Not all indigenous people in New Spain lacked alphabetic writing. Nine of the extant Relaciones Geográficas maps mentioned above, by five different artists, contain longer texts, other than simple toponyms, in Nahuatl.[28] These texts were undoubtedly penned within the communities that made the maps, if not by the same artists. These inscriptions offer us insight into how communities who controlled a written language used it in their self-representations. On these maps, the Nahuatl inscriptions fall into three categories that I have termed (1) nominative, (2) descriptive, and (3) historical (table 9).

The nominative inscriptions are names other than toponyms. They are closely related to the numerous toponyms written in alphabetic Nahuatl in that they replace the kind of information that once would have been written logographically. In the Relación corpus, the nominative inscriptions all name indigenous rulers. In the five maps from the Suchitepec region, picturing Suchitepec (fig. 80), San Bartolomé Tamagazcatepec, San Miguel Macupilco, Tlacotepec, and Zozopastepec, the artist draws the church, looking much like an indigenous tepetl symbol, of each community at or near the center of the map. Near the church he depicts the rulers of each of these towns, who are portrayed much in the manner of pre-Hispanic nobility, some seated on jaguar-skin thrones and one, on the Suchitepec map, clasping a nosegay. Instead of being distinguished by logographic names and

Table 9 Nahuatl Inscriptions on Relaciones Geográficas Maps

Type of Inscription	Maps	Purpose
Nominative (other than toponyms)	Suchitepec group Cempoala	Used to name indigenous rulers depicted on the maps
Descriptive	Suchitepec group Gueguetlan Cholula	Used to explain topographic and demographic information
Historical	Macuilsuchil	Used to narrate regional history and rulership

titles, these men on the five maps are designated with alphabetic ones, if only because their Spanish names—don Francisco Hernández, don Cristóbal, don Martín Hernández, Juan Hernández, Domingo Hernández, Miguel Hernández, don Pedro Hernández, don Luis Hernández, don Bartolomé Pacheco—were not transliterated into logographic script. On the Cempoala map as well, most of the rulers bear Spanish names—don Diego de Mendoza, don Francisco de Guzmán, don Pablo de Aquino, don Juan, and don Andrés—and these are written alphabetically. But the names of those nobles in Cempoala who retained Nahuatl names are for the most part written logographically.

In the Suchitepec group, the inscriptions that are not nominative are descriptive, adding some topographical or demographic information that clarifies or elaborates what the image shows. They, like the rest of the alphabetic text, were written by the artist, because on the map of Tamagazcatepec, he or she writes *çepõ ali yme calli ycha maçehualin nica nictlalia,*[29] or "here I put the houses where the macehualtin live, twenty-two [of them]" next to a drawing of houses. Along one of the roads on the map from Suchitepec, the Nahuatl text reads *vntli çinmatla yhua xalpan yey melio llehuan: ac jztevhuatl: çinmatla,*[30] or "the road to Cimantla and Xalpan, three and half leagues; it arrives at the sea of Cimantla." A descriptive narrative likewise seeps into the Nahuatl texts of the Relación Geográfica map of Gueguetlan (fig. 82). The artist of this picture is not known, but it was almost certainly inscribed by a native principal, don Juan Hernández, who also signed the text of the Relación (Acuña in [illegible] 5: 198–9). While most Spanish [illegible] simply name places, directions, and distances, those by Hernández sometimes provide brief physical descriptions of the places in Nahuatl. San Agustín, a town on the bottom left of the map, is described as *vel tlan, vel atlauhco; vel uvica* [or, to use more conventional spelling, *huel tlani, huel atlauhco, huel ohuihcan*], meaning "very low-lying, in a large canyon, a very dangerous place." Santa Ana, a town at the top center of the map, third from the left edge, is *vel tlani atlauhgo [huel tlani, atlauhco],* or "quite low-lying, in a canyon."

In contrast to the Spanish officials who added to the maps terse inscriptions that were largely descriptive texts, that is, ones that explained the image at hand, those inscriptions added by native authors sometimes went beyond what the image showed. For instance, don Juan Hernández's legends point out topographical details that the map itself does not show. Compare Hernández's words to the Spanish inscriptions that the scribe Hernán García Ruiz adds to the native map of Guaxtepec. García Ruiz explains the symbol of a spring, writing next to it *fuente de agua q[ue] sale junto de la yglesia* ("spring of water that emerges next to the church") even though to the artist and any other native viewer, the symbol for "spring," and its position in relation to the church, were perfectly visible (fig.

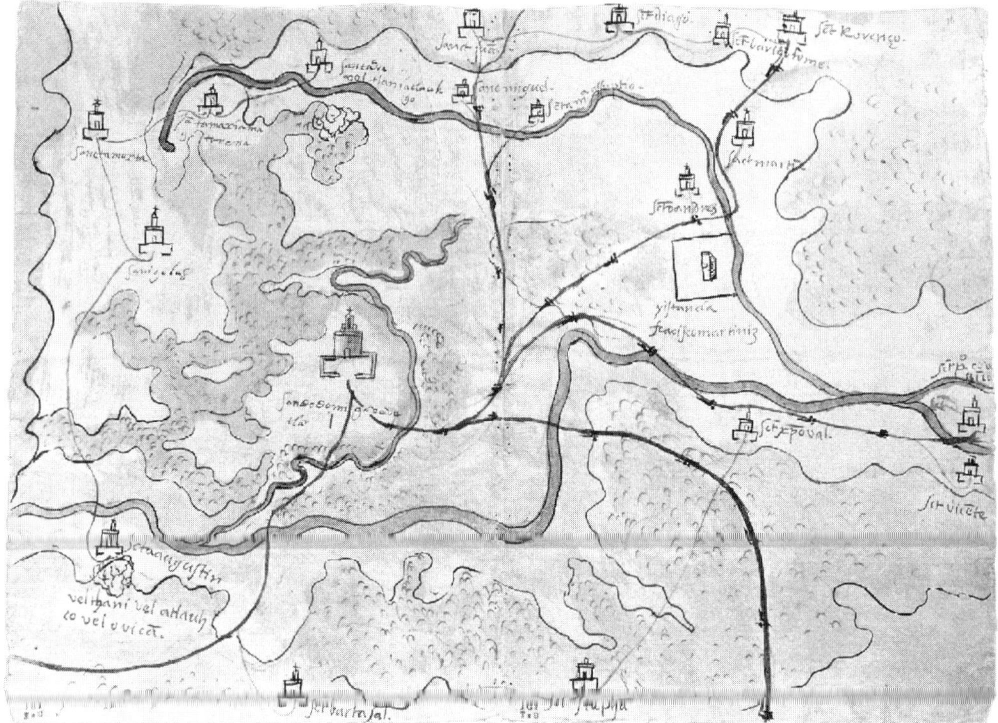

Figure 82. The Relación Geográfica map of Gueguetlan, 1579. The inscriptions on this map were added by an indigenous noble, don Juan Hernández. With them, he names towns and describes regional topography. Size of the original: 31.5 x 43.5 cm. Photograph: B. Mundy; reproduced courtesy of the Benson Latin American Collection, The General Libraries, The University of Texas at Austin (JGI xxiv-6).

30; pl. 2). Understandably, García Ruiz may have added the inscription to clarify the "spring" symbol for a non-indigenous audience, but other corregidores also used inscriptions to reiterate what was clearly shown. In the map of Atlatlauca, the bold inscriptions were added by the region's corregidor, Gaspar de Solís. At the top of the map, where a spring and a stream are clearly depicted, Solís writes *nacimiento de agua* and *corriente de agua.* It is as if Ruiz and Solís never expected viewers to look at these maps, but only to read them. The writers of the Nahuatl texts, on the other hand, clearly expected viewers to look carefully at the image and added words to show the unseen.

In the texts written in Nahuatl, we also sense the continued presence of the oral tradition, but with the alphabetic text now standing in for the reader/reciter, who remained fixed in the community as the map was released into the world, to travel beyond community boundaries to where the community could no longer control what it said. We can glimpse the nature of oral accompaniments to maps when we consider three maps, two of them architectural renderings of palace complexes, that accompanied the Relación Geográfica of Tlaxcala, written by Diego Muñoz Camargo. Its author was a citizen of the region that was the subject of his Relación, a man deeply rooted in Tlaxcala's native community. He had his maps and illustrations in hand before he wrote his text, and used the former as a primary source for his Relación.[31] Muñoz Camargo's written narrative of Tlaxcala's history is closely akin to the oral narrative of history that was kept in preconquest times (Nicholson 1971a: 52–9), thus the relationship between Muñoz Camargo's written narrative and the maps he included in his Relación Geográfica can be used to cast light on the relationship between native maps and the oral tradition that would have once accompanied them.

Muñoz Camargo devoted many pages of his written Relación to describing the palaces of the preconquest ruler Xicotencatl, the royal plaza of Tlaxcala, and the monastic complex of San Francisco (RGS 4: 46–63). These three sites are also described pictorially in the three maps that accompany the text (fig. 83). The maps provide much in the way of layout and detail of palaces, plaza, and monastery. It is left to the narrative of the text to weave these three places together, and it does so by explaining their placement within Tlaxcala, and through this, their spatial relationships to one another. Muñoz Camargo's narrative also colors these illustrations with the hues of historic event. Within the palaces of Xicotencatl, pictured in one of the maps, Hernán Cortés had reached an accord with Tlaxcalan leaders six decades previously (RGS 4: 60). Muñoz Camargo also recalls the songs that once echoed in these same chambers (RGS 4: 61). He brings us through the doors of the royal houses pictured in another of the maps and describes paintings in the inner chambers. He walks among the peach, pear, apple, and olive trees, brought from Spain, that grow in the monastic garden in the upper left of the third of the maps (fig. 83; RGS 4: 54). In the Relación of Tlaxcala and its maps, we see the entangled relationship that the narrative and the map once had within the native community. They were fully complementary, one supporting the other.

It is in the inscriptions of a historical nature that we see most clearly how native writers tried to replace both the histories that community maps traditionally contained and the oral narrative that these histories would inspire. Such memories of pictorial and oral narratives are invoked by the longest of the Nahuatl inscriptions, which appears on the map from the Zapotec-speaking town of Macuilsuchil (fig. 79 [pl. 8]; app. C). The inscription describes how the regional lord of Teozapotlan, or Zaachila, a Mixtec ruler named Ocoñaña (b. 1397), divided lands in the valley into three separate altepetl. While Macuilsuchil cared for its lands, its neighbors did not. The text goes on to give the Zapotec names of three lords who once ruled Macuilsuchil: Coqui Pilla, Coqui Piziatuo, and Yoca Xonaxi Palala, who are depicted on the map, and presumably these are the three who carefully guarded Macuilsuchil's lands. As we understand it, the Nahuatl inscription on the Macuilsuchil map fits the expected parameters of the pictorially written histories that we find as part of cartographic histories. As we saw on both page 36 of the Codex Zouche-Nuttall and on the Relación Geográfica map of Teozacoalco, written narratives concentrate on political legitimacy of the ruling lineage, and in both it is a kind of legitimacy by association. For instance, Teozacoalco's lords traced their connection to Tilantongo's ruling line (figs. 10 [pl. 1] and 52). Likewise, Macuilsuchil's Zapotec rulers seem to be basing their claim to authority over territory on a lands division carried out by

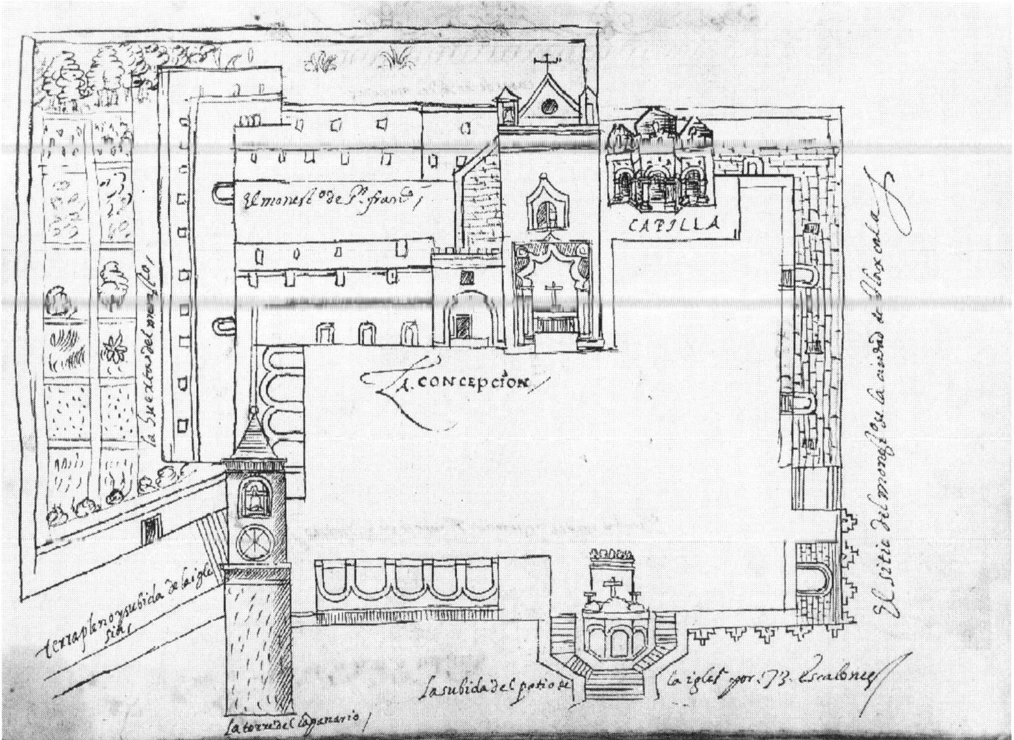

Figure 83. A Relación Geográfica map of Tlaxcala, c. 1584. This map is one of the three maps and 156 illustrations that were drawn to illustrate Diego Muñoz Camargo's Relación of Tlaxcala. It portrays Tlaxcala's Franciscan monastery complex, showing its large patio in front and a small walled garden to the left. Size of the original: 16 x 23 cm. Photograph courtesy of Glasgow University Library, Department of Special Collections (Ms. Hunter 242, fol. 245v).

Ocañaña, a historical Mixtec overlord who reigned in the early part of the fifteenth century. In all three maps, a historical event—ruler migration (as at Teozacoalco), or consecration by an overlord (as at Macuilsuchil)—is recorded to show the basis of the depicted rulers' claim to authority.

The Cempoala map (fig. 42; pl. 5) also offers a brief inscription about regional history: *mexico tlatovani ytzcovatzi ycha,* to label a picture of a house or palace. The text "the house of the Mexica ruler, Itzcoatzin" recalls the Culhua-Mexica's conquest of the region in the late fifteenth century, when Itzcoatl's domain, symbolized by his palace, was established here. Admittedly laconic, this inscription shows how a picture, in this case a house or palace, triggered historical associations on the part of the map's viewer. Here, however, the associations are set down in alphabetic writing.

All in all, in the Nahuatl texts we find native authors trying to replicate or to replace traditional information on the maps; these inscriptions mark that their authors understood the new importance of alphabetic writing. In effect, they seized control of the blank spaces, like those left by the artist of the Guaxtepec map, to try to commandeer the map's meaning. But even as these writers and their communities seized hold of alphabetic writing, they were unable to scale the walls of their ghetto. Yes, they were alphabetically literate, but in a native language, a language unreadable to the ultimate audience for these maps. Native authors may have reinscribed traditional picture-writing, even translating it into a native language (like Nahuatl) that was not their own, but it was still unintelligible to their conquerors.

THE SPANISH INSCRIPTIONS

If in the Nahuatl inscriptions we find a written transcription of what was once the map's spoken accompaniment, in the Spanish inscriptions we find no such thing. Most of them are terse and almost all are devoted to naming, and in this they often are a direct replacement of the logographic words of native maps. In the Relación Geográfica map of Minas de Zumpango, the writer of the inscriptions merely provided the names of each of the settlements and stated the number of taxpayers, or tributaries, in each (fig. 44).

In many cases, the inscriptions run counter to how the indigenous community thought of their local landscapes. Spanish writers often used their inscriptions to establish a hierarchy among the human settlements that the maps depicted. *"Esta es la bocación de la cabecera de S. Juan Evangelista que es Culhuacan,"* reads a central inscription on the map of Culhuacan (fig. 29). While the contours of the political geography that the Spanish established in New Spain often followed the indigenous ones that were in place in the early sixteenth century, the hierarchies that the Spanish established within the altepetl cut against indigenous practice, as described in the two previous chapters. Spanish colonials often designated the first subunit of the altepetl as the cabecera, or head town, and set it above other subunits, which they designated the *sujetos, estancias,* or *barrios* (Gibson 1964: 34; Lockhart 1992: 14–58). We see the importance of these ranks clearly in the Relación Geográfica corpus, where the Spanish writers stress them again and again. In the Acapistla map, for instance, all but one of the named settlements are labeled *estancia,* to distinguish them from Acapistla, the cabecera (fig. 31). The Spanish insistence on internally ranking the altepetl subunits arose from the institutional model they imposed on the New World, where the cabecera, housing a Spanish administrator, would act as a center for collection of tribute from its residents and those of the dependencies, but it disrupted traditional Nahua practice of subunits ordered in rotation, rather than strictly ranked.[32] We see a countering image in the map of Cempoala, where the subunits are all represented equivalently; this native artist imposed no foreign hierarchy upon them (figs. 42 [pl. 5] and 60). In defining the rank of towns within a map like that of Acapistla, the Spanish writers were not creating this hierarchy, but merely recording the hierarchy that had recently been imposed. But in other areas Spaniards were more active agents in shaping the reality that the maps describe.

If we look to what Spanish scribes and local officials were writing upon the maps, we can see not only the immense power they had to arbitrate the meaning of these maps, but also that they tried to make these maps, which native artists saw as unique representations of their communities, fit into a larger picture that they understood New Spain to be. With their inscriptions they often tried to fit the Relaciones Geográficas maps into the great mosaic of New Spain that López de Velasco aspired to amass. To do this, they used a written language that was relational, that described the spatial relationships of one place to another, as we see on the Ameca (Jalisco) map (fig. 84). Pedro de Moras, the scribe who wrote the texts on the map from this region, carefully labeled the six arteries leading out of the town. One was "road of the city of Guadalajara," another was "the road to Mexico and the province of Ávalos," and so on. De Moras, like others, was careful to label "east" and "west" on his map to further orient it. He carefully intended to establish relations among the places pictured in this map, and other nodes outside of it, such as Mexico City and Guadalajara, as well as the cardinal directions. His inscriptions reveal that he saw this map as part of a larger system of spatial relationships, a system applied to, but not particular to, the region of Ameca.

In contrast, many of the native maps offer few of these kinds of orienting inscriptions, perhaps because their nearest antecedents, community maps, offered the singular view-

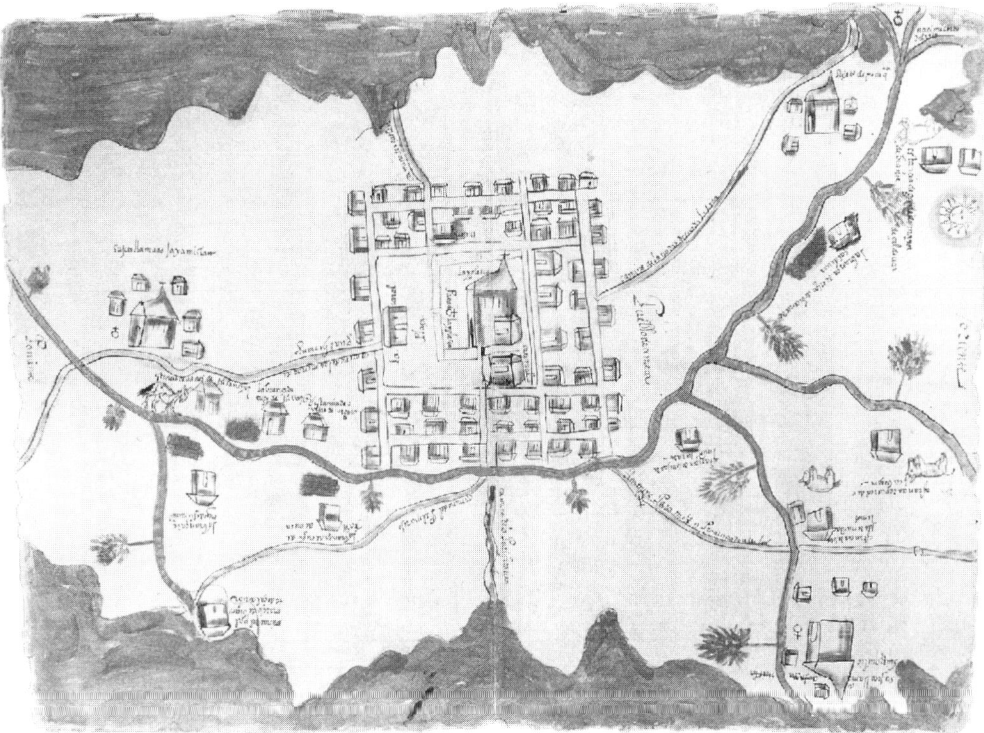

Figure 84. Relación Geográfica map of Ameca, 1579. The inscriptions on this map help orient it within a larger network of roads and settlements. Size of the original: 31.5 x 43.5 cm. Photograph: B. Mundy; reproduced courtesy of the Benson Latin American Collection, The General Libraries, The University of Texas at Austin (JGI xxiii 10).

point of the community that produced it. Community maps saw themselves as the whole, not a small piece of a larger whole. In the map of Amoltepec (fig. 51; pl. 6), which has only one brief inscription in Spanish, the native artist offers us no sense of exterior orientation, does not show us where the sun rises or sets, or where the neighbors' lands lay. Nor on native maps do we find any absolute consistency as to orientation; each mapmaker oriented his or her map as he or she saw fit. Cempoala (fig. 42; pl. 5), likewise, offers little to show relationships to neighboring communities. Pachuca is shown at bottom right flanked by a seated Spaniard, but other than this the reader is given no sense of its relative importance as a mining center in the region (Bakewell 1971; CDI vol. 9: 192–209).

In the biggest break with indigenous community mapping, Spanish writers introduced an entirely new category of information on their inscriptions, one telling of property ownership and individual domain. There seem to be few if any precursors to these individual claims in the pictorial writings of community maps.[33] Such possessive inscriptions not only show what Spanish colonials wanted maps to say, but also prefigure the future of Spanish colonial mapping in New Spain. To turn again to the Ameca map, we see a number of inscriptions that designate tracts of personal property held in the region (fig. 84). One inscription reads "ass ranch of Juan Sanchez," and another, "cattle ranch of Vincente de Çaldivar," and yet another "the wheat farm of Juan Vázquez." These same places are also designated with symbols, and the two ways of encoding information reveal a schism. The symbols show us the general character of the property: Sanchez's ass ranch is flanked by a picture of a donkey, Çaldivar's ranch with a bull, and Vázquez's property with a rectangular symbol to designate a field, but the most important information—names of the owners—is alphabetic.

We can perhaps most clearly see the power of lettering to shape the meaning of a map in returning to the native map of Misquiahuala, with which the chapter began (figs. 62 and 63; pl. 7). In many respects a traditional indigenous boundary map, the Misquiahuala map shows the hills that bound the valley named with tepetl symbols. The landscape is defined by names, shown here through logographic place-names in Nahuatl. Beginning in the sixteenth century, the lands in the region became of interest to Spaniards for sheep raising; by the end of the century about two thirds of all land had passed out of the hands of indigenous communities to become the private property of Spaniards, and in some areas the figure was as high as 99 percent (Cook 1949; Melville 1994).

The land around Misquiahuala was seen one way by the native community that lived there and another by the Spaniards who were coming to own it. We see the schism on the face of the Relación Geográfica map, as the native artist made an indigenous map, carefully naming places within the valley, as well as the boundaries of the valley with logographic names. Francisco Fernández de Córdova, the scribe who wrote the inscriptions, had a different perception of the region, as he alphabetically identifies different parcels of land and notes who owns them (fig. 63); thus we see the "sheep ranch of Martín Seron," and the "sheep ranch of Juan de Cabrera, clerical priest." The native map of Mizantla likewise has its meanings shifted by the inscriptions (fig. 40). The three small identical houses at the top edge of the map take on new meaning via alphabetic inscriptions when we learn that they are cattle ranches owned by Juan de la Cuenca, Juan del Miral, and Sancho Núñez. With such alphabetic inscriptions, the Spanish writers could seize maps that before were the self-imagined landscapes of communities and transform them into writs of private possession.

In examining the writing on the maps, both logographic and alphabetic, and the languages, Nahuatl and Spanish, we see how this corpus was fully colored by the hierarchies of language and writing that came be established in New Spain. The imperialist practices of renaming were carried out by both Aztec and Spanish, but the latter not only robbed the indigenous communities of their proper names, but also imposed new names that were out of the reach of traditional logographic writing. The maps, viewed in this light, tell us a tale of language and loss, as native writers gamely tried to hold on to the meaning of their community representations as they moved into the wider world, only to be outmaneuvered by the Spaniards, who controlled the language that would reveal the meaning of these maps in that final court of authority: the eyes of the Spanish king.

tenocStitla
tenocStitla
tenochtitla
yztacalco
mexicatzinco
ytztapalapa
tenocStitla
colhuaca
tenocStitla
colhuaca
ytztapalapa
colhuaca
ytztapala pa
colhua ca
ytztapa lapa
mexicatzinco
ytztapalapa
colhuaca
yztacalco
mexicatzinco
ytztapalapa
colhuaca
yztacalco
mexi ca tzinco
ytztapa lapa
colhuaca

y estas son suertes de tierras de ytzpalapa y mexicalzingo y culhuacan

cotlaman

tepo

mazatepetl

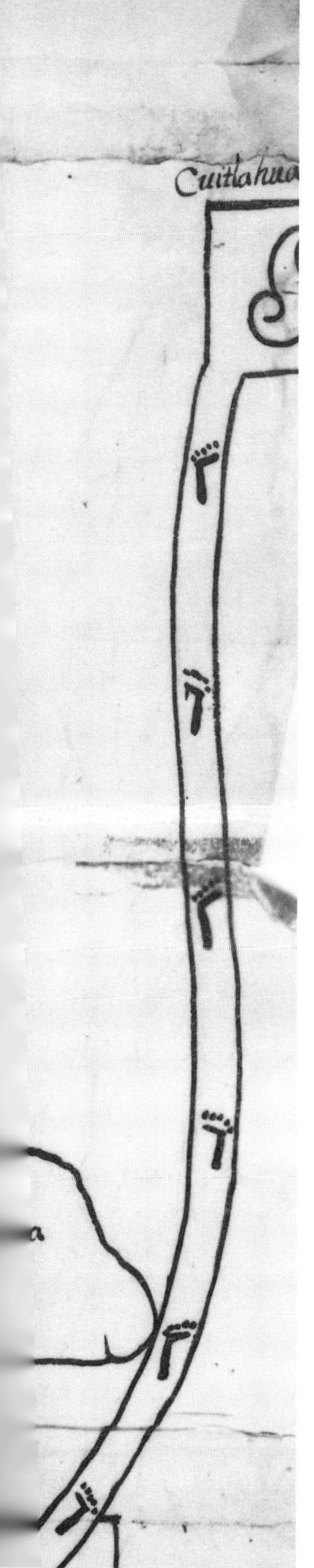

CHAPTER SEVEN

The Relaciones Geográficas and Other Viceregal Maps in New Spain

Many of the Relaciones Geográficas maps were culled from native painters who were called upon by their corregidores. But these maps were not the only ones that these same painters were making for the viceregal government. The immense corpus of maps preserved in Mexico's Archivo General de la Nación (AGN) shows us that scores of native painters made maps, year in and year out, for the hundreds of corregidores across New Spain as part of land grant, or merced, suits. Because mercedes maps were so widespread, and because so many native painters were commissioned by colonial officials to paint them, it would be remiss to discuss the Relación Geográfica corpus without looking at them. Their links to the Relación Geográfica corpus are not merely a matter of conjecture, for at least three of the thirty-two native painters of the extant Relaciones—the artists of Xalapa de la Vera Cruz (fig. 85), Tehuantepec A (fig. 27), and Ixtapalapa (fig. 28)—also painted land grant maps.

In looking at mercedes maps, we will see that they had a magnetic pull on the path of cartography in New Spain, far beyond what one might expect. Their influence stemmed from the land grant process of which they were a part, which was transforming the indigenous landscape. As Spaniards spread across New Spain, they consecrated their occupation through viceregal mercedes, through which they converted once-indigenous land into their personal property (Simpson 1952). From the indigenes' point of view, land was uprooted from its traditional place within

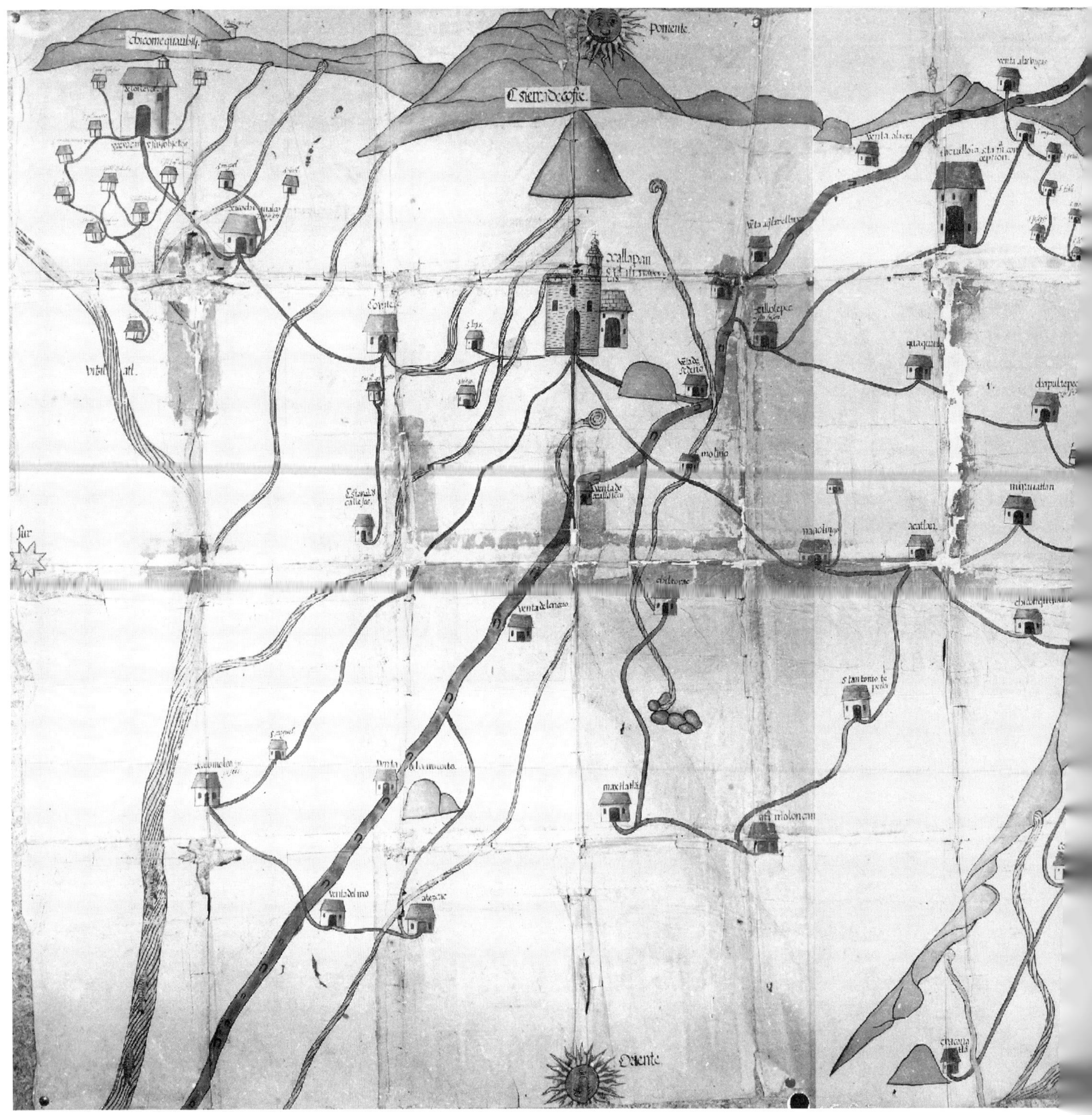
poniente.
xallapan
molino
sur
acatlan
Oriente

networks of social structure and historical rationale. When we find maps by indigenous artists that look like their Spanish counterparts, it is because the model of the deracinated earth that mercedes maps introduced proved to be the authoritative one.

MERCEDES AND LAND GRANT MAPS

The introduction of the merced, or land grant, process came to New Spain early on in the sixteenth century. Not long after Viceroy Antonio de Mendoza came to New Spain in 1535 to establish an orderly government out of the disarray left by the initial conquistadores and the first *Audiencia,* he standardized the granting of lands (Chevalier 1963: 56–8; Gibson 1964: 272–7). As Mendoza envisioned them, land grants were a cheap and convenient way for the crown to reward the clamoring band of ex-conquistadores and their descendants for the services they had rendered to it. By the end of his tenure, Mendoza had firmly established set procedures for granting land, which, like most Spanish legal proceedings, left a considerable paper trail. The written record of the grant began with an *acordado,* in this case a decree stating the beneficiary of the planned land grant and its size and rough position, which was put forth by the viceroy or, in an interregnum, issued without him by the Audiencia (see appendix D for a typical acordado). The acordado was sent to the corregidor or alcalde mayor in charge of the province wherein the grant lay, and the corregidor or alcalde mayor was then to carry out certain proceedings, called *diligencias,* outlined in the acordado. He was to visit the site to make sure that the grant would not impinge on native croplands or on lands owned by Spaniards. He was to announce the site and size of the grant to nearby native communities, making this public notice on a Sunday or feast day when all were gathered in church. Once the impending grant was well publicized, the corregidor or alcalde mayor was to collect testimony from ten witnesses about the status of the lands to be granted—whether the lands were already claimed, or whether the grant would threaten the livelihood of native communities. Lastly, he was to have a map *(pintura)* of these lands drawn as final proof that they could be granted without "harm or damage." The acordado and the written diligencias were then to be sent back to the viceroy or Audiencia in Mexico City for final confirmation.

Once begun, the viceregal prerogative of granting lands swelled far outside the boundaries of its original intent, and by the end of the sixteenth century, land grants had spread like ganglia over the face of New Spain. The government used grants to control the colonial economy, encouraging Spaniards to grow wheat and other needed crops. For Spanish

Figure 85 *(opposite).* The Relación Geográfica map of Xalapa de la Vera Cruz, 1580. This large map covers a great sweep of territory, from the mountains of the Sierra Madre to the Gulf coast. The artist of this map also painted land-grant maps for colonial authorities, maps that covered smaller sections of the same territory. Size of the original: 120 x 126 cm. Photograph courtesy of the Archivo General de Indias, Seville (Mapas y Planos, 18).

colonists, land grants were the quickest way to legally obtain large tracts of land. Lands in the Valley of Mexico were prized, since they allowed Spanish owners to live in Mexico City and still be near their holdings (Gibson 1964: 277). The desirability of other lands shifted from decade to decade. For instance, Spaniards jockeyed for grants around the rich silver mines of Zultepec in the 1540s, while in the early 1560s they aspired for grants around the recently discovered mines of Temazcaltepec (cf. AGN Mercedes vols. 1 [1542], 2 [1543], and 5 [1560–1561]; Gerhard 1986: 275–8; RGS 7: 141). Tehuantepec, with its expanses of land for cattle raising, was at its peak of popularity in the 1590s (Zeitlin 1989: 38).[1] Outside of mining regions, most land grants were for ranches, for Spaniards quickly realized the potential of New Spain's untouched grasslands to fodder ever-growing herds of cattle and other livestock. To do so, they needed to control huge tracts of land, property that they could only gain through royal grants.[2] Land was granted, and herds multiplied like locusts.

From Mendoza's standpoint, the inclusion of a map in the merced process must have been a successful conceit, for it continued to be a standard part of the merced acordado sent out by sixteenth-century viceroys.[3] Such maps allowed him to see towns, fields, and mercedes sites, to ascertain their relative positions without going to the region himself—they were the images of New Spain that allowed him to know it. The maps also allowed him to verify that new grants were the legally mandated distance away from cultivated fields (about two kilometers) and from other ranches (Chevalier 1963: 89–90.) Overall, the map helped eliminate the ambiguity inherent in the acordado's description of the parcel of land, which often stated only that the site lay on the outskirts of such-and-such a town. With the map, the site of the grant could be fixed in space.

Mendoza took his role as protector of New Spain's indigenous population seriously, and it seems that he and his successor, Luis de Velasco, deliberately included the request for a map in the merced process, as well as in the adjudication of land disputes, to protect indigenous communities (for an early example from Velasco's era, see AGN Mercedes, vol. 4, fol. 146). That is, Mendoza recognized that indigenous peoples were more conversant with pictorial rather than alphabetic records, and he wanted to offer them a way of understanding the grants that would be made on all sides of them, so that they could legally protest the ones that worked against their interests. Other facts support this interpretation: in other fora, Mendoza believed native pictographic records to be authoritative and worthy of regard. He may have commissioned the Codex Mendoza; he also made a practice of consulting pictorial accounts in judging native disputes (Borah 1983: 67).

Yet in only a few cases is it perfectly clear that the merced map met this noble intent of acting as a conduit for native education and opinion. To begin with, the land grant process

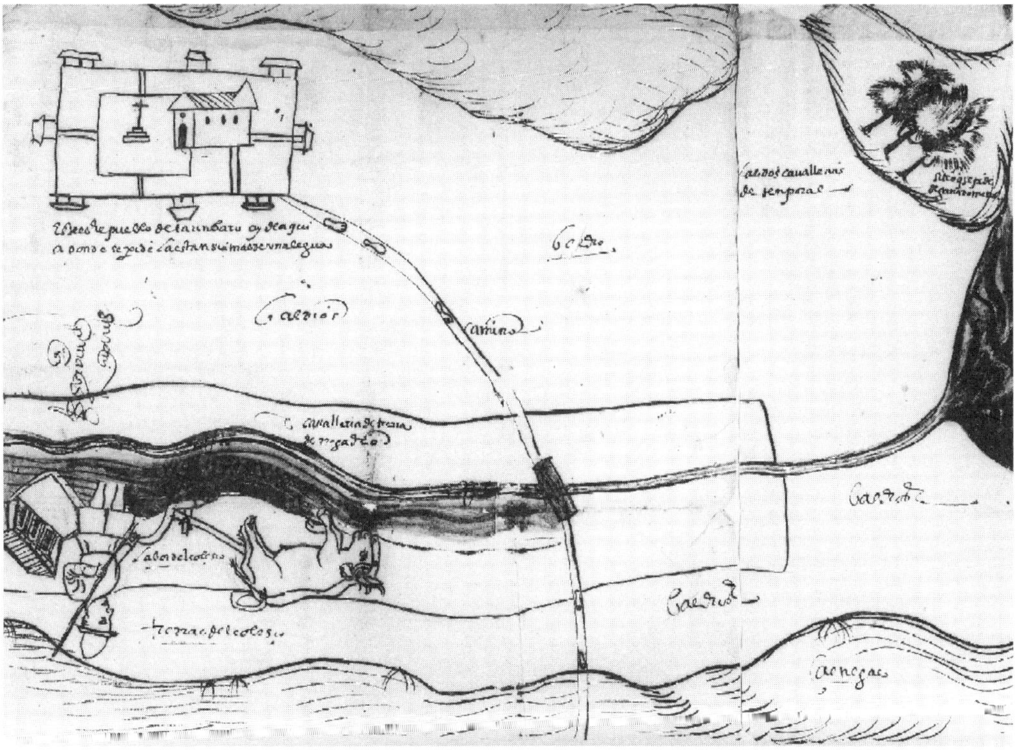

Figure 86. Land grant map, 1585. This map from the region of Tarímbaro, Michoacán, is typical of land grant, or merced, maps. Size of the original: 32 x 43 cm. Photograph courtesy of the Archivo General de la Nación, Mexico (Ramo Tierras, vol. 2721, exp. 38, Mapoteca 1854).

itself was often marked by collusion and fraud on the part of the Spanish official and the grant recipient (Gibson 1964: 276; Zorita 1963: 108–9). It is likely that native communities would not have been properly informed of the specifics of land grants. But one map from Acolman of 1608 states on the front that it was made in the presence of the native community and is signed on the back by members of the native government with a statement that they found it accurate and had nothing more to add (AGN Tierras, vol. 641, exp. 2, Mapoteca 802). This map, though, was an exception.

Ironically, the merced map proved not so much a tool for the protection of native lands as it was an effective conduit for introducing native painters to the kind of map Spanish officials wanted, showing them a conception of territory hitherto foreign to them. The merced map required that new aspects of territory—that is, ones not found on traditional community maps—be entered on the register of the map; it asked the maker to show individual ranches and fields and to designate these, not with toponyms, but by the names of their owners. The acordado's request read: "Have a map [pintura] made of the site of the town on whose outskirts [the grant] would fall as well as the other ranches and lands that would be found [on these outskirts] if this grant should be bestowed. . .and to whom they belong. Also note the lands and untilled fields remaining and state the distance between these and the current request" (my translation; see appendix D). In the merced map, space, or the landscape, was now made visible through the system of ownership and property that now held dominion over it.

A map from Tarímbaro, made in the mid-1580s, is a typical merced map shaped by the demands of the acordado (figs. 86 and 87; AGN Tierras, vol. 2721, exp. 38, Mapoteca 1854). It shows the town of Tarímbaro in a mix of plan and elevation at the top left, using a perspective typical of coeval maps. The desired lands for a cattle ranch fell on Tarímbaro's outskirts and the glosses specify that they lay over a league (about 4.3 kilometers) away. The map also shows—lying at a safe distance from the grant site—a zone of irrigated land being plowed by a man and a team of oxen. These fields are identified as the lands of Colesio; closer to the merced site are the two *caballerías,* lots of about 105 acres, of Senporas (Gibson 1964: 276). Many of the empty spaces on the map are marked as untilled fields *(baldíos)* and a strip at the bottom of the map is labeled a marsh *(ciénagas).* The map-

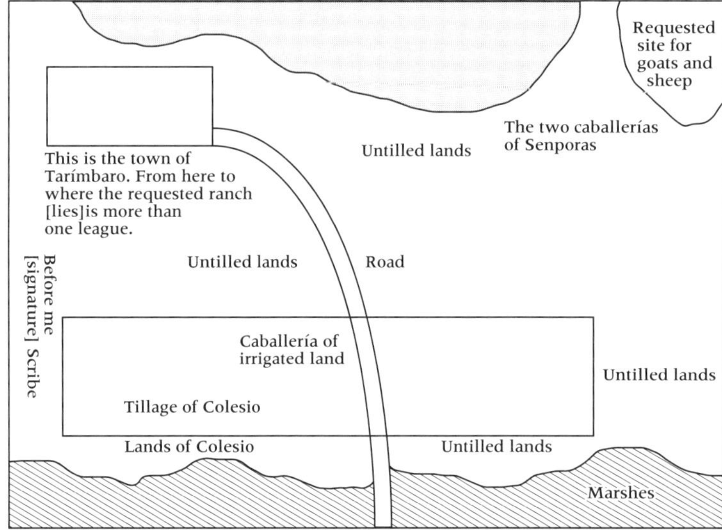

Figure 87. Drawing after the 1585 land grant map from Tarímbaro, Michoacán.

maker includes in his or her map only the kind of information asked for by the viceroy: settlements, types of land, ownership, and placement. The laboring farmer is no mere artistic filler, for he serves to emphasize that this parcel of land is under cultivation and cannot be impinged upon by any new grants.

TRADITIONAL MAPS, LAND TENURE, AND MERCEDES

If we compare the merced map from Tarímbaro to an indigenous community map, one found in the Relación Geográfica corpus, we can see how dislocating this new vision of territory would have been to the native artist. In the Relación Geográfica map from Amoltepec (fig. 51; pl. 6), we see the reaches of the community defined by the regular two-thirds circle of place-names and river; inside lie the region's locii of authority: the church (having replaced the pre-Hispanic temple) and the rulers' palace, wherein the rulers sit. With its pictorial emphasis on boundaries and regional authorities (church and rulers), the map offers us a view of space defined by the limits imposed by the community and, further, offers us a rationale for the existence of that spatial unit—the people within this space shared these rulers and these religious institutions. While the specific boundary names and their rhythmic order were essential to the mapmaker in creating a community map, the parcels of land that the community occupied were not, nor were the specifics of individual property, for these were not elements that defined community. In the Tarímbaro map, on the other hand, we see an artist who is very concerned with spatial layouts of territory; he or she names individual parcels of land with inscriptions, showing how they were used, and is very careful to show us their spatial relationships to one another. The basic difference is clear: the Amoltepec mapmaker was picturing community territory, which he or she envisioned as a contiguous, harmoniously perfect whole, whereas the Tarímbaro mapmaker was envisioning the landscape as a collection of land parcels to be differentiated by ownership and use.

This is not to say that indigenous communities in New Spain were like communitarian utopias, eschewing private property and holding all land collectively. There is significant ethnohistorical evidence that many, if not all, indigenous peoples in New Spain held private property, property that they bought, sold, and left to heirs (Cline 1986: 125–59, esp. 150–9; Lockhart 1992: 149–55; Spores 1967: 9–14, 164–71). While a number of large-scale indigenous maps from Mexico document lands held in private hands, this genre was a separate one from community mapping, and on the whole indigenous community maps like the ones made for the Relación Geográfica corpus do not register the property held by indigenes. Rather, they are concerned with the spatial substrate of collective identity:

community property, as defined by its boundaries, and often its relationships to society and history. The request for a map outlined by the acordado, in addition to asking for a map of property, also requested that the mapmaker show nearby settlements and the placement of the grant request, so the map's scale and scope were akin to a community map, which covered a wide swath of territory. In the end, it seems that native artists must have drawn upon the idea of the community map in making the merced map.

This being the case, the merced map proved as corrosive to the community map as, perhaps, the land grant did to the native community itself. In painting the merced map, a native artist was asked to abandon his or her traditional view of an autonomous community and to adopt instead a view of his or her community as a collection of land parcels, some to be given away at the whim of the viceroy. In addition, given the new role that the alphabetically written inscription played, the artist was also being asked to create a map that was a mere backdrop; other people would add the text that defined its meaning. In the next section we will turn to the artists who we know to have painted Relaciones Geográficas maps as well as mercedes maps, to see individual cases of how the merced, and the land grant process that accompanied it, as well as the introduction of alphabetic script, eroded the hold that indigenous community maps once had across New Spain.

THE XALAPA RELACIÓN GEOGRÁFICA ARTIST

The alcaldía mayor of Xalapa, stretching out from the spine of the Sierra Madre Oriental toward the sands of the Gulf coast, embraced some of the first lands encountered by Cortés and his men in 1519. Along with that initial incursion came the diseases that were to ravage the New World. Their toll was highest in these hot, wet lands along the Gulf coast and on the cooler slopes to the west, reducing the richly peopled landscape to a scattering of hamlets (Gerhard 1986: 385; RGS 5: 343–71). Although the alcaldía mayor of Xalapa was not a place where many Spaniards chose to live, it was where they wanted land, and they replaced the abandoned native towns with livestock ranches. Pasturage was plentiful, and the Camino Real, linking the alcaldía to Mexico City and to the Gulf ports, was a conduit to markets for meat and leather. In addition, the profitable endeavor of sugar raising was just beginning (Chevalier 1963: 77; RGS 5: 365).

It is this region that the Xalapa artist portrayed (fig. 85). The Relación Geográfica map is large—it measures 120 centimeters by about 126 centimeters—to show fifty-seven towns, most of which had a tenth of the population they did before the conquest (RGS 5: 343–71). This depiction of the region is straightforward, with few details and flourishes. A scalloped half-frame on the top and right side of the map (west and north) represents the Sierra Madre. The little houses that stand for the towns of the region are so alike they could have been cut from the same die. Two-story houses of a similar pattern show the towns of Ixhuacán de los Reyes and Tlacuiloia (today Tlacolula). Xalapa, the cabecera, is central, shown by a church of cut stone topped with a belfry. The highway linking coastal Veracruz to Mexico City (the Camino Real) neatly bisects the map along the diagonal from top right to bottom left. This highway is distinguished from minor roads by the prints of horse hooves.

The Xalapa mapmaker did have a few defining traits of his or her style, some of which lead me to believe he (or she) was a native painter.[4] Certain hills of the map—notably the "Sierra del Cofre" (today the Cofre de Perote) lying behind Xalapa—have an individuality and solidity reminiscent of the tepetl symbol. In addition, this painter used a bold outline—a hallmark of native style—to define all the forms on the map. Distinctive to his or her hand are the many rivers that carry water from the mountains to the Gulf. They are composed of twisting strands and look like ropes of hemp.

The Xalapa artist eschewed the bright palette of most of the other polychrome maps and instead used mostly opaque earth tones to fill in the forms first outlined in black. Grey washes add a shadowing effect to the houses. The roads are painted with a flat yellow ocher, as are the roofs of the buildings. A pale, even green covers the half-frame of mountains and blue, close in shade to the green, colors the rope-like rivers. A bright cherry red, perhaps derived from cochineal, is used sparingly to outline the sun and stars of the cardinal directions and to fill in the red orbs of east and west.

The style of the rivers, the forms of the buildings, and the restrained palette link the hand of the Xalapa artist to two land grant maps that survive in the AGN (fig. 88), one from 1578 (fig. 89; AGN Tierras, vol. 2688, exp. 10, Mapoteca 1681) and the other from 1587 (fig. 90; AGN Tierras, vol. 2680, exp. 5, Mapoteca 1583). Both are from the Xalapa region and are dated within a decade of each other and of the Xalapa map, whose accompanying Relación is dated 1580. The dates signal that the Xalapa artist had been commissioned to paint mercedes maps both before and after he or she was chosen to paint the map for the Relación Geográfica. The AGN maps are monochrome, done only in washes of grey; even the limited pigments used in the Xalapa map may have been too precious for these common maps. The map of 1578 shares the solid and distinct hills of the Relación map of two years later. In the later map of 1587 (fig. 90), the artist has attempted to blend the hills into the background, in the style of landscape painting, but the hills still have a somewhat abrupt quality.

In all three maps, the artist reveals a similar conception of the landscape, and no doubt the mercedes maps that he or she was commissioned to paint colored his (or her) Relación

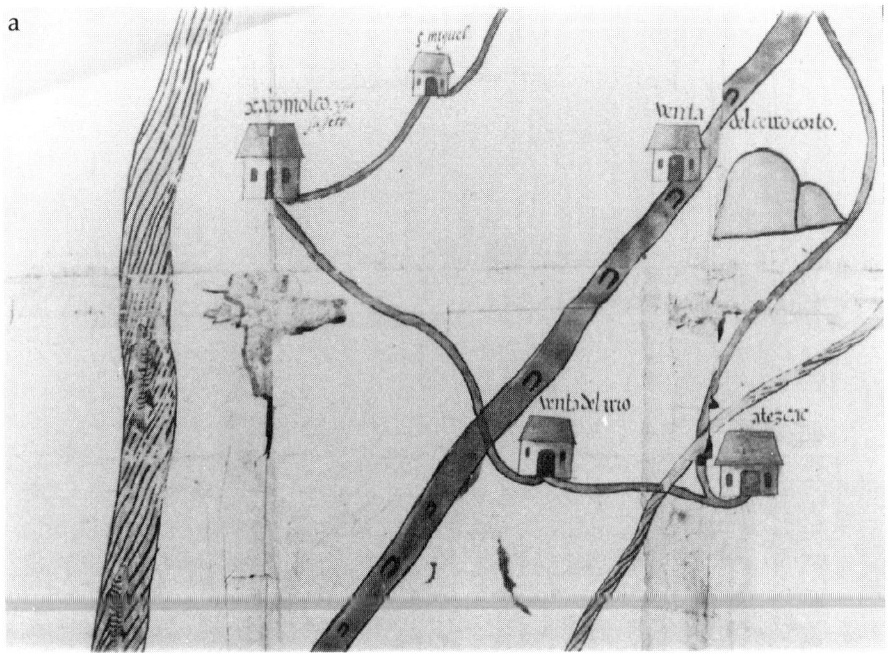

Figure 88. Details of maps made by the Xalapa artist. The forms of houses and the pattern marking rivers are unmistakably similar in all three maps. The inscriptions, however, are all by different hands. (a) the Relación Geográfica of Xalapa de la Vera Cruz, detail; (b) merced map from Actopan of 1578, detail; (c) merced map from Atezca of 1587, detail. Photographs: (a)courtesy of the Archivo General de Indias, Seville (Mapas y Planos, 18); (b) B. Mundy, reproduced courtesy of the Archivo General de la Nación, Mexico (Tierras, vol. 2688, exp. 40, Mapoteca 1681); (c) B. Mundy, reproduced courtesy of the Archivo General de la Nación, Mexico (Tierras, vol. 2680, exp. 5, Mapoteca 1583).

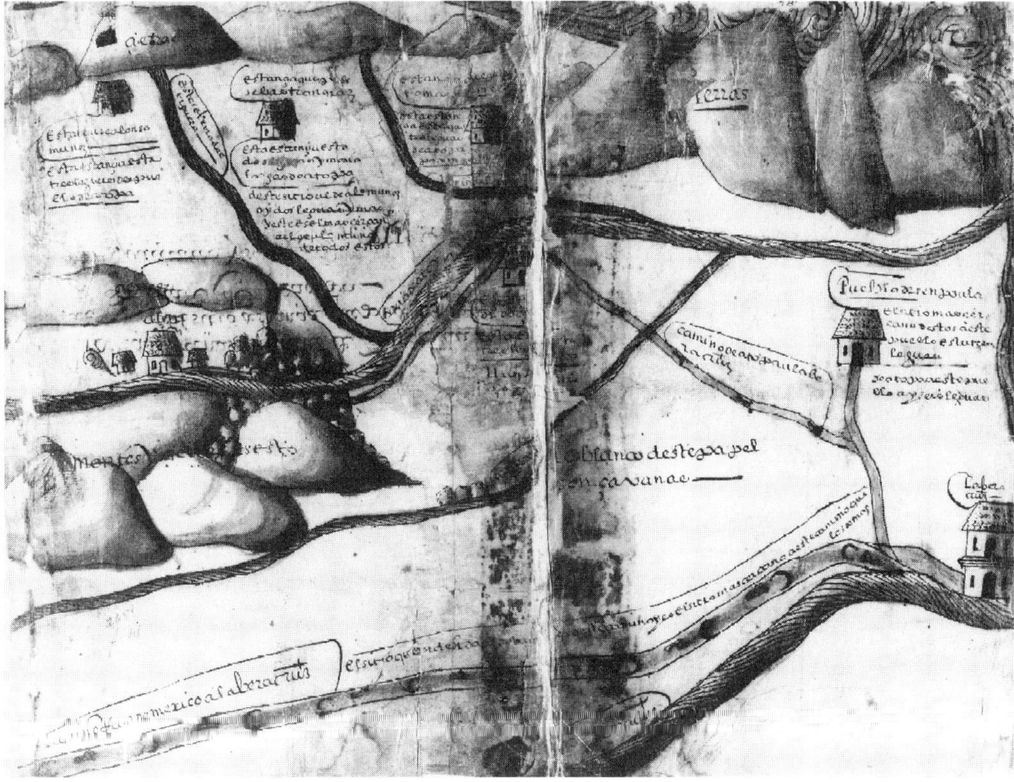

Figure 89. Land grant map from Actopan, 1578. This map, as comparison of details shows, was made by the artist of the Xalapa Relación Geográfica map. The inscriptions describe the landscape and the land to be given away. Size of the original: 31 x 40 cm. Photograph: B. Mundy, reproduced courtesy of the Archivo General de la Nación, Mexico (Tierras, vol. 2688, exp. 40, Mapoteca 1681).

Geográfica work. On all three maps, the artist represents the space with frameworks of rivers and roads with scattered houses to stand for towns and villages, adding an occasional group of rocks or hills; there is no indication of a communicentric viewpoint. The mercedes maps cover areas of smaller scope and provide somewhat greater detail than the Xalapa Relación Geográfica map. The smaller map of 1578 covers a piece of territory also shown on the larger Relación map of 1580. The map of 1587 shows lands, which fall outside Xalapa's jurisdiction in the neighboring district of Vera Cruz Vieja, in a similar fashion to the Relación map (fig. 91).[5]

Like the Xalapa map, the map of 1587 is bisected by the Camino Real running from Veracruz to Mexico City; it is hung with rope-like rivers. In the 1587 map, the artist shows two settlements, which inscriptions identify as the *"Venta Lencero"* and the *"Pu[ebl]o de Atezca"* and a few hills. Clustering small trees blur boundaries between map and landscape. There is little else. The earlier map of 1578 is much the same: rope-like rivers, winding roads, solid hills, and seven settlements, the most prominent being the towns identified by inscriptions as Actopan, Cempoala, and Veracruz. Veracruz is represented by a two-story house, as are Ixhuacán and Tlacolula in the Xalapa Relación Geográfica map. If these maps were tesserae, it seems one could piece together the mosaic and produce parts of the Xalapa Relación Geográfica map. The mercedes maps and Relación Geográfica map differ in the inscriptions. Those of the former are more detailed as to distances, ranches, and landowners, the specifics asked for in the acordado.

Unfortunately, the texts of the Relación and the mercedes give no clue to the identity of the mapmaker working in and around Xalapa in this decade; it was not common practice

Figure 90. Land grant map from Atezca, 1587. This map was also made by the artist of the Xalapa Relación Geográfica map, some years later. Size of the original: 31 x 42 cm. Photograph courtesy of the Archivo General de la Nación, Mexico (Tierras, vol. 2680, exp. 5, Mapoteca 1583).

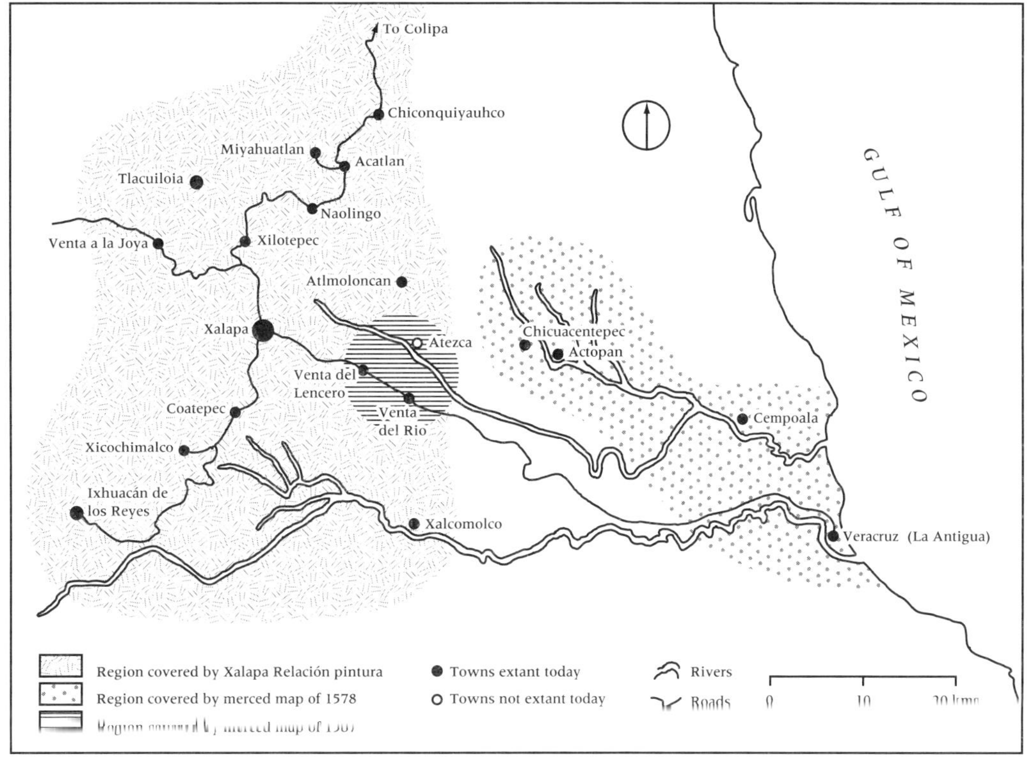

Figure 91. Map of the Xalapa region showing areas covered by Relación Geográfica map and two land grant maps of 1578 and 1587.

to name the painter of the map in either Relación or merced.[6] What is certain is that the Xalapa painter created maps that held visions of territory that were in line with Spanish expectations because he (or perhaps she) found consistent employ with the Spanish colonial government of Xalapa, producing maps for at least three different officeholders.[7] Clearly, this mapmaker was called upon when Spanish colonial officials needed a representation of the region, be it a land grant map or a Relación Geográfica map. This painter knew the region so well as to be able to draw in his or her Relación map the small subject villages clustering around Ixhuacán de los Reyes, a town about thirty kilometers from Xalapa, and to map areas in the mercedes maps that fell some distance from the seat of Xalapa (fig. 91). He or she may have gleaned this topographic knowledge while traveling with local officials when they visited these and other grant sites. But most important was that he or she shared with these patrons a view of the landscape that was a mosaic of individual properties, set among the rivers, roads, settlements, and mountain ranges that further defined the space.

The Xalapa mapmaker also accepted the map's role as backdrop to the written inscriptions, for on all of the maps it was the scribe, not the artist, who tailored the map to its final usage; a different hand glossed each of the mercedes maps and neither was that of the Xalapa artist.[8] The scribe provided the place-names, marked the distances, and described the type of land. Since the different scribes added the inscriptions to the map, not the painter, it seems that the Xalapa painter would submit maps, bare of text, and leave inscriptions to the scribes. Whether the artist had a hand in the composition of all of the inscriptions, we simply do not know. But it is clear that he or she was not their author.

When we look at the inscriptions on the mercedes maps, we see clearly how the mapmaker lost control of the ultimate meaning of the map, for the inscriptions on both maps extend the oral testimony taken in the case, allowing the alcalde mayor, through the scribe, to set down either the majority opinion put forth by witnesses or to advance his own view. Since the function of the testimony was usually to predict that no damage would come to either neighboring lands or to the native estate by the granting of new lands, the two mercedes maps convey to the viewer that the land consisted of empty savannas, which the viceregal government would be free to give away. The depiction of space here is meant to suggest the possibility of its conversion to private property. In the merced map of 1578 (fig. 89), the inscriptions insist that the four ranches *(estancias)* pictured (one extant and the others desired as mercedes) lie at a safe enough distance from the native town of Actopan so as not to infringe on the fields and property of that community. These inscriptions follow like a loyal dog the statements of witnesses who agreed that the land was three leagues from Actopan, well away from native plots and fields. Similarly, in the map of 1587 (fig. 90), inscriptions closely follow the arguments initiated by the text of the merced. Witnesses said that the nearest native town, Atezca, had only about thirty tributaries and plenty of nearby land for its needs. In addition, they said, the land to be granted lay a safe four leagues away.

In both of these maps, the inscriptions carry most of the information needed for the grant. It is the inscriptions that tell us how far the desired ranches lie from the native town and that the land is either untilled *(baldías)* or empty grasslands *(zavanas)*. For the purpose of the merced, the image created by the Xalapa mapmaker was overshadowed by the inscriptions by other hands, who controlled the meaning of the map as far as the grant was concerned. The Xalapa mapmaker's maps gave those others the opportunity to make whatever argument they wished. Thus, the noble intent of the viceregal government, to allow a way to represent indigenous interests by requesting a map to complement the merced process, was foiled. In both the land grant procedures and the accompanying merced map, land that once may have been named and defined by native communities now became property available to Spaniards; this vision of the landscape penetrated both the native community and native mapmaking, and thus appears on works like the Xalapa Relación Geográfica map.

THE TEHUANTEPEC RELACIÓN GEOGRÁFICA ARTIST

When we look at the Tehuantepec region, we find that these new ways of understanding territory were not merely whims of Spaniards, but were the result of both a property-owning class who usually had few ties to the local community and to the economics of ranching, which needed vast tracts of territory and little manpower. Labor-intensive agriculture, in contrast, had been the economic foundation of Mesoamerica. With ranching, land in New Spain became valuable as real estate. The region of Tehuantepec had many of the same qualities as the Xalapa region: vast expanses of land suitable for raising sheep and cattle, nearby roads, and ports giving access to markets for meat and leather in New Spain and Peru. By the 1580s, land in Tehuantepec began to be ceded to Spaniards at an astonishing rate, estimated by one scholar as 89 to 94 percent "of the total surface area of the coastal plain and adjoining piedmont zones" (Zeitlin 1989: 39). At the time the Relación Geográfica was painted, the new economics of livestock, the new business of real estate, and the new enterprise of transcontinental commerce was booming in Tehuantepec.

The indigenous artist of the Relación Geográfica map from Tehuantepec supplies a vision of the region that reflects the new understanding of the landscape brought about both by the land grab and maritime commerce (fig. 27). The map's native authorship is revealed by the logographic place-name that shows a jaguar (*tecuani* in Nahuatl) climbing a hill (tepetl) behind the main church (fig. 92a), but otherwise the map embraces a new vision of Tehuantepec's space. The map covers an area of about 16,000 square kilometers (figs. 27 and 93), and the shape of the long coastline was no doubt influenced if not drawn from nautical charts of this important zone of commerce.[9] Such nautical charts would have been commonplace. Juan de Torre de Lagunas, the alcalde mayor of Tehuantepec who

Figure 92. Details of logographs made by the Tehuantepec artist. Tehuantepec is a Nahuatl name meaning "hill of the jaguar." At left (a), is the place-name from the Relación Geográfica map of Tehuantepec A, and at right (b) is the same place-name, appearing in the land grant map from Tehuantepec of 1573. Photographs: B. Mundy; (a) reproduced courtesy of the Benson Latin American Collection, The General Libraries, The University of Texas at Austin (JGI xxv-4); (b) reproduced courtesy of the Archivo General de la Nación, Mexico (Tierras, vol. 3343, exp. 4, Mapoteca 2378).

compiled the Relación, understood them as another way to map the region, for he included a nautical chart with the native map illustrating the Relación (fig. 25).[10] The native map emphasizes this important coastline, the topography and the number and position of towns and villages in the area that defined the space, and the tracts of empty land between them, these tracts having a new, unforeseen value in New Spain.

A close analysis of the Tehuantepec artist's hand has allowed me to identify the larger corpus of his or her works. As we can see in this Relación Geográfica, this artist had a very distinct way of drawing houses, which appear frequently on the map to indicate settlements (fig. 94). The body of the house is a rectangle and it holds one oversized black doorway. Rising from the body of the house is a steep-pitched roof in the conical shape of a dunce cap. This artist also had a hallmark style for rendering hills. They are filled with color; their solid outline marks their shape but then this profile is softened by stippling to give the impression of wooded areas. No fewer than six mercedes maps by this same Tehuantepec artist survive today at the AGN, dating from 1573, 1580, 1583, 1585, 1586, and 1598 (figs. 94–96).[11]

The earliest map, painted some seven years before the Relación Geográfica, was for a grant of lands near Tehuantepec, and in it the artist marks this town by the same logographic place-name as he or she uses on the Relación Geográfica map: a jaguar scaling the hill (fig. 92b). In both maps, the jaguar climbs up the right side of the hill, the lower part of his body following its curve. In both, his mouth is open and tongue protrudes.[12] In both, the church of Tehuantepec sits at the foot of the hill, in a blending of Christian and native symbolism so often found in native maps. In time, this artist was called upon to paint other mercedes maps, from which he or she omitted logographic place-names like

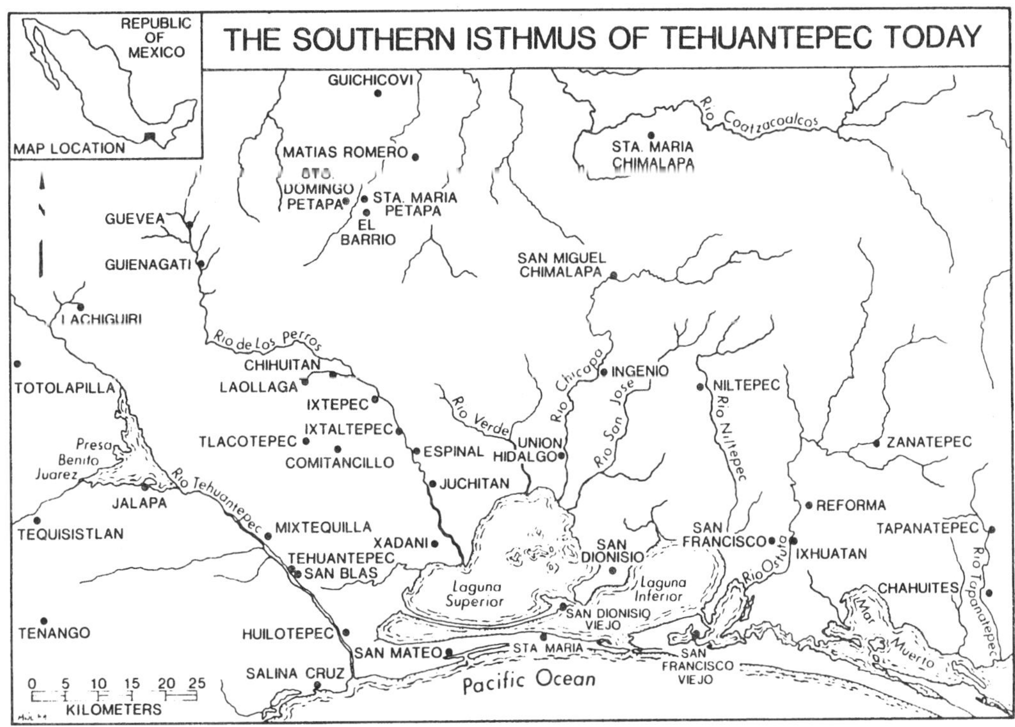

Figure 93. Map of the Tehuantepec region. (After Zeitlin 1989: fig.1.)

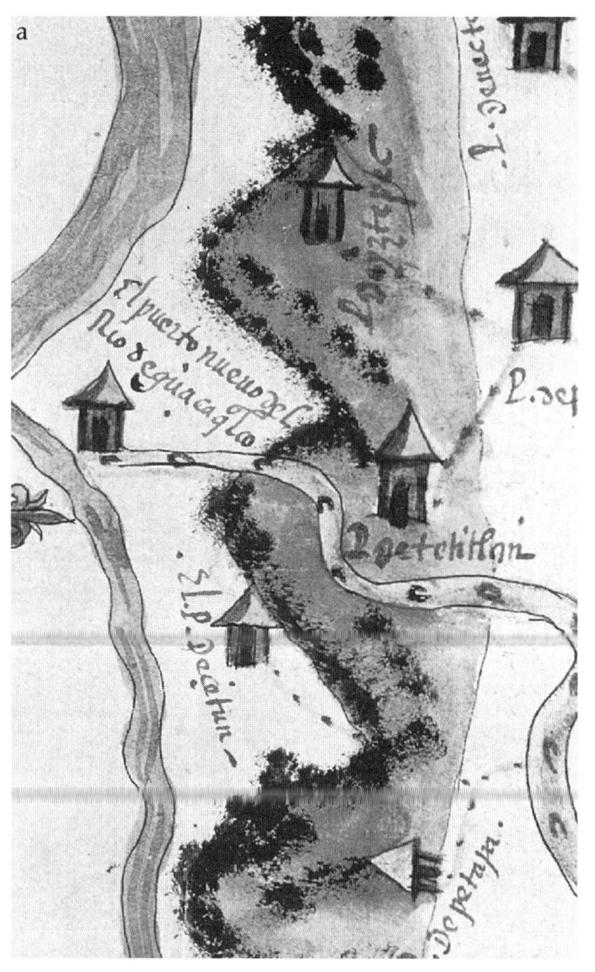
a

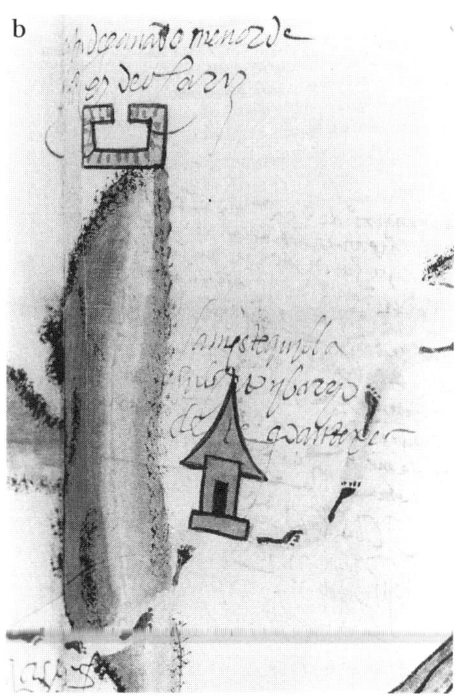
b

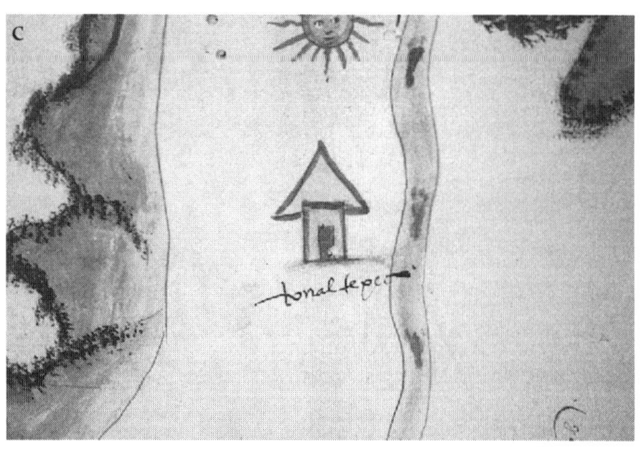
c
tonaltepec

d

the one of Tehuantepec. However, other stylistic features unite the group (fig. 94). All settlements are shown with the same wide-doored house with a dunce-cap roof. In all but two of the maps, the peak of this steep roof is adorned with a small cross. The Tehuantepec artist is consistent in his or her representation of hills as solid shapes softened by stippling. The style of these hills grows looser and the brush strokes more impressionistic as

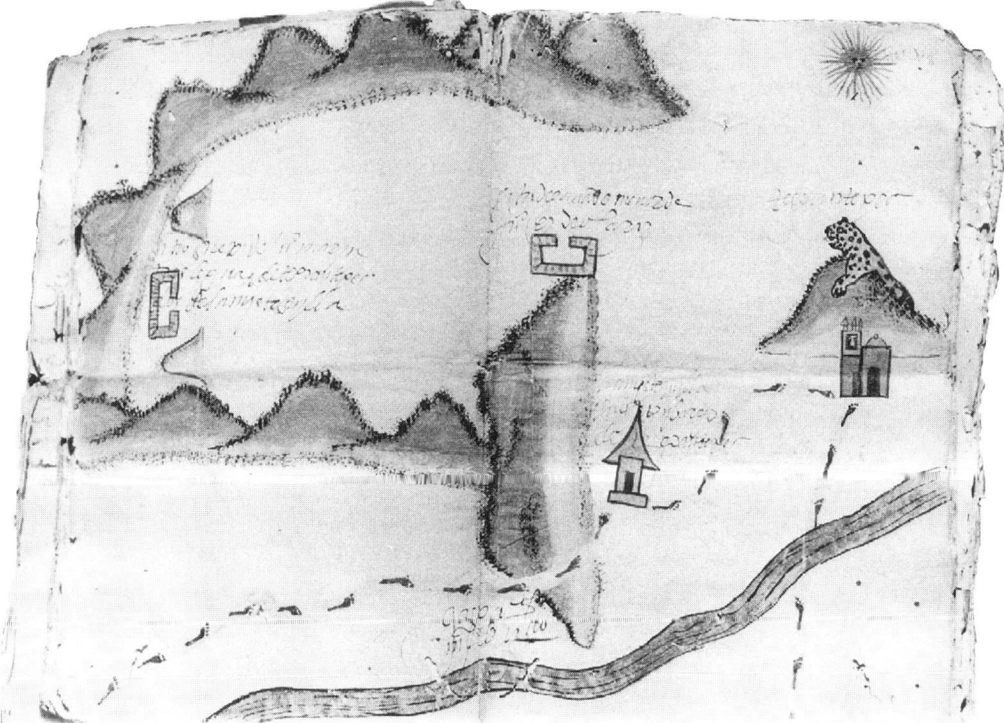

Figure 95. Land grant map by the artist of the Relación Geográfica map of Tehuantepec A, 1573. Size of the original: 32 x 40 cm. Photograph: B. Mundy; reproduced courtesy of the Archivo General de la Nación, Mexico (Tierras, vol. 3343, exp. 4, Mapoteca 2378).

Figure 94 *(opposite)*. Details of maps made by the Tehuantepec artist. The similarity of house-forms and paintwork on hills shows these maps to have been created by the same hand. (a) the Relación Geográfica of Tehuantepec, detail; (b) land grant map of 1573, detail; (c) land grant map of 1580, detail; (d) land grant map of 1598, detail. Photographs a–c: B. Mundy; (a) reproduced courtesy of the Benson Latin American Collection, The General Libraries, The University of Texas at Austin (JGI xxv-4); (b) reproduced courtesy of the Archivo General de la Nación, Mexico (Tierras, vol. 3343, exp. 4, Mapoteca 2378); (c) reproduced courtesy of the Archivo General de la Nación, Mexico (Tierras, vol. 2729, exp. 4, Mapoteca 1903); (d) photograph courtesy of the Archivo General de la Nación, Mexico (Tierras, vol. 2764, exp. 26, Mapoteca 2084).

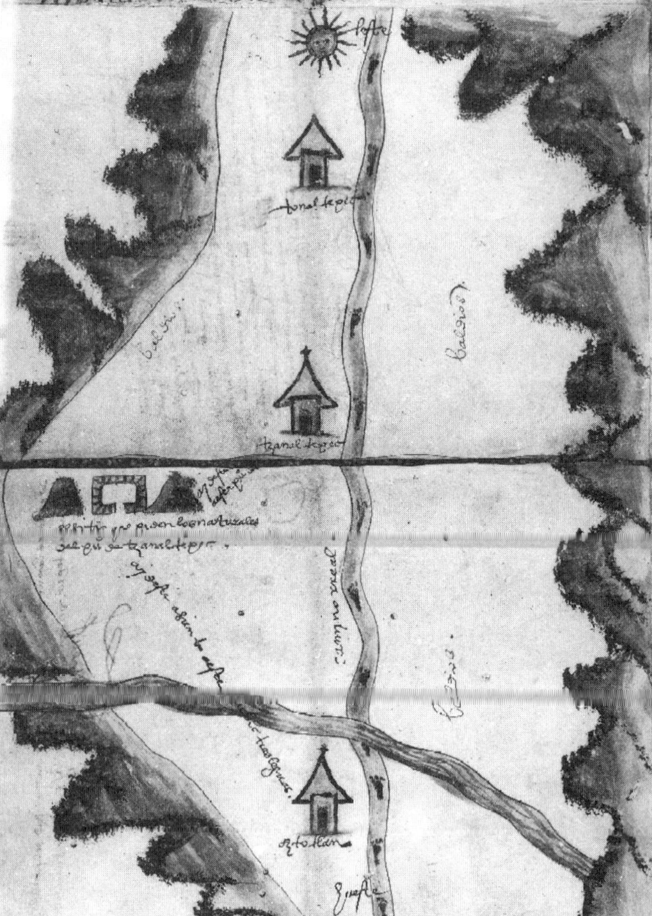

Figure 96. Land grant map by the artist of the Relación Geográfica map of Tehuantepec A, 1580. Size of the original: 40 x 31 cm. Photograph courtesy of the Archivo General de la Nación, Mexico (Tierras, vol. 2729, exp. 4, Mapoteca 1903).

the artist advances in years. In the mercedes maps, he or she also presents ranches *(estancias)* marked by a rectangle open on one side that seems to suggest a corral (this being common to maps of this period). In the two earliest land grant maps, those of 1573 and 1580 (figs. 95 and 96), the artist marks the roads with footprints; in the other maps footprints give way to hoofprints.

Like the Xalapa artist, the Tehuantepec artist provides us with a bare-bones picture of the landscape in this Relación map, showing it as an armature of roads and rivers connecting identical settlements, with the occasional range of stipple-topped hills. This same view is also put forth in his (or her) mercedes maps, which, like the Xalapa map, could be seen as fragments ripped from the larger whole of the Relación Geográfica map. They emphasize the expanse of territory, at the expense of a community-centric view.

We know that the Tehuantepec artist, like the Xalapa artist, was successful in creating maps that met with Spanish expectation, for his or her relationship with the Spanish official community endured at least a quarter of a century, during which time he (or she) worked for five alcaldes mayores.[13] The Tehuantepec artist painted wherever in the expansive alcaldía mayor that the grants were being made (fig. 97). These six maps are the

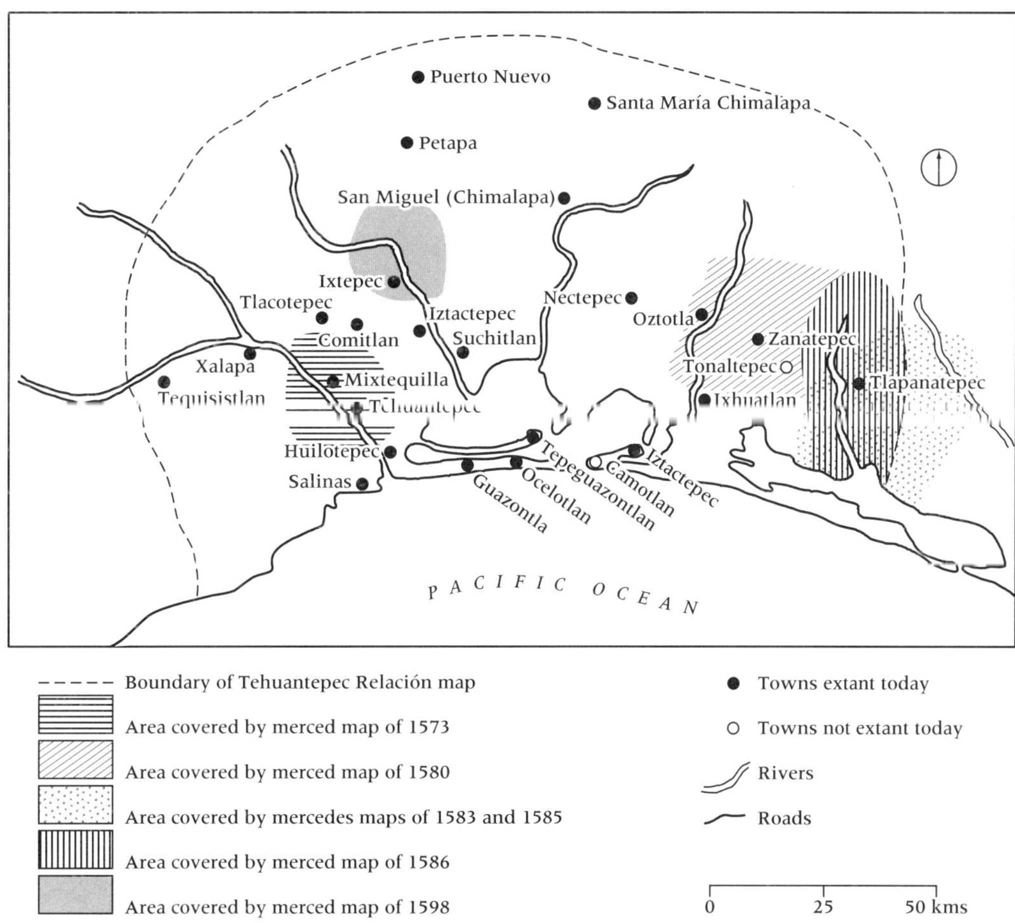

Figure 97. Map of the Tehuantepec region showing areas covered by the Relación Geográfica map and the six land grant maps made by the Tehuantepec artist.

only ones I found in the AGN, but they may have been part of a larger corpus of land grant maps by the Tehuantepec artist, maps that are either lost or in a different repository.

While the Tehuantepec artist was a prolific mapmaker, his or her mercedes maps were the backdrop for texts written by others: all were inscribed by the hands of Spanish officials.[14] Most of the inscriptions on the maps—following the demands of the acordado—establish the distance between the desired land and any nearby native community. In general, these glosses follow the general opinions put forth by witnesses in the cases, showing that land grants would not endanger nearby Spanish ranches or native communities. However, in one case, the native community contested the grant (AGN Tierras, vol. 2737, exp. 25). On one of the maps corresponding to the contested merced, the glosses state that the town of Tlapanaltepec is "more or less four leagues" away from the site of the grant. The images on the map also attested to a distance sufficient to isolate the native community from the dangers posed by herds. Nonetheless, the native community of Tlapanaltepec complained that the herds would indeed be dangerously close. While we are unable to assume that all native peoples—that is, the community of Tlapanaltepec and the Tehuantepec artist—shared the same interests, or that the Spanish text contradicted the native mapmaker's intent, we can conclude that on this and other maps, the Tehuantepec artist had little control over how his (or her) work would eventually be interpreted.

Thus, the Tehuantepec and Xalapa mapmakers both provided very basic, unadorned maps of whatever part of their region was of interest to Spanish colonial officials. The conception of the region that they were to present in their Relaciones Geográficas maps varied little from what they were asked to paint in land grant suits, where they supplied a rendering of the territorial expanse—the potential real estate—that their regions offered. Their mercedes maps often look like small fragments of the Relaciones Geográficas maps. For both mapmakers, the merced map was probably the starting point for the Relación Geográfica map. And both may have realized that they were not the final arbiter of meaning of the maps: in their mercedes maps, most of the information in the map was supplied by a different hand, the literate hand, whose glosses brought the map in line with the demands of the colonial world.

THE IXTAPALAPA RELACIÓN GEOGRÁFICA ARTIST, MARTÍN CANO

Perhaps the most interesting overlap between a merced and a Relación map is found on two maps of Ixtapalapa, both painted by the indigenous artist identified in the Relación Geográfica text as Martín Cano, the "oficial de pintor" whose training was discussed chapter 4 (fig. 28; RGS 7: 41). Cano was a sophisticated artist, and he seems to have been well aware that the picture of Ixtapalapa needed for a merced map—a display of space as property—was quite different from the picture of Ixtapalapa to be sent back to the Spanish king; on the latter, he shows Ixtapalapa as a ordered collection of Christian communities. In addition, written documents accompanying the merced give us rare insight into this native painter's traffic with the Spanish colonial authorities.

Again, a close scrutiny of the hands allows us to see that Martín Cano created the merced map as well as the Relación Geográfica. The Ixtapalapa map is done with a firm and careful line and is filled with details that mark the native hand at work. The bottom of the map is filled with the two most important buildings of Ixtapalapa and Ixtapalapa's logographic place-name. The town-government buildings *(casas de comunidad),* where the native cabildo would meet, are portrayed frontally with an emphasized post-and-lintel and a frieze of circles decorating the roofline, in the same manner as palaces were portrayed in preconquest manuscripts. The church is a simple rectangle with additional narrow rectangles marking its base and top. The wide door of this church is delineated by double lines; the bell in the belfry is carefully drawn. The Relación text declares the name Ixtapalapa to mean "town situated in the place of flagstones and of water" (RGS 7: 37), from the Nahuatl *itztapalli* ("flagstone") and *apantli* ("river"). On the map, below the church of San Lucas Ixtapalapa, a current of water surrounds a stone symbol, drawn as if its customary volutes have been chiseled off to shape it into a hexagon. The roads leading out from the church and to the *casas de la comunidad* of Ixtapalapa are marked with distinct footprints, each with five toes. In the upper half of the map the space is compressed to show Ixtapalapa's outlying subjects and the shore of Lake Tetzcoco in the upper left corner.

Cano's rendering of the church and place-name of Ixtapalapa ties his Relación map to a land grant map of 1589 (fig. 98; AGN Tierras, vol. 2809, exp. 4, Mapoteca 2206). On the two maps, the churches are similar: their bodies are simple rectangles with narrow bases and tops. The wide door arches are outlined with double lines and the bells in the belfries are carefully rendered. Emerging directly from the base of the church in the merced map is the same logographic symbol as in the Relación map, a hexagonal stone symbol surrounded by a flow of water. Footprints also mark roads on the merced map, and they are as distinct as on the other map, with rounded heels and each of the five toes carefully painted.

Unlike the work of the Xalapa and Tehuantepec artists, whose Relaciones Geográficas maps closely resembled their mercedes maps, the native painter Martín Cano presents us with two wholly different views of Ixtapalapa, even though he covers much the same ground in both Relación map and merced map (fig. 99). He seems fully aware that even though both Relación questionnaire and merced acordado asked for a map, they expected different information. Since the Relación questionnaire was concerned with human set-

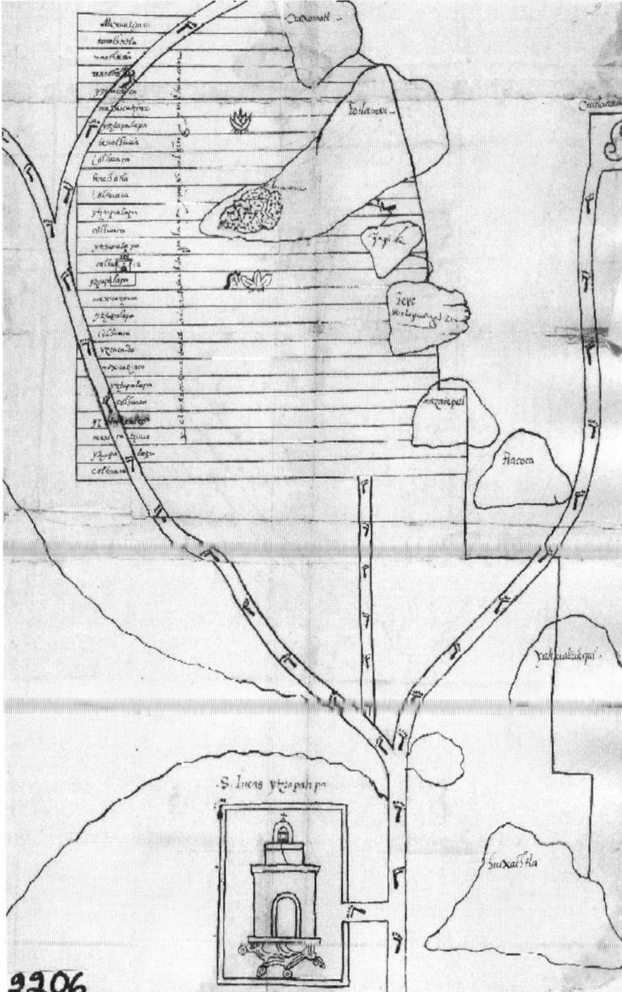

Figure 98. Land grant map from Ixtapalapa by Martín Cano, 1589. Martín Cano is the only identified artist of the Relación corpus who we know to have painted a land grant map. The logograph for the name of Ixtapalapa, below, under the church, is quite similar to his Relación map. However, his land grant map covers a larger reach of territory and offers a more detailed description of topography. Size of the original: 65 x 41 cm. Photograph courtesy of the Archivo General de la Nación, Mexico (Tierras, vol. 2809, exp. 4, Mapoteca 2206).

tlements, Martín Cano designates the main town of Ixtapalapa with its church. Likewise, he uses churches to indicate subject towns that lay to the east: San Cristóbal, San Juan, San Phelipe, Santa Cruz (Meyehualco), and Santa María (Aztahuacan). In the map he made in response to the Relación Geográfica questionnaire, a map that was to represent his community before the Spanish king, Cano calculated that the king's interest would be in the landscape colonized by religion. He shows native settlements being one and the same as Christian communities (Leibsohn, n.d.: 1). This Christian landscape is one of great order: Cano casts the entire region of Ixtapalapa onto a rectilinear grid: the roads cross at right angles, the community houses stand directly perpendicular to the church. Along the vertical axis of the map, the main church is prominent; along the horizontal axis, the seat of the native government dominates.[15] Lined up behind the central church of San Lucas Ixtapalapa are the subject chapels. Even water flows in a sharply angled canal. Only in the great lake of Tetzcoco at the top left corner of the map do the natural contours of landscape reassert themselves.

In Martín Cano's land grant map, by contrast, space is defined, not by the settlements of Christians, but by topography and agricultural fields (fig. 98). His map covers much of the same area as the Relación map, but runs farther to the south and east in the area between Cuitlahuac (today Tlahuac) and Ixtapalapa (figs. 99 and 100). Ixtapalapa sat on

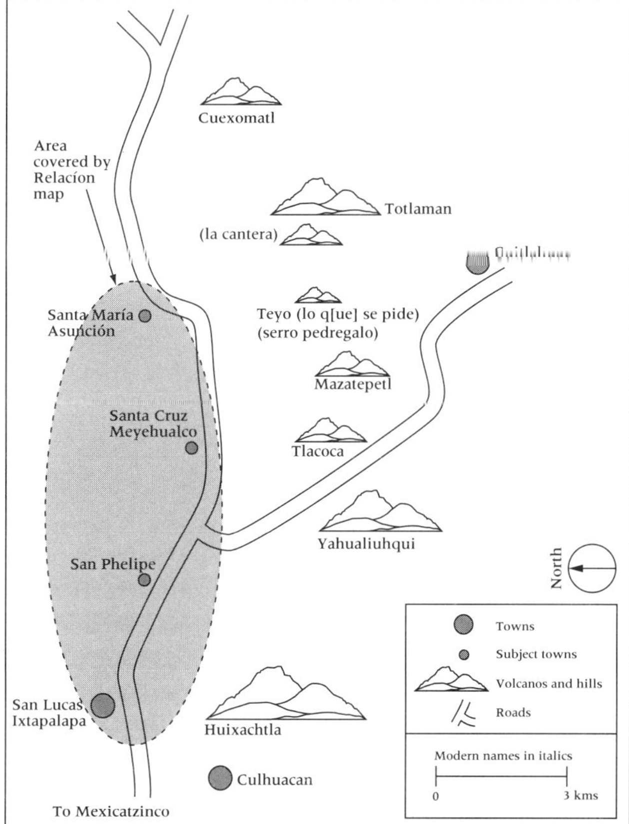

Figure 99. Map of Ixtapalapa region. This map shows the region Martín Cano covered in his land grant map and that of his Relación Geográfica map shaded within.

the narrow band of land running between the lakes of Xochimilco and Tetzcoco, while Cuitlahuac was set within the swampy Lake Xochimilco. The edge of Lake Tetzcoco is marked on the left side of the merced map by the same undulating line that Cano used to mark the shore on his previous map. The right side of the map is cut by a rectilinear line, an artificial border marking the boundary of the lands of Cuitlahuac. This line runs around and through eight amorphous closed shapes representing the small string of volcanos that rose near the marshy lake bed. The bottom four seem to be influenced by the shape of the tepetl symbol, but as we move up the map, the hills become more like topographic projections of hills. All are glossed, most with Nahuatl names. They read, from the bottom: Huixachtla, Yahualiuhqui, Tlacoca, Mazatepetl, Teyo *serro pedregalo, lo q~[ue] se pide, la cantera,* Totlaman, Cuexomatl.[16] Some of these names can be correlated to modern

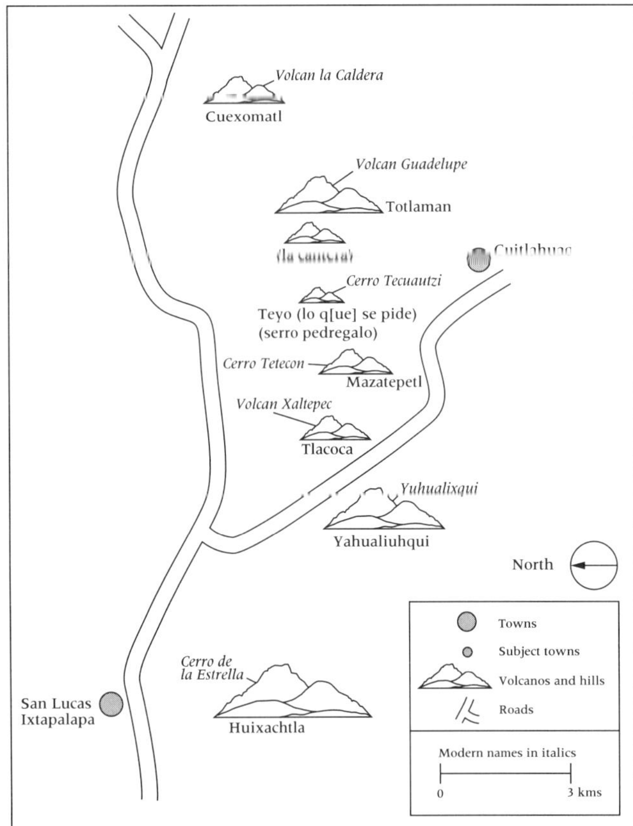

Figure 100. Topography and inscriptions of Martín Cano's land grant map of Ixtapalapa.

names; all of the hill shapes can be correlated on a modern map to a string of volcanic cones (fig. 100). The first of the hills, beginning at the bottom of the merced map, bears a version of the Nahuatl name of the Cerro de la Estrella, where the New Fire ceremonies were held every fifty-two years, the same hill that is pictured on the top left corner of the Culhuacan map (fig. 29).[17] The second hill, identified on the map as Yahualiuhqui, is a small volcano, which today is called Yuhualixqui. The modern-day names of the other volcanos in this string have changed from the days of the merced map: Tlacoca is now Volcan Xaltepec; Mazatepetl is Cerro Tetecon; Teyo and *"lo q~[ue] se pide"* are the peaks of Cerro Tecuautzi; Totlaman is now Volcan Guadelupe (also known as El Borrego and Cerro Santa Catarina), at whose northern edge lies a number of outcrops that were marked as a quarry *(cantera)* on the merced map; Cuexomatl is now Volcan La Caldera.

Not only does Cano concentrate on rendering space as topography, he also provides a detailed account of the landowners in this zone of alluvial soil and rocky volcanic tuff (Sanders et al. 1979: map 3, geology). He draws twenty-seven narrow-strip fields that are marked with the names of five different towns: Ixtapalapa; Culhuacan; Mexicatzinco; Itztacalco; and Tenochtitlan. Presumably these lands were held by these different communities and were worked for their benefit.[18] Pictographs on two of the fields suggest that this is also a record of how the land was used; these pictographs fall on the lands of Ixtapalapa, the lands the resident Martín Cano would know best. In one of the upper fields, he draws a maguey plant, probably to show the crop of this land. Farther below, he paints a woman's head alongside a broad-leafed plant. This land may have been the *cihuatlalli,* the land controlled by women (Cline 1986: 141, 145, 149). The trilobed plant beside the woman's head may indicate the edible greens *(quilitl)* that were a staple of the Valley diet. Most of the glosses on this map were done by one hand, and I suggest, since the glosses seem so well integrated into this merced map, that the hand is that of Martín Cano. Who better to know the elaborate division of lands and the Nahuatl names of the volcanos?

The merced map from Ixtapalapa shows a debt to a kind of cadastral map made in the Valley, a map of ownership not known to have been made in other regions of Mesoamerica. The Humboldt Fragment II (fig. 101) is another such map, being a similar record of strip-fields (Seler 1904: 154–76). The Humboldt, probably the earlier of the two colonial maps, records the owners of the fields by using heads that bear both name logographs and alphabetic inscriptions. Cano, on the other hand, depends almost entirely on alphabetic inscriptions to convey information about ownership; his ability to write in alphabetic script allowed him a control over the meaning of his maps that other native mapmakers lacked.

That Martín Cano came up with two wholly different ways to map the same piece of land not only testifies to his ability, it also shows that he had a nuanced understanding of

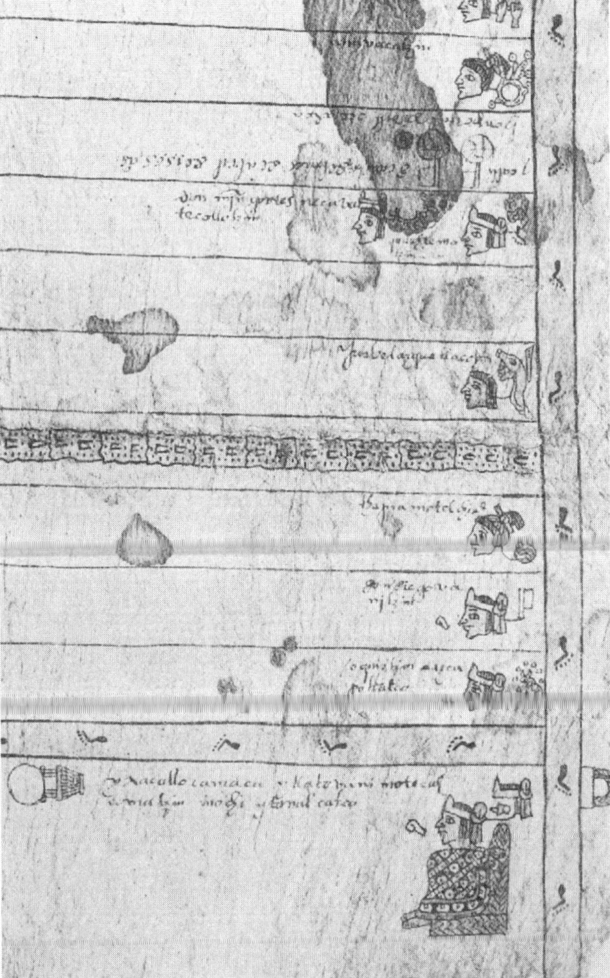

Figure 101. The Humboldt Fragment II, detail of lower right corner, after 1565. This map shows long rectangular fields, and identifies them with figures and profiles of the pre-Hispanic nobility who once controlled them. The emperor Moteuczoma held the largest plot, and he appears in the lower right corner. Size of the original: 68 x 40 cm. Photograph reprinted from Seler 1892. (Deutsche Staatsbibliothek, Berlin, MS. Amer. 1).

what the Relación questionnaire was looking for and what the merced map should show, and that they were different. Not only did he have this understanding, he also had different pictorial vehicles to convey these separate concepts. He knew what was in Spanish minds. And we know that he was accorded great respect by local Spanish officials, who not only named him in their written Relación, but added his title, "oficial de pintor."

The written records of the Ixtapalapa land grant tell us even more about Martín Cano's relationship to both the Spanish officials representing the royal government and to the native government of the town (AGN Tierras, vol. 2809, exp. 4). The grant was for lands that Juana Ximénez de Bohorquez desired for a sheep ranch. The ranch would fall on one of the volcanic protrusions from the Valley floor (marked *"lo q~[ue] se pide"* on the map), land of limited agricultural value that abutted native fields. The corregidor's lieutenant, who was in charge of carrying out the proceedings for the grant, was in favor of Ximénez de Bohorquez's suit, as were a number of Spanish and native residents. The proceedings required that testimony of ten people be taken, five chosen by the government *(de oficio)* and five put forth by the party desiring the grant *(de parte).* Martín Cano, the painter, identified only as a resident Indian of Ixtapalapa in the merced, was put forth by Ximénez de Borhorquez as a witness in her favor. In his testimony he promoted the land grant, saying the land was not under cultivation by Ixtapalapa or other native communities.

With this testimony on behalf of a Spaniard, Martín Cano set himself against the native cabildo of Ixtapalapa. Headed by don Agustín Bonifacio and Luis de Mendoza and representing all the nobles and the commoners, the native government protested the grant. These men put forth their opposing *contradicción* claiming that herds of sheep would endanger their fields and obstruct their path to the market at Cuitlahuac. Their fears were not unfounded, for Spanish herds frequently ravaged native fields (Chevalier 1963: 92–102). In addition, the natives of Ixtapalapa may have desired to use those lands themselves.

In the case of the map of Ixtapalapa, we have clear evidence that its painter, Martín Cano, was also at work painting land grant maps. His views in the one merced for which we know he painted a map were at odds with those of the native community of which he was a part. His sympathies lay with Juana Ximénez de Bohorquez and the local Spanish official, who supported the grant. In the case of Martín Cano, not only did his map uphold the interests of his patron and oppose those of the community in which he lived, so did his recorded testimony.

Given Martín Cano's training, his employ by Spanish patrons, and the closeness of Ixtapalapa to Mexico City, the center of Spanish influence in the colony, we must wonder why he, unlike so many other native artists, had reason to preserve native style, using logographic place-names, footprints, and indigenous map forms. Given his perspicacity in constructing two wholly different images of the same place, one contoured to the merced demand, the other to the Relación Geográfica questionnaire, Cano's decision to use indigenous style was no doubt a conscious one. I can only imagine that native style had a more authentic ring, a useful timbre to strike in arguing to Spanish authorities that the native community was free from threat.

CONCLUSIONS

Mercedes maps were widespread, the dominant kind of map made by native artists in the late sixteenth century. These maps made for viceregal land grants allow us to see the wider context within which the Relaciones Geográficas maps were painted. They also show how land grants became a means through which Spanish notions of territory came to infiltrate the indigenous world of New Spain (Melville 1994). As it was solicited in the merced request, the map was not asked to show the spatial substrate of a community self-portrait. Rather, it was meant to be a picture of parcels of property for the use of acquisitive Spaniards. The mercedes maps offer certain explanations for the trajectory of native mapping toward the close of the sixteenth century—and the nadir of native population. Spaniards wanted maps that documented possessions, and asked for them through the merced acordado; in this arena, native maps shifted to conform to the contours of Spanish expectations. In addition, the emphasis the merced process put on texts to carry meaning came to overshadow native expression on the map, which depended on pictorial form and symbol as vehicles for meaning. On mercedes maps, it was the glosses that stated distances, ownership, and other specifics asked for in the acordado. Native artists, like those of Tehuantepec and Xalapa, would provide bare maps while other hands added these all-important glosses. Their images of territory became the backdrop for alphabetic text.

The implantation of the Spanish view of the landscape was both indirect and subtle. Native mapmakers were employed by Spanish corregidores and alcaldes mayores; in order to continue working, these painters needed to make maps that met Spanish norms. In the years that the Xalapa artist (1578–1587) and the Tehuantepec artist (1573–1598) were at work for colonial authorities, the native hallmarks of each artist's style diminished. Logographic place-names mark only the earlier of the Tehuantepec artist's maps; the native convention of footprints marking roads is soon replaced with hoofprints. With both artists, the solidity of the hills, reminiscent of the native tepetl symbol, gradually gives way to more impressionistic renderings. In light of this, Martín Cano's use of native style seems calculated, a way of concealing that the vision of the territory it presented differed considerably from that of native community leaders.

The influence of mercedes maps is not confined to these three regions, or these three Relaciones Geográficas maps. On other Relaciones maps from other regions, especially regions where ranching was supplanting agriculture, we find the same kind of information as we would on mercedes maps. For example, the map of Mizantla (fig. 40) plots the bare layout of roads, rivers, hills, and settlements, showing the town of Mizantla as a kind of rudimentary gridiron. The native painter reveals himself through the footprints marking roads and the tepetl shapes standing for hills.[19] Along the top of the map, the north, the artist shows three ranches, names their owners, and marks the distances between them and the sea, as well as marking a half league of savanna outside Mizantla. Likewise, the map of Misquiahuala (figs. 62 [pl. 7] and 63) also shows sites of sheep ranches in the upper half of the map, marking them with rectangles to indicate corrals. This kind of information, as well as the implicit view of space as a patchwork of individual possessions, was asked for on a merced map, not in the Relación questionnaire.

As more lands were granted, more mercedes maps were made of New Spain's land. In the eyes of the viceregal government, these documents of private property and landholdings, with their written inscriptions, crowded out indigenous community maps, which defined space by community norms and presented it with logographic place-names. It is mercedes maps, given their official importance, that have survived by the scores to the present day in Mexico's AGN. It was these maps that enjoyed official patronage, and in the reflection of the country that they offer, we see a landscape of New Spain that was no longer primarily envisioned through the eyes of the indigenous peoples who populated it.

CHAPTER EIGHT

Conclusion

If the current physical state of the Relaciones Geográficas maps is any indication, then we can see that their reception in Spain was not one of celebration: they were folded into small rectangles, tied up in bundles, neglected in archives. Over the years, the corpus was split up, much of it leaving the royal hold, some to wend its way into private hands (Cline 1972d). Fittingly, perhaps, for these maps commissioned by the first foreign empire to enter the New World, many of the Relaciones Geográficas now lie in the United States, the New World's most recent imperial power.

Their cool reception in Spain is understandable. The Spanish cosmographer Juan López de Velasco, who engendered the project, thought that with these maps he would be able to create an atlas showing New Spain, perhaps like the ones Pedro de Esquivel or Anton van den Wyngaerde had made of Spain. In particular, López de Velasco's Relación Geográfica project calls out to be compared to the representational practices inherent in Philip II's commission of the Esquivel map; just as Philip wanted to register his politically unified Spain in map form, so López de Velasco wanted to register New Spain with a planimetric, scale model map, as if to declare its reality. But just as Spain needed to be imagined before it could be mapped, so did New Spain (O'Gorman 1961).

To López de Velasco, New Spain was a geographic expanse, a region of the globe needing to be reined in by the best of man's rational powers, yet blank because it was a synthetic place, one designated as "New Spain" from the Iberian side of the globe by men who had no knowledge of it. From this vantage across the Atlantic, "New Spain" could be cut out of whole cloth, unrent by the factional loyalties and parochial vantages that so undermined the prospect of a national unity for Spain itself.

When López de Velasco received the maps, some of them, like the ones from Teozacoalco (fig. 10; pl. 1), Amoltepec (fig. 51; pl. 6), and Cempoala (fig. 42; pl. 5), revealed a country even more balkanized than his own. The "indios" who painted them showed New Spain to be an archipelago of individual and separate communities, each with a unique sense of identity that they saw fit to express in their maps, thereby helping to derail López de Velasco's aspirations for the creation of "New Spain." Far from showing the successful imposition of New Spain in the New World, many of the indigenous maps of the corpus showed him space that was still defined and represented by the communities of its indigenous inhabitants.

It is hardly surprising that few of the indigenous residents of New Spain had any sense of a "New Spanish" identity that would knit their communities into the larger fabric of New Spain. Pre-Hispanic rule had done little to pave the way for a national, or even a nonlocal, sensibility among the residents of New Spain; although the Aztec empire covered much of the same territory as New Spain, this "empire" was really a very loose coalition of independent communities. Their common bonds were little more than their shared obligation to pay tribute to Culhua-Mexica lords residing in the Valley of Mexico (Collier et al. 1982). If each community in New Spain were to make maps that showed how it fit into the larger whole of New Spain, it would have meant that each altepetl or cacicazgo would have primarily seen itself as a small part of some larger whole. But maps like those of Teozacoalco and Amoltepec resist doing this—their artists, fiercely loyal to their communities, never saw themselves as the small parts of the whole colony. Rather, their maps reflect their sense of themselves as autonomous regions. Their allegiances were steadfastly local, and the community map was the form in which their self-understanding found an expression.

For New Spain to be mapped, it needed to be imagined. How did this act of imagining come about? On the level of representation, how did maps made within that expanse we know to be New Spain register themselves as part of this larger whole? In looking at the Relaciones Geográficas, I have found that "New Spain," no matter how dismal the prospects for it looked to López de Velasco, was beginning to make itself felt in the indigenous imagination. We may not see much of it on the maps of Cempoala and Teozacoalco, but we do on other maps, like those from Xalapa and Tehuantepec (figs. 85 and 27). "New Spain" filtered through to its indigenous inhabitants in the ways that Spanish colonists looked at the landscape and through the exercise of power by the viceregal government, the embodiment of New Spain. Official power made itself felt in possessing the landscape, at least implicitly, and then giving it away with land grants. It also asserted itself in renaming communities with names that were represented in alphabetic script, thereby undermining local representations of the landscape, which formerly were able to capture space with a net of their own toponyms.

If we look at the Relación Geográfica corpus and concentrate not just on the indigenous maps that must have so disappointed López de Velasco, we can see an erosion of the power of indigenous communities to represent themselves to the colonial government with maps of their own imagining and making. Why else would the towns around Teotihuacan, once subject to Tetzcoco and holders of rich powers of self-representation, find themselves represented by the scrawly pen of Francisco de Miranda? Why else would the region of Los Peñoles, whose manuscript tradition was among the most fecund in the New World, have found itself portrayed by with a bare itinerary (fig. 9)? It is little wonder that earlier scholars who looked at the corpus took from it a history of European style overwhelming the indigenous, or compared the history of indigenous representation to a shipwreck with few survivors (Robertson 1972a; Kubler 1961).

But in seeing indigenous communities lose the power to represent themselves, we must also remember that they were representing themselves to a foreign king, and in sending these maps off into the hands of the colonial government, they delivered maps to men who did not want them. For it is clear from the Relación questionnaire, which avoided aiming a map question to indigenous communities, that Iberian administrators like López de Velasco felt strongly that mapping was their prerogative: they would be the ones to control the representation of New Spain, not its indigenous people.

For New Spain to be mapped as López de Velasco envisioned, it meant that indigenous self-representations had to be suppressed. In the fate of the Relaciones Geográficas after their arrival in Spain, we see the process of indigenous maps being gradually pushed out of the official colonial record. These maps show us that indigenous representations were barred from the dialogue shaping "New Spain." The Spanish government unwittingly solicited indigenous maps, apprised them, and then shut them away.

But the maps' reception in Spain, and history in the seventeenth and eighteenth century, are overly negative notes with which to end a study of the corpus. It was remarkable then and now that indigenous maps were created as part of this notable project. In addition, the circumstances of their production and patronage allows them to stand in for a much larger set of indigenous representations in the New World. For these maps were painted to enter the Spanish sphere; no doubt even the painter of Suchitepec, a town so remote that it appears on few maps today, knew he was painting for the Spanish king (fig. 80). Thus they are maps that anticipate: from their inception they were colored by indigenous expectations of a Spanish audience. In this respect they show us the double-consciousness of the colonized artist: working to satisfy an immediate local audience and laboring with a set of expectations about the colonizers; this artistic double-consciousness marks a much larger set of images from the New World.

In addition, the evidence that so many indigenous artists across New Spain were able to create maps of great richness and variety is testament to the resilience of indigenous self-conceptions and the inventiveness of the artists at work. Although their Relación Geográfica maps ended up in Spain and survive today, perhaps the most important work they did was other maps, maybe not the product of such an exalted commission, that remained in New Spain. Here, at home, these artists and their maps bolstered and fostered local traditions of self-representation. The maps, now nameless and unknown, that stayed behind kept tradition alive so that indigenous maps, ones that look like the Lienzo of Zacatepec and the Relación Geográfica of Teozacoalco (figs. 49 and 10 [pl. 1]), could be and are still made and used in Mexico today, largely within indigenous communities (Oettinger 1993; Parmenter 1982). As Oettinger and Horcasitas showed in their study of the Lienzo of Petlacala, a community map whose antecedents are to be found in the sixteenth century, indigenous Mexican maps are reused, repainted, and reinterpreted as need be (Oettinger and Horcasitas 1982).

While the Relaciones Geográficas maps may have had little impact on the shaping of "New Spain," this was not the end of their history. Not only did they, and the maps that remained, play a part in the continuance of native mapping in the sixteenth century, they now offer modern Mexico, both nationally and locally, a chance to repossess and reinterpret its tortured colonial past. In 1989, I visited the town of Macuilxochitl, Oaxaca, where the Relación Geográfica map of Macuilsuchil had been painted some four centuries before (fig. 79; pl. 8). I happened to arrive midmorning, just as a town meeting was breaking up, and was welcomed into a large, sparsely furnished room in the Ayuntamiento, where about twenty men had gathered. On a wall in the room was a framed photograph of their Relación Geográfica map. These townsmen, speakers of Zapotec and Spanish, easily identified the topography pictured on the map as their own. Ironically, they could shed no light upon their map's long Nahuatl text, the means through which their colonial forebear had valiantly tried to make town history legible, nor upon the Spanish translation that I put forth. Instead, the men who gathered that morning offered me their own interpretations of what the map meant. Above all, they held the map to be an enduring testament of Macuilxochitl's long and illustrious history, evidence that they, the people of Macuilxochitl, had a land and a legacy that reached back in time. Far from being locked up in Philip's dusty archives, the Relaciones Geográficas are recirculating in the world, and the map of Macuilsuchil, like a homing pigeon, has found its way back.

Appendix A

Catalogue of Maps Studied

NAME (modern name of town, state)
Group name, if any
Date of accompanying Relación
Repository and number [Abbreviations used: AGI = Archivo General de Indias, Seville; RAH = Real Academia de la Historia, Madrid; UTX = Benson Latin American Collection, University of Texas at Austin]
Robertson (1972a) number; Cline (1972a) number
Style, after Robertson (1972a); Native language(s) of town, after Harvey (1972)
Artist, if known
Publication of text in RGS series [pages relating directly to town in brackets]; placement of map

1. ACAMBARO (Acambaro, Guanajuato)
6/15/1580
RAH 9-25-4/ 4663-x
Robertson number 1; Cline number 18a
European; Tarascan, Otomi, Chichimec, Mazahua
artist unknown
RGS 9: 47–72 [59–68]; map follows p. 58

2. ACAPISTLA (Yecapixtla, Morelos)
10/10/1580
UTX, JGI xxiii–8
Robertson number 2, Cline number 1
Mixed; Nahuatl
artist unknown
RGS 6: 177–223 [212–23]; map follows page 222

3. AMECA (Ameca, Jalisco)
10/2–12/15/1579
UTX, JGI xxiii-10
Robertson number 3, Cline number 4
European; Cazcan, Totonac
artist unknown
RGS 10: 23–50; map follows p. 30

4. AMOLTEPEC (Santiago Amoltepec, Oaxaca)
1/21/1580
UTX, JGI xxv-3
Robertson number 4; Cline number 108a
Native; Mixtec
artist unknown
RGS 3: 147–151, map follows p. 150

5. ATLATLAUCA AND MALINALTEPEC (San Juan Bautista Atatlahuca and Maninaltepec, Oaxaca)
9/8/1580
RAH 9-25-4/ 4663-xxvi
Robertson number 6; Cline numbers 11 and 11a
Mixed; Cuicatec, Chinantec
artist unknown
RGS 2: 43–59; map follows p. 46

6. ATLATLAUCA AND SUCHIACA (Atlatlahuca, Mexico)
[illegible]
UTX, JGI xxiii-13
Robertson number 7; Cline numbers 10 and 10a
European; Nahuatl, Matlatzinca
artist unknown
RGS 6: 39–52; map follows p. 46

7. CEMPOALA (Zempoala, Hidalgo)
11/1/1580
UTX, JGI xxv-10
Robertson number 10; Cline number 19
Mixed; Nahuatl, Otomi, Chichimec
artist unknown
RGS 6: 67–93; map follows p. 78

8. CHICOALAPA (Chicoloapan de Juarez, Mexico)
12/3/1579
AGI, 12
Robertson number 11; Cline number 29b
Mixed; Nahuatl
artist unknown
RGS 6: 123–76 [169–76]; map follows p. 174

9. CHIMALHUACAN ATENGO (Santa María Chimalhuacan, Mexico)
12/1/1579
AGI, 11
Robertson number 13; Cline number 29a
Mixed; Nahuatl
artist unknown
RGS 6: 123–76 [155–68]; map follows p. 166

10. CHOLULA (Cholula de Rivadabia, Puebla)
1581
UTX, JGI xxiv-1
Robertson number 14; Cline number 25
Mixed; Nahuatl
artist unknown
RGS 5: 121–45; map follows p. 126

11. COATEPEC CHALCO (Coatepec, Mexico)
AGI, 10
11/16/1579
Robertson number 15; Cline number 29
Mixed; Nahuatl
artist unknown
RGS 6: 123–76 [129–68]; map follows p. 150

12. COATZOCOALCO, VILLA DE ESPIRITU SANTO (Coatzacoalcos River; town near Tuzandepetl, Veracruz)
4/29/1580
UTX, JGI xxiv-2

Robertson number 16; Cline number 30
European; Nahuatl
Francisco Stroza Gali
RGS 2: 111–26; map follows p. 126

13. CUAHUITLAN (Cahuitan, Oaxaca)
8/14/1580
RAH 9-25-4/ 4663-xxxi
Robertson number 18; Cline number 33
Mixed; Mixtec
artist unknown
RGS 2: 127–36; map follows p. 128

14. CULHUACAN (Culhuacan, Distrito Federal, Mexico)
1/17/1580
UTX, JGI xxiii-14
Robertson number 19; Cline number 41
Mixed; Nahuatl
Pedro de San Agustín
RGS 7: 23–47 [31–35]; map follows p. 30

15. CUZCATLAN A (Coxcatlan, Puebla)
Cuzcatlan group
10/26/1580
AGI, 19
Robertson number 20, Cline number 42
Mixed; Nahuatl, Chocho, Mazatec
artist unknown: painter of Cuzcatlan B
RGS 5: 87–103; map A follows p. 94.

16. CUZCATLAN B (Coxcatlan, Puebla)
Cuzcatlan group
10/26/1580
UTX, JGI xxiii-15
Robertson number 21; Cline number 43
Mixed; Nahuatl, Chocho, Mazatec
artist unknown: painter of Cuzcatlan A
RGS 5: 87–103; map B follows map A

17. EPAZOYUCA (Epazoyucan, Hidalgo)
11/1/1580
UTX, JGI xxv-10
Robertson number 22; Cline number 19a
Mixed; Nahuatl, Otomi, Chichimec
artist unknown
RGS 6: 67–93; map follows p. 86

18. GUAXTEPEC (Oaxtepec, Morelos)
9/24/1580
UTX, JGI xxiv-3
Robertson number 23; Cline number 47
Mixed; Nahuatl
artist unknown
RGS 6: 177–223 [196–212]; map follows p. 206

19. GUEGUETLAN (Santo Domingo Huehuetlan, Puebla)
9/15/1579
UTX, JGI xxiv-6
Robertson number 24; Cline number 85a
European; Nahuatl, Mixtec
artist unknown
RGS 5: 195–213; second map following p. 214

20. GUEYTLALPA (Hueytlalpan, Puebla)
Gueytlalpa group
5/30–7/20/1581
UTX, JGI xxiv-5
Robertson number 25; Cline number 49
Mixed; Totonac, Nahuatl
artist unknown: painter of Gueytlalpa group
RGS 5: 147–180; first map following p. 158

21. HUEXOTLA (Huejutla de Reyes, Hidalgo)
2/3-4/1580

AGI, 16
Robertson number 26; Cline number 51
European; Nahuatl, Tepehua
artist unknown
RGS 6: 241–54; map follows p. 246; detail follows p. 254

22. IXCATLAN A, SANTA MARÍA (Santa María Ixcatlan, Oaxaca)
Ixcatlan group
10/13/1579
UTX, JGI xxiv-7
Robertson number 27; Cline number 54
European; Chocho, Nahuatl
Gonzalo Velázquez de Lara (?)
RGS 2: 223–41; map follows p. 226

23. IXCATLAN B, SANTA MARÍA (Santa María Ixcatlan, Oaxaca)
10/13/1579
UTX, JGI xxiv-7
Robertson number 28; Cline number 54
European; Chocho, Nahuatl
Gonzalo Velázquez de Lara (?)
RGS 2: 223–41; map follows p. 236

24. IXTAPALAPA (Ixtapalapa, Distrito Federal, Mexico)
1/31/1580
UTX, JGI xxiv-8
Robertson number 29; Cline number 56
Mixed; Nahuatl
Martín Cano
RGS 7: 23–47 [36–42]; map follows p. 38

25. IXTEPEXIC (Santa Catarina Ixtepeji, Oaxaca)
8/27–30/1579
RAH 9-25-4/4663-xiv
Robertson number 30; Cline number 57
Mixed; Zapotec
artist unknown
RGS 2: 243–64; map follows p. 264

26. JUJUPANGO (Jojupango, Puebla)
Gueytlalpa group
5/30–7/20/1581
UTX, JGI xxiv-5
Robertson number 31; Cline number 49b
Mixed; Totonac, Nahuatl
artist unknown: painter of Gueytlalpa group
RGS 5: 147–80; third map following p. 158

27. MACUILSUCHIL (San Mateo Macuilxochitl, Oaxaca)
4/9/1580
RAH 9-25-4/4663-xix
Robertson number 32; Cline number 62
Mixed; Zapotec
artist unknown
RGS 2: 325–40; map follows p. 264

28. MACUPILCO, SAN MIGUEL (not known; vicinity of Santa María Xadan, Oaxaca)
Suchitepec group
8/23–29/1579
AGI, 25
Robertson number 33; Cline number 88d
Mixed; Chontal (de Oaxaca)
artist unknown: painter of Suchitepec group
RGS 3: 55–72; second map following p. 62

29. MATLATLAN AND CHILA (Chila, Puebla)
Gueytlalpa group
5/30–7/20/1581
UTX, JGI xxiv-5
Robertson number 34; Cline number 49c

Mixed; Totonac
artist unknown: painter of Gueytlalpa group
RGS 5: 147–80; fourth map following p. 158

30. MEZTITLAN (Metztitlan, Hidalgo)
10/1/1579
UTX, JGI xxiv-12
Robertson number 37; Cline number 66
European; Nahuatl
Gabriel de Chávez
RGS 7: 49–75; map follows p. 70

31. MISQUIAHUALA (Mixquihuala, Hidalgo)
10/8/1579
UTX, JGI xxiii-12
Robertson number 38; Cline number 8b
Mixed; Otomi, Nahuatl
artist unknown
RGS 6: 25–38; map and diagram follow p. 38

32. MIZANTLA (Misantla, Veracruz)
10/1/1579
UTX, JGI xxiv-13
Robertson number 39; Cline number 67
Mixed; Totonac
artist unknown
RGS 5: 181–94; map following p. 190

33. MUCHITLAN (Mochitlan, Guerrero)
3/7/1582
UTX, JGI xxv-13
Robertson number 41; Cline number 132a
Mixed; Nahuatl, Tuztec, Matlatzinca
artist unknown
RGS 5: 261–77; map following p. 278

34. NOCHIZTLAN (Asunción Nochixtlan, Oaxaca)
4/9–11/1581
RAH 9-25-4/ 4663-xxxiii
Robertson number 43; Cline number 74
Mixed; Mixtec
artist unknown
RGS 2: 361–72; map following p. 372

35. PAPANTLA (Papantla de Olarte, Veracruz)
Gueytlalpa group
5/30–7/20/1581
UTX, JGI xxiv-5
Robertson number 44; Cline number 49d
Mixed; Totonac, Nahuatl
artist unknown: painter of Gueytlalpa group
RGS 5: 147–80; fifth map following p. 158

36. LOS PEÑOLES (Santa Catarina Estetla, San Antonio Huitepec, Santa María Peñoles, Santiago Huajolotipac, San Juan Elotepec, San Pedro Totomachapan, Oaxaca)
8/20–10/3/1579
UTX, JGI xxiv-15
Robertson number 45; Cline number 80
European; Mixtec, Zapotec
Diosdado Treviño
RGS 3: 41–53; map follows p. 46

37. QUATLATLAUCA (Huatlatlauca, Puebla)
9/2–5/1579
UTX, JGI xxiv-6
Robertson number 46; Cline number 85
European; Nahuatl
artist unknown
RGS 5: 195–213; map follows p. 214

38. SUCHITEPEC (Santa María Xadan, Oaxaca)
Suchitepec group
8/23–29/1579
AGI, 29
Robertson number 49; Cline number 88
Mixed; Zapotec
artist unknown: painter of Suchitepec group
RGS 3: 55–72; second map following p. 70

39. TAMAGAZCATEPEC, SAN BARTOLOMÉ (not known, vicinity of Santa María Xadan, Oaxaca)
Suchitepec group
8/23–29/1579
AGI, 26
Robertson number 51; Cline number 88c
Mixed; Chontal (de Oaxaca)
artist unknown: painter of Suchitepec group
RGS 3: 55–72; third map following p. 70

40. TECOLUTLA (Tecolutla, Veracruz)
Gueytlalpa group
5/30–7/20/1581
UTX, JGI xxiv-5
Robertson number 53; Cline number 49e
Mixed; Totonac
artist unknown: painter of Gueytlalpa group
RGS 5: 147–80; sixth map following p. 158

41. TECUICUILCO (Teococuilco, Oaxaca)
10/2/1580
UTX, JGI xxiv-19
Robertson number 54; Cline number 101
European; Zapotec
artist unknown
RGS 3: 83–102; map follows p. 94

42. TEHUANTEPEC A (Santo Domingo Tehuantepec, Oaxaca)
9/20–10/5/1580
UTX, JGI xxv-4
Robertson number 55; Cline number 102
Mixed; Zapotec, Mixe
artist unknown
RGS 3: 103–28; second map following p. 126

43. TEHUANTEPEC B (Santo Domingo Tehuantepec, Oaxaca)
9/20–10/5/1580
UTX, JGI xxv-4
Robertson number 56; Cline number 102
European; Zapotec, Mixe
Francisco Stroza Gali
RGS 3: 103–28; map follows p. 126

44. TEMAZCALTEPEC, MINAS DE (Real de Arriba, Mexico)
Temazcaltepec group
12/1/1579–1/1/1580
AGI, 20
Robertson number 57; Cline number 103
European; Nahuatl, Matlatzinca
Melchior Núñez de la Cerda (?): painter of Temazcaltepec group
RGS 7: 131–61; map faces p. 142

45. TEMAZCALTEPEC (Temascaltepec, Mexico)
Temazcaltepec group
12/1/1579–1/1/1580
AGI, 23
Robertson number 57 bis; Cline number 103c
European; Matlatzinca, Mazahua, Nahuatl
Melchior Núñez de la Cerda (?); painter of Temazcaltepec group
RGS 7: 131–61; map faces p. 151

46. TENANPULCO AND MATLACTONATICO (Tenampulco, Puebla)
Gueytlalpa group
5/30–7/20/1581
UTX, JGI xxiv-5
Robertson number 58; Cline number 49f
Mixed; Totonac
artist unknown: painter of Gueytlalpa group
RGS 5: 147–80; seventh map following p. 158

47. TEOTITLAN DEL CAMINO (Teotitlan del Camino, Oaxaca)
9/15–22/1581
RAH 9-25-4/ 4663-xxvii
Robertson number 59; Cline number 107
European; Nahuatl, Mazatec
Francisco de Miranda(?); also paints map from Tequizistlan
RGS 3: 191–213; map follows p. 198

48. TEOZACOALCO (San Pedro Teozacoalco, Oaxaca)
1/9/1580
UTX, JGI xxv-3
Robertson number 60; Cline number 108
Mixed; Mixtec
artist unknown
RGS 3: 129–51; map follows p. 134, with details after pp. 138, 140, 142, 144, and 146.

49. TEQUIZISTLAN, TEPECHPAN, ACOLMAN, AND SAN JUAN TEOTIHUACAN (Tequisistlan, Tepexpan, El Calvario Acolman and San Juan Teotihuacan, Mexico)
2/22–3/1/1580
AGI, 17
Robertson number 64; Cline number 116a-c
European; Nahuatl, Otomi, Popoloca
Francisco de Miranda(?); also paints map from Teotitlan del Camino
RGS 7: 211–51; map follows p. 214

50. TESCALTITLAN (Santiago Texcaltitlan, Mexico)
Temazcaltepec group
12/1/1579–1/1/1580
AGI, 21
Robertson number 65; Cline number 103a
European; Matlatzinca, Nahuatl
Melchior Núñez de la Cerda (?): painter of Temazcaltepec group
RGS 7: 131–61; map faces p. 143

51. TETELA (Tetetla de Ocampo, Puebla)
10/29/1581
AGI, 31
Robertson number 66; Cline number 118
European; Nahuatl
artist unknown
RGS 5: 375–436; first map follows p. 430

52. TETLISTACA (San Tomás, Hidalgo)
11/15/1581
UTX, JGI xxv-12
Robertson number 67; Cline number 19b
Mixed; Otomi, Nahuatl
artist unknown
RGS 6: 67–93 [91–93]; map follows p. 94

53. TEUTENANGO (Tenango de Arista, Mexico)
3/12/1582
AGI, 33
Robertson number 68; Cline number 122
Mixed; Nahuatl, Matlatzinca
artist unknown
RGS 7: 273–83; second map following p. 278

54. TEXUPA (Santiago Tejúpan, Oaxaca)
10/20/1579
RAH 9-25-4/ 4663-xvii
Robertson number 69; Cline number 124
Mixed; Mixtec, Chocho
artist unknown
RGS 3: 215–222; map follows p. 222

55. TEXUPILCO (Tejupilco de Hidalgo, Mexico)
Temazcaltepec group
12/1/1579–1/1/1580
AGI, 22
Robertson number 70; Cline number 103b
European; Matlatzinca, Nahuatl
Melchior Núñez de la Cerda (?): painter of Temazcaltepec group
RGS 7: 131–61; map faces p. 150

56. TLACOTALPA (Tlacotalpan, Veracruz)
2/18–22/1580
RAH 9-25-4/ 4663-xxxvii
Robertson number 73; Cline number 134
European; Nahuatl
Francisco Stroza Gali
RGS 5: 279–97; map follows p. 288

57. TLACOTEPEC (not known, vicinity of Santa María Xadan, Oaxaca)
Suchitepec group
8/23–29/1579
AGI, 28
Robertson number 74; Cline number 88a
Mixed; Chontal (de Oaxaca)
artist unknown: painter of Suchitepec group
RGS 3: 55–72; first map following p. 62

58. TLAXCALA (Tlaxcala, Tlaxcala)
Tlaxcala group
c. 1584
Hunter collection, University of Glasgow
No Robertson number; no Cline number
Mixed; Nahuatl
Scribe of Diego Muñoz Camargo
RGS 4: 23–286; map is illustration 16

59. TLAXCALA (Tlaxcala, Tlaxcala)
Tlaxcala group
c. 1584
Hunter collection, University of Glasgow
No Robertson number; no Cline number
Mixed; Nahuatl
Scribe of Diego Muñoz Camargo
RGS 4: 23–286; map is illustration 17

60. TLAXCALA (Tlaxcala, Tlaxcala)
Tlaxcala group
c. 1584
Hunter collection, University of Glasgow
No Robertson number; no Cline number
Mixed; Nahuatl
Scribe of Diego Muñoz Camargo
RGS 4: 23–286; map is illustration 18

61. TUZANTLA (Tuzantla, Michoacan)
Temazcaltepec group
10/20/1579
AGI, 23 bis
Robertson number 76; Cline number 103d
European; Tarascan, Mazahua
Melchior Nunez de la Cerda (?): painter of Temazcaltepec group
RGS 7: 131–61 [154–61]; map faces p. 158

62. XALAPA DE LA VERA CRUZ (Jalapa Enríquez, Veracruz)
10/20/1580
AGI, 18
Robertson number 83; Cline number 141
Mixed; Nahuatl, Totonac
artist unknown
RGS 5: 337–74; map (in two parts) follows p. 374

63. XONOTLA (Jonotla, Puebla)
10/20/1581
AGI, 32
Robertson number 84; Cline number 118bis
European; Nahuatl, Totonac
artist unknown
RGS 5: 375–436; second map following p. 430

64. YURIRPÚNDARO A (Yuriria, Guanajuato)
6/15/1580
AGI, 24
Robertson number 85; Cline number 18b
European; Tarascan, Chichimec
artist unknown
RGS 9: 47–72 [68–72]; map following p. 66

65. YURIRPÚNDARO B (verso) (Yuriria, Guanajuato)
6/15/1580
AGI, 24
Robertson number 86; Cline number 18b
European; Tarascan, Chichimec
artist unknown
RGS 9: 47–72 [68–72]; map unpublished

66. ZACATLAN (Zacatlan, Puebla)
Gueytlalpa group
5/30–7/20/1581
UTX, JGI xxiv-5
Robertson number 87; Cline number 49a
Mixed; Nahuatl, Totonac
artist unknown: painter of Gueytlalpa group
RGS 5: 147–80; second map following p. 158

67. ZIMAPAN, MINAS DE (Zimapan, Hidalgo)
8/11/1579
AGI, 13
Robertson number 89; Cline number 155
European; Otomi, Nahuatl, Chichimec
artist unknown
RGS 6: 95–104; map follows p. 102

68. ZOZOPASTEPEC (not known, vicinity of Santa María Xadan, Oaxaca)
Suchitepec group
8/23–29/1579
AGI
Robertson number 90; Cline number 88b
Mixed; Chontal (de Oaxaca)
artist unknown: painter of Suchitepec group
RGS 3: 55–72; first map following p. 70

69. ZUMPANGO, MINAS DE (Zumpango del Río, Guerrero)
3/10/1582
RAH 9-25-4/ 4663-xxxvi
Robertson number 91; Cline number 164
Mixed; Nahuatl
artist unknown
RGS 8: 189–202; map follows p. 198

ITEMS FROM ROBERTSON CATALOGUE OMITTED IN PRESENT STUDY

1. ATITLAN (Santiago Atitlan, Guatemala)
Robertson number 5; Cline number 9
In gobierno of Guatemala

2. COMPOSTELA (Compostela, Nayarit)
Robertson number 17; Cline number 31
In gobierno of Nueva Galicia

3. MOTUL (Motul de Felipe Carrillo Puerto, Yucatan)
Robertson number 40; Cline number 69
In gobierno of Yucatan

4. TABASCO (not located in detail)
   Robertson number 50; Cline number 90
   In gobierno of Yucatan

5. VALLADOLID (Valadolid, Yucatan)
   Robertson number 79; Cline number 139
   In gobierno of Yucatan

6. VALLADOLID (Valadolid, Yucatan)
   Robertson number 80; Cline number 139
   In gobierno of Yucatan

7. VERACRUZ (La Antigua Veracruz, Veracruz)
   Robertson number 81; Cline number 140
   outside of corpus; painted by Alonso de Santa Cruz c. 1540

8. VERACRUZ (La Antigua Veracruz, Veracruz)
   Robertson number 82; Cline number 140
   outside of corpus; painted by Alonso de Santa Cruz c. 1540

9. ZAPOTITLAN (Areas in Departamentos of Suchitepequez and Solola, Guatemala)
   Robertson number 88; Cline number 152
   In gobierno of Guatemala

LOST MAPS

1. ATLITLALAQUIA
   Robertson number: 8

2. CELAYA, VILLA DE
   Robertson number: 9

3. CHILAPA
   Robertson number: 12

4. MEXICATZINCO
   Robertson number: 35

5. MEXICO
   Robertson number: 36

6. NEXAPA, SANTIAGO DEL VALLE DE
   Robertson number: 42

7. QUERETARO
   Robertson number: 47

8. SAN MIGUEL Y SAN FELIPE DE LOS CHICHIMECAS[*]
   Robertson number: 48

9. TAMAZULA
   Robertson number: 52

10. TEPEAPULCO
    Robertson number: 61

11. TEPUZTLAN
    Robertson number: 62

12. TEQUALTICHE
    Robertson number: 63

13. TILANTONGO
    Robertson number: 71

14. TILANTONGO
    Robertson number: 72

15. TOTOLTEPEC
    Robertson number: 75

16. USILA
    Robertson number: 77

17. VALLADOLID
    Robertson number: 78

[*]This map was missing when I visited the Real Academia de la Historia in Madrid to see it in 1989. Thus it was not one of the maps I included in this study. However, a photograph of it is reproduced in RGS 9: following p. 370.

Appendix B
The Questionnaire of the Relaciones Geográficas

This translation is based on the printed version of the 1577 questionnaire. The Spanish version of this text is included in all the RGS volumes. See also the translation of the same text by Clinton R. Edwards (Cline 1972d: 234–7). Edwards notes the handwritten additions to some questionnaires as well as the variations between the 1577 questionnaire and the 1584 version.

MEMORANDUM OF THE THINGS THAT ARE TO BE ANSWERED, AND OF THAT WHICH SHALL BE TAKEN INTO ACCOUNT.

1. Firstly: For towns of Spaniards, state the name of the district or province in which it lies. What does this name mean in the native language, and why is it so called?

2. Who was the discoverer and conqueror of this province? By whose order was it discovered? In what year was it discovered and conquered, as far as is readily known?

3. What is the general climate and character of the province or district? Is it very cold or hot, humid or dry? Does it have much or little rain, and when, approximately, does it fall? How violently and from where does the wind blow, and at what times of the year?

4. Is the terrain flat or rugged, clear or wooded, with many or few rivers or springs, abundant in or lacking water? Is it fertile or without pasture, abundant or lacking in fruits and sustenance?

5. Are there many or few Indians? Were there more or fewer at other times, and what are the known causes of this? Are they presently settled in planned and permanent towns? Describe the degree and quality of their intelligence, inclinations, and way of life. Are there different languages in the province or a general language that all speak?

6. What is the latititude, or the altitude of the Pole Star, at each Spanish town, if it has been taken and is known, or if anyone knows how to observe it? Or on which days of the year does the sun not cast a shadow at midday?

7. What are the league distances and the direction of each Spanish city or town in the

district to the city of its *Audiencia,* or to the town where the governor to whom it is subject resides?

8. Likewise, what are the league distances and the direction of each Spanish city or town to adjacent ones? Are these leagues long or short, through flat or hilly land, over straight or winding roads, easy or difficult to travel?

9. What are the present and former names and surnames of each city or town? Why are they so called, if known? Who gave each its name, who was its founder, by whose order was it settled? What year was it founded? How many residents did the settlement first have, and how many does it have now?

10. Describe the sites upon which each town is established. Is each upon a height, or low-lying, or on a plain? Make a map of the layout of the town, its streets, plazas and other features, noting the monasteries, as well as can be sketched easily on paper. On it show which part of the town faces south or north.

11. For native towns, state only how far they are from the town in whose *corregimiento* or jurisdiction they are, and how far they lie from their *cabecera de doctrina.*

12. In addition, state how far the native towns lie from surrounding native or Spanish towns. Declare for all their direction from these other towns. Are the leagues long or short, the roads through level or hilly land, straight or winding?

13. What does the name of this [native] town mean in the indigenous language? Why is it called this, if known? What is the name of the language spoken by the natives of this town?

14. Who were their rulers in heathen times? What rights did their former lords have over them? What did they pay in tribute? What forms of worship, rites, and good or evil customs did they practice?

15. How were they governed? With whom did they wage war? How did they battle? What was their battle dress and clothing like, both formerly and now? What was their former and is their present means of subsistence? Were they more or less healthy than now and what are reasons for this that you may know?

16. For all towns, both Spanish and native, describe the sites where they are established. Are they in the mountains, in valleys, or on open flat land? Give the names of the mountains, valleys, and districts, and for each, tell what the name means in the indigenous language.

17. Is the land or site healthy or unhealthy? If unhealthy, why (if it is known)? What illnesses commonly occur, and what cures are commonly used for them?

18. How far or near, and in what direction does each town lie from a nearby prominent mountain or range? Supply its name.

19. What major river or rivers flow nearby? How distant and in what direction do they

lie? How great is their flow? Is there anything notable known about their sources, water, orchards and other growth along their banks? Are there or could there be irrigated lands of value?

20. What are the significant lakes, lagoons, or springs within the town boundaries? Note anything remarkable about them.

21. What are the volcanoes, caves, and all other notable and remarkable aspects of nature in the district worthy of being known?

22. Which wild trees and their fruits are commonly found in the district? What are the uses of them and their woods, and to what good are they or could they be put?

23. Which cultivated trees and fruit orchards are in the region? Which were brought from Spain and elsewhere? Do these grow well?

24. What are the grains and seed plants, and other garden plants and vegetables that are or have been used as sustenance for the natives?

25. Which were brought from Spain? Does the land yield wheat, barley, wine, or olive oil, and in what quantities? Is there silk or cochineal in the region, and in what quantities?

26. What are the herbs or aromatic plants that the natives use for healing? What are their medicinal or poisonous properties?

27. What animals and birds, both wild and domestic, are in the region? Which ones were brought from Spain, and how well have they bred and multiplied?

28. In the region or within the town lands, are there any gold or silver mines or sources of other metals, or black or colored pigments?

29. Are there any quarries of precious stone, jasper, marble, or other notable ones? What value might they have?

30. Are there sources of salt in the town or nearby? If [residents] lack salt and other items necessary for their sustenance or clothing, where do they procure all these things?

31. How are the houses built and what is their form? What building materials are found in the town, or brought from elsewhere?

32. What are the towns' fortifications? Are there barracks, or any fortified or impregnable places within the boundaries of the district?

33. Through what dealings, trade, and profits do both Spaniards and Indians live and sustain themselves? What are the items involved and with what do they pay their tribute?

34. In which diocese of the archbishopric, bishopric, or abbey does each town lie? In which district does each lie? How many leagues and in which direction does each town lie from the town of the cathedral and the *cabecera* of the district? Are the leagues long or short, along straight or winding roads, and through flat or hilly land?

35. In each town, what are the cathedral and parish church or churches? What is the

number of endowed church offices and allotments for clergymen's salaries in each? Do any have a chapel or significant endowment, and, if so, whose it is and who established it?

36. What are the monasteries or convents of each Order? By whom and when were they founded? How many notable things do they contain? And what is the number of religious?

37. What are the hospitals, schools, and charitable institutions in the towns? By whom and when were they founded?

38. If the towns are on the seacoast, in addition to the above, report in your account the nature of the sea in the vicinity, whether it is calm or stormy, the nature of the storms and other dangers. What times, approximately, are they frequent?

39. Is there beach or cliffs along the coast? What are the prominent reefs and other dangers to navigation along it?

40. How great are the tides and tidal ranges? On which days and at which hours do they occur? In which season are they greater or lesser?

41. What are the notable capes, points, bights, and bays in the district? Note their names and sizes, as well as can be determined.

42. What are the ports and landings along the coast? Make a map showing their shape and layout as can be drawn on a sheet of paper, in which form and proportion can be seen.

43. What is their size and capacity? Note their approximate width and length in leagues and paces, as well as can be determined. Also note how may ships they will accommodate.

44. What is the depth in fathoms of each? Is the bottom clear? Are there any shallows and shoals? Indicate their locations, and whether the port is free of shipworm and other impediments.

45. What directions do their entrances and exits face? With which winds must one enter and depart?

46. What are their advantages or lack of them in the way of firewood, water, and provisions? What are the favorable or unfavorable considerations for entering and remaining?

47. What are the names of the islands along the coast? Why are they so named? Make a map, if possible, of their form and shape, showing their length, width, and lay of the land. Note the soil, pastures, trees, and resources they may have, their birds and animals, and the notable rivers and springs.

48. Where are the abandoned Spanish towns located in the region? When were they settled and abandoned? What is known about the reasons for abandonment?

49. Describe any other of the notable aspects of nature, and any notable qualities of the soil, air, and sky in any part of the region.

50. Having completed the account, the persons who have collaborated on it will sign it. It must be returned without delay, along with these directions, to the person from whom it was received.

Appendix C

The Nahuatl Inscriptions of the Macuilsuchil Map

TRANSCRIPTION OF THE MAIN TEXT OF THE MAP

1. yoqui yni[n] ymotenehua macuilsuchil
2. ynpa[n]pa ytechcopa. tlatoani teotzapotla[n]
3. oquiçexeloque tlali oquimaca yn çeçe
4. tlatohuani çeçe altepetl. ça yx´qu´j
5. ch itlatoani nica macuilsuchitl
6. quipiya tlali yhua cuasuchit[l]
7. yhua chinamic. manel teoti
8. tlan. tlacochabaya
9. [ni]ca ytlali macuilsuchi
10. ypapa ycuac yohua
11. ya[n] oconana tlatua
12. ni macuilsochitl tlali
13. ytuca tzapatecatl co
14. quipilla yva coqui pi
15. ziatuo yhua çee civa
16. pili ytlaca yoca xo
17. naxi palala ça nica
18. ca qui yeytin tlato
19. huani macuisu
20. chitl

AUTHOR'S TRANSLATION

This, this was called Macuilxochitl. Because of the tlatoani of Teotzapotlan (i.e., Ocoñaña), he divided up the lands, he gave some to each tlatoani in each altepetl, a certain amount [to each]. Here, Macuilxochitl's tlatoani, he protects the lands, boundaries, fields, although Teotitlan and Tlacochihuaya [do not.] Here [on the map] are Macuilxochitl's lands, because of when [yohuaya?] Ocoñaña [divided them]. The lord of Macuilxochitl lands, his name, this Zapotec person, [is] Coqui Pilla, and [there is] Coqui Piziatuo and one noble lady, called Yoca(?) Xonaxi Palala, here she is (?). There were three lords of Macuilxochitl.

PADDOCK TRANSLATION (1982)

1–4. So this is called Macuilxochitl because on behalf of the lord of Teotzapotlan they were divided, there was given to each lord just one town. 5. Lord here Macuilxochitl 6. has lands and boundaries 7. and there is a barrio Teotitlan 8. Tlacochahuaya 9. is land of Macuilxochitl 10. because long ago 11. Ocoñaña the lord 12. of the Macuilxochitl lands 13. who in Zapoteco is called 14. Lord Pilla and Lord 15. Piziatuo and a woman 16. noble called Yoca 17. Lady Palala here are 18. the three lords 19. of Macuilxo 20. chitl.

ACUÑA TRANSLATION (RGS 2: 339–40)

Éste, su nombre Macuilsucil. Por causa del señor de Teotzapotlan dividieron la tierra. Entrególes a cada señor cada un pueblo. Solamente toda, el señor de aquí, de Macuilsuchil, cuidaba de la tierra y mojones y lindes; aunque Teozapotlan, Tlacochahuaya, aquí su tierra de Macuilsuchil; por eso, al anochecer, le tomaron al señor de Macuilsuchil la tierra el llamado zapoteca Coqui Pilla, y Coqui Piziatuo y una señora noble del nombre Yoci Xonaga P[e]la La[a]. Por eso, aquí están los tres señores de Macuilsuchil.

Appendix D

A Typical Viceregal Acordado

From AGN Tierras, vol. 2688, exp. 40, fol. 1

AUTHOR'S TRANSLATION

Don Martín Enríquez, Viceroy and Governor and Captain General for your Majesty in this New Spain, and President of the *Audiencia,* which in it resides. Know ye, *alcalde mayor* of the town of Actopan, that Tomás de Herrera has asked me in the name of his Majesty that he be granted a site for a cattle ranch on the outskirts of towns of Chicuacentepec and the aforementioned Actopan. Before making the grant it is proper that [the site] be viewed and that the necessary procedures be carried out. Because of this, we order you, upon receiving these orders, to go see the place where the aforementioned [Tomás de Herrera] has asked [for a grant]. Also inform those affected [by the grant]: the natives of the town on whose borders the lands [of the proposed grant] fall, as well as the other persons who have sites, ranches, or lands nearby, who might be harmed or aggrieved. Make your announcement on a Sunday or a day of observance when all people are gathered together at High Mass. After the priest has finished the celebration, you are to find out if the aforesaid grant should be made, whether it will be harmful and how. For greater surety, you are to gather information from ten witnesses, five chosen by the interested party [Herrera] and five by you. They may be Spaniards or Indians. Have a map made of the site of the town on whose outskirts [the grant] would fall as well as the other ranches and lands that would be found [on these outskirts] if this grant should be bestowed and made, and to whom they belong. Also note the lands and untilled fields remaining and state the distance between these and the current request. Having made the drawing and signed it with your name, send all this together with your sworn opinion to my attention [so that I may] approve what is fitting. Dated in Mexico the 16 of March, 1578. [Signed,] Martín Enríquez.

TRANSCRIPTION OF THE SPANISH TEXT

Don Martín Enriquez, visorrey y gobernador e capitan general por su mag[esdad] en esta nueva españa e presidente de la audiencia rreal que en ella rreside hago saber a vos el alcalde mayor del pueblo de atocpa que tomas de herrera me a pedida que en nombre de su mag[esdad] le haga m[erce]d de un sitio de estancia para ganado mayor en terminos del pueblo de chicacentepec e del d[ic]ho pueblo de atocpa e porque primero que se le haga la m[erce]d conbiene que se bea y se hagan las diligencias necesarias por la presente os mando que luego queste mi mandamiento beare bare a la parte y lugar donde el susod[ic]ho pide y atados para ello los naturales del pueblo en cuyos terminos cayere e las demas personas que cerca tengan otros sitios de estancias o tierras o en alguna manera puedan rrecibir dano o prejuysio la qual citacion hacere en un domingo o fiesta de guardar estando toda la gente junta e congregada en misa mayor despues que el sacerdote aya echado las fiestas os informarere sabiere e aberigarere si de se le hacer la d[ic]ha m[erce]d les biene el d[ic]ho dano e perjuysio y en que e para mas justification recibiere ynformacion con diez testigos, cinco de parte e cinco de oficio que sean espanoles e indios e harere pintar el asiento del pueblo en cuyos terminos cayere e las demas estancias o tierras que en el d[ic]ho termino estubieren proveyos y h[ech]a m[erce]d y cuyas son e las tierras y baldios que quedan con declaracion de la distancia que ello ay a lo que agora se pide y h[ech]a la d[ic]ha pintura y firmada de v[uest]ro nombre con todo lo demas juntamente con v[uest]ro parecer jurado en forma lo enbiarere ante mi para que visto yo probea lo que conbenga fecha en mexio a xvi dias del mes de marzo de mil y quinientos y setenta y ocho anos. [firma] Martín Enríquez

Notes

NOTES TO THE PREFACE

1. The term "map" used in this book follows the inclusive definition given by Harley and Woodward: "Maps are graphic representations that facilitate a spatial understanding of things, concepts, conditions, processes, or events in the human world" (1987: xvi).

2. As used within, "reality" refers to the way a society constitutes, interprets, and understands the real. See Gruzinski (1993: 285, n. 1).

3. For discussion of the idea that the changes in visual representation can be connected to changes in the social and political matrix, see Barrell (1987), and Prown (1980, 1982). For a critique of this approach, see Bryson (1985: 13–35).

4. In these descriptions of projective systems, I follow those discussed in Robinson and Petchenik (1976), except that they call my Albertian space "projective," a term I have avoided because it is too easily confused with "projection."

5. On the history and content of this manuscript, see Codex Mendoza (1992).

6. "Planimetric" is used here and throughout to denote a map that seeks to capture spatial relationships through a scale model.

7. I have greatly simplified the complexities of this page. The basic relationship of the Mendoza page to calpolli was set out by Alfonso Caso (1956). Van Zantwijk (1985: 57–93) gives a somewhat different interpretation. Following Lockhart (1992: 16), I use "calpolli" as both a singular and a plural noun, as I do the related "altepetl" discussed in chapter 5.

8. I use "humanistic" here, to convey the socially based projection of Mesoamerican maps, in its sense of "of or pertaining to humanity" rather than "of or pertaining to Humanism."

9. In treating Relaciones Geográficas maps from only the gobierno of New Spain, I have left out seven maps from the other gobiernos that formed the larger political entity of the Viceroyalty of New Spain. I have done this because it was within the gobierno of New Spain that the largest corpus of native maps was concentrated (see appendix A).

10. The Relación Geográfica corpus of New Spain is discussed and catalogued by Cline (1964, 1972a, 1972d) and an annotated bibliography can be found in Cline (1972e). The maps are discussed and cat-

alogued by Robertson (1959b, 1972a, 1972b). More recent studies of the corpus of texts or maps and its members not cited by Cline (1972e) include Bailey (1972); Cook de Leonard (1974); Kubler (1985a); Montêquin (1974); and Gruzinski (1993: 70–97). A new publication, edited by René Acuña, and referred to as RGS throughout, reproduces all the known texts and maps of the Relaciones Geográficas from New Spain.

NOTES TO CHAPTER TWO

1. For a survey of attempts to measure the earth, see Lafuente (1984). My thanks to Anthony Aveni for bringing this work to my attention.

2. Since the Pole Star is not precisely situated at the Pole, its altitude is an approximation of the position of the Pole. More precise latitude measurement can be made by averaging the elevations of the Pole Star from observations made twelve hours apart. Anthony Aveni (1992: personal communication) clarified this point.

3. At the equator, the sun passes directly overhead on March 21 and September 23 (the equinoxes). At the tropic of Cancer, the zenith passage falls on June 22 (the summer solstice), while at the tropic of Capricorn, it falls on December 22 (the winter solstice). At the lines of latitude within the tropic lines, zenith passage is equally predictable. For more on the measurement of latitude and longitude see Aveni (1980: 56–67), Brown (1979), and Lafuente (1984).

4. See Goodman 1988: 53–65. The problem of longitude was connected to another scientific deficit Europe shared: the lack of a timepiece that would keep time accurately over long distances. Had Europe possessed such a chronometer, Santa Cruz could have simply found out when the sun set in Manila, Mexico City, and Seville on a certain day, from this figured out the different time zones in which they lay, and then determined where they lay in relation to one another on the face of the globe.

5. Pietro Martire d'Anghiera, *De Novo Orbe Decades* (1964–1965) [written in 1493–1525 and published 1530]. Gonzalo Fernández de Oviedo y Valdés, *Historia general y natural de las Indias* (1944–45) [written 1520– c. 1535 and published in 1535 and 1557]. A later chronicler not mentioned by Santa Cruz but working in the same vein was Antonio Herrera y Tordesillas, author of *Historia general de los hechos . . .* (1944?; first published 1601–1615). See Ballesteros Gaibrois (1973), Burrus (1973), and Warren (1973).

6. The post was called both *cosmógrafo-cronista* and *cronista-cosmógrafo* in the Ovantine laws that established it (Jiménez de la Espada 1965: 43–6). I follow the usage of Jiménez de la Espada.

7. For a summary of López de Velasco's life and work, see González Muñoz (1971).

8. When López de Velasco sent out his questionnaires to the colonies, another questionnaire, also initiated by Ovando, was circulating in regions of New Castile (Cline 1972d: 188; Jiménez de la Espada 1965: 5–10; López Piñero 1979; Viñas y Mey and Paz 1949–1971). While a number of questions of the Castilian Relaciones Geográficas asked about local geography, none of them asked for maps. Two versions of the Castilian questionnaire and the responses have been published by Viñas y Mey and Paz (1949–1971).

9. Using this inventory, scholars have begun to correctly identify and attribute Santa Cruz's works and to assess his importance. See Cuesta Domingo (1983); Dahlgren (1892); Elsasser (1974?); Latorre y Setien (1913); Saralegui y Medina (1914); Schuller (1912); and Wieser (1908).

10. González Muñoz draws this point from the comments of Justo Zaragoza in the first edition of López de Velasco's *Geografía y descripción . . .* (1894).

11. The following exegesis of López de Velasco's method is based on the instructions the cosmographer sent out which are printed in translation in Edwards (1969: 20–1). A transcription of a Spanish original (from a second attempt at observation launched in 1581) is found in CDI, vol. 18: 129–36.

12. Antonio de León Pinelo, an early seventeenth-century bibliographer, mentions an unpublished description of New Spain by Domínguez, now lost (cited in Somolinos D'Ardois 1960: 258).

13. The Relación Geográfica of Papaloticpac mentions the visit of this royal physician (*protomédico*) to collect botanical information (RGS 3: 31), as do the Relaciones of Los Peñoles (RGS 3: 50), Teozacoalco (RGS 3: 146), the Mines of Tasco (RGS 7: 129), Tetzcoco (RGS 8: 109, 111), and Tiripitio (RGS 9: 355). There may be a reference to his visit in the Relación from the Minas de Zimapan (RGS 6: 103). A pamphlet written by Hernández on the use of plants to treat disease is mentioned in Relación of Querétaro (RGS 9: 224).

14. Somolinos D'Ardois wrote that the last report from Domínguez was a letter of complaint written to the king from New Spain in 1581. However, we find Domínguez measuring eclipses in Mexico City in 1584 (Edwards 1969: 21), and there exists a 1594 petition to the royal government on behalf of Domínguez, who died a few years later (Goodman 1988: 66–7, 83 fn. 33).

15. Edwards reports that eclipse instructions were sent for eclipses of 1577 and 1578, 1581, 1582, and finally in 1584 (1969: 18–9).

NOTES TO CHAPTER THREE

1. In the Middle Ages, "there were no precedents for the idea of drawing maps at all" outside of certain traditions (Harvey 1987: 484). These traditions were, according to Harvey, those of maps of Palestine, of Italian city and district maps, and of the *isolarii*, descriptive atlases of islands (1987: 464–84). The awkwardness and indifference toward the map-image were not atypical of Iberian Spanish officials as well. When a series of similar questionnaires, also sponsored by Ovando, was sent out to various provinces within Spain about the same time as the New World Relaciones Geográficas, they did not ask for maps. Even so, the graphic material that local Spaniards sent back to the crown was even more idiosyncratic than that from New Spain. See Viñas y Mey and Paz (1949–1971)

2. Given that their official functions were much the same, local officials—comprising alcaldes mayores and corregidores—are uniformly referred to as corregidores, unless alcaldes mayores are specifically meant.

3. It is clear that many local officials evaded these laws, from the viceroy down. See Haring (1975: 128–38); also McAlister (1984: 203–7).

4. *"Todo esta tierra son sierras y cordilleras, tan bravas y terribles, q[ue] a ninguna podemos poner particular nombre porque son innumerables, y, aunq[ue] en su antigüedad (y aún ahora) tienen los indios puesto nombre a todas, suena en n[uest]ro castellano tan mal y tratamos tan pocas veces de ellas, que parece excusado para este lugar, por no ser cosa importante sus nombres."*

5. The influence of prints on art in New Spain is well established. See Kelemen (1951: 200–3); Kubler and Soría (1959: 303–7); Manrique (1982); Peterson (1993); Robertson (1959a: 187–8); Robertson (1972a: 259, fn. 25); Soría (1956: 15–43).

6. As Robertson points out (1972a: 246, fn. 10), the Tlacotalpa Relación points to its map to answer items 11, 20, 38, 41 to 45, and 47 (RGS 5: 283, 284, 286); the writer of the Tepoztlan Relación says its map, now missing, answers items 11, 16, 18, and 19 (RGS 6: 184, 189, 191).

7. Soría writes of how colonists failed to be inspired by the landscape: "few regions boast landscape, people, flora, and fauna more exciting than Latin America. Yet Colonial painting knows no landscapes" (Kubler and Soría 1959: 303).

8. Here the map differs, showing the distance between Quauxoloticpac and Huiztepec to be two leagues, not the one figured by the text.

9. Gerhard gives the date for the establishment of this presidio as 1581 (1986: 63), but it must have been in existence some time earlier to appear on this Relación map of 1579.

10. See, for example, Codex Xolotl (1980) and Mapa de Quinantzin (1983), fol. 1.

11. Native painters after the conquest continued these visual distinctions between the civilized and the Chichimecs, even when adopting features of European style, as in the fresco paintings of Ixmiquilpan, a monastery across the Sierra de Pachuca from Meztitlan (Pierce 1981).

12. Most trained European painters, Kubler and Soria suggest, would have been located in Mexico City and their primary works would have been religious images (1959: 303-4).

13. This recopying is evident by the report from Tuzantla, dated first but set last in the text.

14. The map that is the second to appear in the text looks as if it were executed first, for it is less carefully carried out, with buildings more schematic and smudgy, and with lettering less carefully done.

15. In 1585 Stroza Gali was sent by the viceroy of New Spain on an expedition to explore the Philippines, accounts of which were published in 1596 and 1598 (Acuña in RGS 5: 282).

16. I contend that Stroza Gali used a magnetic compass because most mariners of his time used them. The maps, as discussed below, deviate from true north as a compass might, but this cannot be considered evidence of compass use because there are, as Aveni points out, "many long-term magnetic variations of indeterminate origin and magnitude" (1980: 119–20).

17. He wrote his record of these latitudes in the center of the map, and by some error in transcription or in instrumentation, his readings were off by about a degree (by between 56" and 1' 21"), errors he avoided in the later Coatzocoalco map. The mouth of Rio Alvarado is actually 18'48" so Stroza Gali's reading for the Tlacotalpa map of 17'52" is 56" off; the point of Anton Lizardo is 19'4" and his reading of 18'8" is again off by 56"; Tuxtla is 18'27" and Stroza Gali's reading of 17'15" is at variance by 1'12"; Roca Partida is 18'43" and his reading of 17'22" leads to a difference of 1'21".

18. López de Velasco's laconic inventory makes no note of two maps accompanying this Relación (see Cline 1972d, 230).

19. Notable exceptions are the survey maps of Bavaria made by Philipp Apian, whose methods were discussed in chapter 2. His *Bairische Landtafeln* was published in Ingolstadt in 1568.

20. In the large collection that includes hundreds of sixteenth-century maps housed in the Archivo General de la Nación (AGN) in Mexico City, evidence of survey mapping is equally rare. See *Catálogo de Ilustraciones* (1979–: vols. 2–5, 11). Also, the group of colonists who answered the questionnaire took no note of Domínguez, although a number of them mentioned Francisco Hernández, the scientist with whom he traveled. See chapter 2, fn. 13.

21. The 1572 inventory in *Códice Mendieta* (1971: 255–7) lists one as Cosmografia Camponi, which I have yet to identify. Another listed is a work likely to be by Peter Apian ("Apiano de Beliz"). The book in question would have been either his *Cosmographiae introdvctio* or *Cosmographicus liber Petri Apiani*, both of which went through numerous editions in the sixteenth century, including at least one Spanish edition of the latter of 1548 (see Ortroy 1963).

22. Survey maps were used for these ends from Roman times onward, as well as to levy taxes. See Dilke (1971, 1987).

23. Haggard notes that in Spain, a paso measured about 1 1/2 yards, or 1 2/3 varas (1941: 82). In New Spain, the paso is sometimes equated to the vara (AGN Tierras, vol. 2683, exp. 11). Gibson tries to sort out measures used in the Valley of Mexico (1964: 257), especially the measurement of the braza, while Pezzat Arzave makes no such effort, calling the system of mensuration "anarchy" (1990: 60–1).

24. An early desagüe map was mentioned in the acts of the town council of Mexico City in 1551. See

Actas de Cabildo . . . (1889, vol. 6: 202). Some desagüe maps have been published by Apenas (1947), and the seventeenth-century projects are discussed by Hoberman (1972).

25. An eighteenth-century Spanish dictionary gives the following entries for "pintura": "[1] A liberal art, an imitation of the proportions of nature. It is an image or imitation of what is visible, delineated on a plane surface, not only in that which is the form, but also in the color and other attributes . . . [2] Also refers to the boards, plate, or canvas upon which something is painted . . . [3] Metaphorically, it is used for the description or narration that is made in writing or in speech of something, referring minutely to its circumstances and qualities, as in the *pintura* [portrait] of a city, of a lady, etc." (Academia española 1726–1739, vol. 5: 278).

26. For example, the posthumous inventory of Santa Cruz's works, many of them cartographic, never uses mapa, referring to all maps, be they world maps or island charts, as *descripciones* (Jiménez de la Espada 1885–1897, vol. 2: xxx–xxxvii), while López de Velasco's 1583 inventory of the Relaciones Geográficas refers to the maps therein as "pinturas" (Cline 1972d: 237–40).

NOTES TO CHAPTER FOUR

1. "Ethnic" and the ethnic terms that I use mean the self-defined affiliation of peoples in the colony; Spaniards are those born in Spain, who act as part of the Spanish community; Creoles are those born in the New World who are members of the Spanish-Creole community; "natives" or "indigenes" are those who counted themselves as members of the native community.

2. Two of the Relaciones texts are written by men whom we know to be mestizos—don Juan Bautista de Pomar wrote the one from Tetzcoco, and Diego Muñoz Camargo wrote the one from Tlaxcala. It seems that the pictures they used to accompany their texts were drawn from native manuscripts.

3. The Spanish text reads: *Don Diego de Mendoza, principal de Cempoala, y a Don Francisco de Guzmán, gobernador del pueblo de Tzaquala, y a Don Pablo de Aquino, gobernador de Tecpilpan, y a Martín de Ircio, gobernador de Tlaquilpa, y a los alcaldes de los dichos pueblos y a otros muchos indios viejos y ancianos de los dichos pueblos* . . .

4. The mill has recently been excavated under the auspices of INAH. See Montellano (1988).

5. All the identifiable artists at work in the corpus are men, and historical evidence strongly suggests that indigenous professional painters would have been male. But given that we cannot positively determine the gender of all the artists and thereby rule out that women may have painted maps, I have used both pronouns to reflect this uncertainty.

6. For the most part, the pigments used in the Relaciones Geográficas maps seem to have been water-based, consisting of earth and mineral pigments. Applying these kinds of pigments evenly took practice and skill, since the medium was thin and liquid by nature, and pigments were not permanently suspended. Laid on by the hand of a novice, colors would puddle and smudge. See Anderson (1963).

7. The friars were just part of a larger Spanish colonial incursion into Texupa. Once the center of silk production in the region, Texupa enjoyed great commercial success (Borah 1943; Spores 1967: 81). On the map, see Bailey (1972).

8. Reyes-Valerio (1989) makes a similar argument about the quality of native fresco painting in monasteries, attributing it to a convergence of the training native painters would receive in large monastic establishments and the training they had received in the traditional native school of the *calmecac.* I disagree that the particular institution of the *calmecac* (described in the Valley of Mexico by writers such as Sahagún) proliferated all over preconquest Mexico. I think other, more varied systems of native education existed in Mexico and many would have continued after the conquest when not supplanted by monastic schools. The continuing influence of the native manuscript tradition into the colonial peri-

od is registered in the Relación Geográfica corpus as a whole, where regions like the Tarascan zones to the northwest of the Valley of Mexico, which seem to have had little in the way of a pre-Hispanic painting tradition, produced no native maps.

9. Early writers on the subject held that much of the painting of monasteries was done by Spanish painters, who would use native charges for purely rudimentary tasks (Kubler 1948: 367–8; Toussaint 1965). After it was discovered that Juan Gerson, the "European" painter of Tecamachalco, was in truth a native painter, other scholars have been attributing many early colonial mural paintings to native hands (Peterson 1993; Pierce 1981; Reyes-Valerio 1989).

10. Peterson points out that trained native painters were highly mobile. In light of this, she argues that the Augustinian convent of Malinalco may have been painted by artists sent from Franciscan Tlatelolco around 1571 (1993: 50–6). However, the Relaciones Geográficas maps show that trained painters were to be found throughout New Spain, not just in Tlatelolco or Mexico City.

11. For a more detailed discussion of the sources of monastic church fresco cycles—book illustration, church retables, Spanish mural paintings, and tapestries—see Kubler (1948: 372–8); McAndrew (1965: 137–9); Peterson (1993: 57–82); Reyes-Valerio (1989: 97). For the varied subject matter of monastic church frescos painted by native painters, see Artigas Hernández (1979); Cordero (n.d.); Edwards (1966); Peterson (1993); Pierce (1981).

12. Europeans did devise pictographic records drawing upon native conventions (but not duplicating them) in order to teach Catholic catechism to native peoples. These manuscripts, called Testerian after their inventor, have been studied by Normann (1985).

13. An exception may be when roads are shown marked with hoofprints. This feature occurs in many early colonial maps that seem to have been drawn by native and European alike, and is a transformation of the preconquest convention of using a string of footprints to indicate travel or movement. Another example is to be found in the Relaciones Geográficas maps of Meztitlan and Tecaltitlan, both discussed in chapter 3, in their representation of native warriors.

14. In an earlier work (1959a), Robertson had defined the stylistic traits that were the basis of a pure "native" painting style that reigned in Mesoamerica before the Spanish conquest. Among its characteristics were a perfectly even, steadfast pen stroke that preconquest artists cultivated (Robertson 1959a: 65). Color was another diagnostic of the trained hand; skilled painters applied it in solid blocks (Robertson 1959a: 66). A balanced composition was another quality that marked native style; preconquest-style painting left an effect "similar to that of an oriental rug or wallpaper" (Robertson 1959a: 15).

15. In an earlier work on the Relaciones Geográficas maps, Robertson asserted that the corpus was "mainly the work of native painters" (1959b: 540).

16. Only one extant map is classified as "native," that from Amoltepec. In all other cases, Robertson applies the label "native?" to maps that are presently lost (Robertson 1972a: 265–78).

NOTES TO CHAPTER FIVE

1. For a survey of the wider topic of indigenous cartography, see Boone (1992b); Smith (1973a); and Mundy (1998).

2. I define regions with a strong pre-Hispanic tradition of painting as ones where we have both some documentary evidence of this tradition and a surviving colonial corpus of native manuscripts.

3. The Tetzcocan manuscript style resonated far from Tetzcoco: the Lienzos de Tuxpan (1970), made in distant Veracruz, are exquisite manuscripts and unusual products for a region a great distance from a major urban center. Their quality can be attributed to Tetzcocan control of the Tuxpan region in the preconquest period (Ixtlilxóchitl 1985: ii, 107).

4. Robertson (1959a: 151–3) discusses only the Tequisistlan and the Teotitlan del Camino Relación maps as products of this Tetzcoco school, an unfortunate choice of examples because these maps are crudely drawn sketches, of uncontrolled line and with a chaotic organization, that almost certainly were done by Miranda.

5. Nicholson (1966: 154) says a "fair case can be made for a Cholula provenience."

6. As Andrews notes about Nahuatl compound nouns, the "matrix stem" is preceded by modifying words, much as it is in English (1975: 156). Consider the example of Coatepec, a place-name meaning "hill of the snake." In this example, *coatl* ("snake") modifies *tepetl* ("hill"), thus the hill has on it, or is shaped like, a snake. If we reverse the order of the words, we change the meaning. *Tepecoatl,* for instance, means a snake living on a hill. The difference is analogous to that between the terms "doghouse" and "house dog."

7. All Nahuatl nouns include an absolutive suffix, usually *-tl* or *-tli* or *-tla,* that drops off a root word when that word is combined or possessed, or when a locative suffix is added. Thus, in the name *Xilotepec,* the component *xilotl* is the root word *xilo* with the suffix *-tl* attached.

8. This hill shown on the map probably corresponds to the "conical mountain" that contained remnants of terraces and structures to the north of the town (Spores 1967: 40).

9. The Zouche-Nuttall, like other Mixtec screenfolds, was read in boustrophedon fashion. Beginning at the bottom register, the viewer read the first register from right to left, the second from left to right, and the top from right to left. The reading order of various Mixtec manuscripts is described by Smith (1973a: 10, fig. 1).

10. In many ways, I find the term "cartographic history" less than ideal, since it fails to adequately emphasize that these maps express overlapping aspects of one community's identity. However, it is the common currency of contemporary scholarship (including my own), and so I use it here to avoid the confusion of too many neologisms.

11. The dynastic foundation is recounted in the Codex Vindobonensis. See Furst (1978) and [illegible] (1982).

12. We have little evidence that would allow us to evaluate whether there was a schism between these notions of community identity that elites disseminated and those of commoners, but given the longevity of cartographic histories, it seems that the histories they depict were accepted by their viewers as the community history. For an alternate view, see Marcus (1992).

13. The Lienzo de Zacatepec I has been studied by M. E. Smith (1973a: 89–121), and her analysis is the basis for the following discussion. See also Caso (1977–1979, vol. 1: 137–44) and Rauh (1972).

14. The historical narrative overlying the map may have been drawn from a screenfold manuscript, for it is read from upper left to upper right, then continues in a zig-zag, or boustrophedon, pattern shared by some Mixtec screenfolds. This point is suggested by Smith (1973a: 93). In addition, one of the figures in the historical narrative, a man named 4 Wind, is an actor in four of the pre-Columbian–style Mixtec screenfolds. See Caso (1955) and Caso (1977–1979: vol. 1, 137–44).

15. On the numerous disputes within the Mixtec, especially over boundaries, see Spores (1984: 209–25).

16. A work by Anders et al. (1992) was brought to my attention just as this book was going to press; in it they provide the most detailed analysis of the geographic reaches of the Teozacoalco map available.

17. Anders et al. (1992) reach a similar conclusion independently.

18. Smith also points out an example closer to home, the Mixtec Aubin Manuscript no. 20, which shows the four directions surrounding a circular center (personal communication 1995).

19. The prototype of the Teozacoalco Relación map, if made when Elotepec was still part of its domain, may have also shown its boundaries in a perfect circle, with the boundary revision added on an arc inside the circle. When drawing up his (or her) map for the Relaciones Geográficas, the Teozacoalco artist arranged Teozacoalco's present boundaries along a perfect circle and relegated its former ones to the exterior arc.

20. This differed from the Mixteca, where communities were discrete, a situation that was reinforced by geography. In the Mixteca, the community history was the preferred map.

21. Molina (1977: fol. 11v) defines calpolli as *"casa o sala grande, o barrio"* ("house or large hall, or neighborhood").

22. This image of Cholula is prefigured by a map on fols. 9v–10r of the Historia Tolteca-Chichimeca which shows the historic organization of the Cholula region when it was divided into ten polities under the control of Olmeca-Xicalanca rulers, who were later usurped by the Tolteca-Chichimeca. On this page, each one of the ten blocks shows an identifying place-name and a ruler within his palace, or tecpan. Kirchhoff et al. (1976) have identified many of these Olmeca polities as having once clustered around Cholula (see their map 5) and they postulate that their order on this page corresponds to social ranking rather than planimetry (p. 150).

23. The presence of fifteen rulers may have been numerologically important; van Zantwijk discusses the importance of the number fifteen for Tenochtitlan in van Zantwijk (1985).

24. See Historia Tolteca-Chichimeca (Kirchhoff et al. 1976: paragraph 265, pp. 180–1).

25. The order (and existence) of the twelve calpolli follows Kirchhoff et al., who analyze the various calpolli lists appearing in the text (Kirchhoff et al. 1976: 148 fn. 3, 240–6, 258).

26. The other boxes framing the city center name not calpolli, but other rulers.

27. The Historia Tolteca-Chichimeca identifies the turkey on a hill as the place-name of Amaquemecan in folio 25r. In the Historia version, the turkey wears a paper *(amatl)* collar not shown on the Map of Chichimec History.

28. *Teccalco* derives from *teuctli* ("lord," "noble"), *calli* ("house"), and the locative *-co*. Molina defines the term slightly differently as *casa, o audiencia real* ("house," or "royal courtroom") (1977: fol. 92r).

29. *Tecuhtli* is a variant spelling of *teuctli*. Chimalpahin relates that the other lord of Tecuanipan bore the title *teohua teuctli* or *tzompahuaca teuctli* but does not attest to him ever being called *motlahuaçcoma* (Schroeder 1991: 61–5).

30. The arrow shooting is described in Ixtlilxóchitl (1985: vol. 1: 295–6); Angel García Zambrano postulates that arrow-shooting was widely used throughout Mexico as part of the foundation ritual (García Zambrano 1992: 239–96).

31. Only eleven of these calpolli are named in the Historia Tolteca-Chichimeca, but the editors argue that twelve actually existed, one being unnamed in the text. See Kirchhoff et al. (1976: 148, fn. 3).

32. RGS 6: 74; the order of the last two is reversed in *Códice Franciscano* (1941: 14), where a description of Cempoala appears.

33. As registered in the Relación Geográfica text, each of the four altepetl of Cempoala had either three or four calpolli, or subunits. It is unclear whether the larger altepetl was just an umbrella for the subunits, or whether it too had a representative calpolli. If so, most altepetl in Cempoala would have had five parts, not four.

34. While the Epazoyuca map offers little in the way of the social divisions of territory, what we can glean from the Cempoala map and Epazoyuca's own Relación Geográfica text presents us with an equally complex society. Epazoyuca seems to have been divided in two—the four parts of one of these moieties described as *estancias* in the Relación Geográfica text, and the four parts of the other as *barrios*. This

two-part organization is further suggested by the two rulers named in the text—don Juan de Austria and don Bernardino de Tolentino—as well as four alcaldes. (The Relación Geográfica text is ambiguous on the office held by Tolentino, but as he bears the title "don," it is likely that he is not merely an alcalde.) Just who don Andrés (who appears on the map next to one of the estancias of Epazoyuca) is is unclear—perhaps he is a moiety leader.

35. Another gloss, *intlalehual cenpohualteca,* marks "the earthen mound of the Cempoallans" (Louise Burkhart, personal communication, 1993; cf. Acuña in RGS 6: 71). This enigmatic mound is shown by a circle studded with the U-shapes that mark cultivated land.

NOTES TO CHAPTER SIX

1. For a discussion of different writing systems, see Coe (1993: 13–45) and Boone and Mignolo (1994: 3–26). Central Mexican "picture writing" was a combination of logograms—written signs, often pictures, standing for entire words or smaller word-parts called morphemes—and phonetic signs, standing for spoken sounds. For the sake of convenience, it is called "logographic" writing herein, although it did have phonetic elements. Overviews are provided by Lockhart (1992: 326–73) and Marcus (1992: 153–89).

2. The relationship between naming and power in the conquest of indigenous America has been explored by Derrida (1976); Owens (1986); and Todorov (1984).

3. Two early studies of Nahuatl place-names are Peñafiel (1885) and Peñafiel (1897). For a sampling of more recent discussions of both place-names and writing, see Dibble (1971) and Galarza (1979); see also Prem (1992).

4. For works showing the point of view of the former camp, see, for example, Galarza (1972) and Galarza (1979). For the latter, see Boone and Mignolo (1994: 3–26).

5. This is not to say that texts can be fully interpreted without native language. As Smith has argued (1973a), the Mixtec language is tonal, lending itself easily to puns, and manuscript writers followed suit, often rendering words with homophones, or puns. See, for example, the discussion of the Apoala place-name, "river of the lineages," in chapter 5.

6. The other is the fragmentary Matrícula de Tributos, which is also a tribute list, undoubtedly a close relation of the one copied in the Codex Mendoza. See Matrícula de Tributos (1980). The relationship between Matrícula and Mendoza is discussed by Robertson (1959a), as well as in the Codex Mendoza (1992) and Barlow (1949b).

7. The source for the Nahuatl translations and for the etymologies of Nahuatl words in this and other chapters is Karttunen (1983). I follow her spellings wherever possible, but have omitted diacritical marks; for these, Karttunen (1983) can be consulted. Other secondary sources used for translations of Nahuatl words are Campbell (1985); Molina (1977); and Siméon (1986). The translations are the author's unless otherwise noted.

8. For discussions of Mendoza place-names, see Barlow and McAfee (1949); Berdan (1992); Nicholson (1973); and Nowotny (1959). The transliteration of names in the Codex Mendoza is discussed by Nicholson (1973: 14–18).

9. Its botanical name is *Leucaena diversifolia, Crescentia alata* (Karttunen 1983: 82).

10. Perhaps to show that the tepetl symbol of Tepexocotlan alone had a logographic role, the artist colored half of it yellow and half blue.

11. The limited phoneticism of this example is characteristic of Nahuatl writing. For discussions of the phoneticism in these place-names, see Berdan (1992); Dibble (1973); Galarza (1979); Nicholson (1973); Nowotny (1959); and Prem (1970).

12. The two other sites on the Jujupango map have somewhat anomalous logographic place-names.

Tecpatlan, "place of flints," is marked by a *tecomatl,* or clay pot; the town of Coayango is marked with a black-and-white circle of unknown meaning.

13. According to Karttunen, the *chicuahtototl* "resembles a barn owl" (1983: 50). However, the description of the *chicuahtototl* bird in the Relación and its portrayal in the accompanying map do not resemble that of an owl. Acuña holds that the town's name comes from another bird, the *chiqualotl,* mentioned by a sixteenth-century source (RGS 6: 169–70, fn. 47).

14. This name comes from the Nahuatl *zoyatl,* "palm," and *quiyotl,* "shoot of a plant." *Zoyatl* may be a homophonic substitution for *cihuatl;* in many Nahuatl dialects, *cihuatl* is pronounced *zohuatl.* This substitution may have saved the artist from the difficult task of representing *cihuayo,* meaning "womanhood" or "female genitalia," with a picture. The substitution of *zoyatl* for *cihuatl* was done by other Nahuatl-speaking communities; see Lockhart (1991: 47).

15. See also chapter 5, fn. 8.

16. RGS 3: 148; *ahmolli* is the Nahuatl word for both soap and the plant root from which it was derived; *nama* is the Mixtec word for "soap" (Alvarado 1962: 87, fol. 203r).

17. Spanish documents, such as the colonial list of Teozacoalco's subject towns, which includes Amoltepec (listed as "Examoltepec"), favor the town's Nahuatl name (PNE 1: 284).

18. M. E. Smith has shown that this phenomenon of double-naming occurred among a number of Mixtec towns, among them Teozacoalco (1973a: 58).

19. M. E. Smith pointed out that the correct Mixtec name of this white four-petaled flower would be *yuhu* (known in Nahuatl as *izquixochitl*) rather than the more standard Mixtec word for flower, *ita* (M. E. Smith, personal communication 1991; see also Alvarado 1962: fol. 111v).

20. The form of the flower *yuhu* (Mixtec) in Teozacoalco's pictograph of "plain of flowers" is different from the buds in the pictograph shown on the Amoltepec map. However, they seem to be the same plant, but at different stages. Their botanical name is *Bourreria huanita.* In the *Badianus Manuscript,* (Emmart 1940) a sixteenth century catalogue of the plants of Mexico, the *yuhu* flower, which resembles that on the Teozacoalco map, is coupled with its bud, which matches that of the Amoltepec place-name. The representations of flower and bud in the *Badianus Manuscript* and the botanical name of *yuhu* were pointed out to me by M. E. Smith (personal communication, 1991).

21. The Lienzo is now in the American Museum of Natural History in New York City.

22. On the Lienzo de Yolotepec, near the "plain of flowers," lies another logographic place-name, a stepped-fret frieze—to stand for "town"—attached to a shield fringed with bells. A similar fringed shield, linked to a plain rather than a frieze, is prominently displayed on the Amoltepec Relación map and may refer to the same place as is mentioned on the Lienzo of Yolotepec.

23. Tehuantepec was not listed in the Codex Mendoza as a tributary state but is shown as conquered by the Culhua-Mexica, who used it as a land bridge to the southeastern province of Soconusco (Xoconochco). See Barlow (1949a: 99). Paddock gives the Zapotec name of Tehuantepec as "gui zii" or "guíe zíj" (1983: 100, fn. 17).

24. Classic-period Zapotec writing is much better documented. See Marcus (1992) and Urcid (1992).

25. Galarza (1967) has found some evidence of Spanish names being transliterated into Nahuatl and then rendered logographically, but this practice seems to have been a relatively isolated one.

26. The map has an inscription on the hill that reads "This is where the town used to be." Tommasi de Magrelli (1978: 136) provides archeological evidence to show that the hilltop ceremonial zone was already partially abandoned by the time that Teutenango was conquered by the Culhua-Mexica in 1474; Cortés, in his third letter, seems to suggest that the Teutenango hill was abandoned at the time of the Spanish conquest (Cortés 1986: 245–6).

27. In the map, the pictograph for *Comulco* was added as an afterthought by a different hand; the symbol for *Axochitlan* lies at some distance from the settlement named as such and it is ambiguous whether the logograph pertains to the place. For other details and difficulties of transcription, see Cook de Leonard (1974), Mundy (1993: 26–8, 223), and Musset (1989).

28. While the same person may have painted the map and written the inscriptions, as was the case with the Suchitepec group, few of the native artists are identified and none signed his or her map, so any correlations are difficult to establish. But it seems clear that the writers were part of the indigenous community: there was no reason for a Spaniard or European to write in Nahuatl on a map that was destined for the Spanish king.

29. I interpret this inscription to read *Cempohualli ome calli ychan macehualtin nica nictlalia.*

30. I read this inscription as *ohtli Cinmatlan yhuan Xalpan yei medio lleguas: ahciz teohuatl [?] Cinmantl.* "Sea" is Acuña's translation of what he believes to be *teohua atl* (priest-water), but I have not found this term attested in other sources.

31. Muñoz Camargo possessed, and added as part of his Relación, a copy of the famous Lienzo de Tlaxcala. The relationship of Muñoz Camargo's work to the Lienzo is discussed in Brotherston and Gallegos (1990) and Martínez (1990).

32. In the Mixteca, this Spanish model was closer to indigenous political arrangements (Spores 1967). Towns that were held in *encomienda*—by an individual rather than the crown—owed their tribute to their *encomendero,* but by the 1570s, it was collected through the same networks as crown tribute (McAlister 1984: 162–3, Zavala 1935).

33. This is not to argue that Mesoamericans did not possess personal property or map it; see the discussion in chapter 7.

NOTES TO CHAPTER SEVEN

1. Mercedes maps would have exerted less influence in the fief of Hernán Cortés and his heirs, the marqueses del Valle. Within the marquesado, whose domain included Guaxtepec and Acapistla, where Relaciones Geográficas maps were made, viceregal grants were limited. Although the Cortés family leased lands with procedures mirroring those of crown grants, they did not seem to ask for maps as a regular part of the process (cf. AGN Hospital de Jesús; Chevalier 1963: 131–3; Martin 1985: 17, 25–6).

2. The native elite was quick to catch on, and soon they sought land grants, often to confirm their rights to the lands they already occupied or to receive license to use the lands to raise livestock (Gibson 1964: 262–3). Usually they requested the grants for entire native communities, but sometimes they requested them for themselves alone (Gibson 1964: 265–6). By and large, however, land grants were made to Spaniards out of native lands, thus diminishing indigenous holdings (Melville 1994).

3. There are some exceptions when the request for a merced map was omitted. See, for example, AGN Tierras, vol. 2713, exp. 7, fol. 3.

4. Robertson (1972a: 275) is of a different opinion, classifying this map as European.

5. Normally, the process of the land grant would have been handled by the alcalde mayor or corregidor of that region (and in fact, the initial order of the viceroy was sent to the neighboring alcalde mayor). This merced fell near the border of Xalapa's lands, so the matter was handled by the alcalde mayor of Xalapa, who may have carried out the grant process during an interregnum in the seat of Vera Cruz Vieja's alcalde mayor.

6. In contrast, the naming of the scribe was an elaborate procedure in the merced process. The scribes' names were almost always noted in Relaciones texts, reflecting their important legal roles, which were akin to those of lawyers today (Pezzat Arzave 1990: 113–5).

7. In 1578, the Xalapa painter worked for Francisco Vasco de Andrada, who was then the alcalde mayor of Xalapa, to draw a map of the lands to be granted to Tomás de Herrera. In 1580, he worked for Andrada's successor, Constantino Bravo de Lagunas, to paint the Relación map of Xalapa. In 1587, he was at work again for the alcalde mayor of Xalapa, who at this time was Bernabé Salmerón, to map a land grant for Hernando de Godoy. These dates correspond to the normal term in office for alcaldes mayores and corregidores, which was three years (Haring 1975: 130).

8. The maps of 1578 and 1587 appear to have been written on by the scribes carrying out the diligencias, Gerónimo de Benavides and Melchor de Torres, respectively. The text of the Relación of Xalapa is missing from the Archivo General de Indias in Seville, so it is impossible to say who wrote on the map.

9. While Tehuantepec's shelterless shores made for a less than ideal port, it was one of the few routes linking the Pacific Ocean to the Gulf of Mexico. Radiating from Tehuantepec were roads leading out to Guatemala and inland towards Oaxaca.

10. I proposed in chapter 3 that the artist of this second map of Tehuantepec was the mariner Francisco Stroza Gali, author of the Relaciones Geográficas maps of Tlacotalpa and Coatzocoalco. Even though the Stroza Gali map was probably a "scientific map" drawn with the help of measuring devices, it is far less accurate than the more impressionistic native map of Tehuantepec. The Stroza Gali map understates the size of the lagoon and incorrectly shows the Tehuantepec river emptying into the lagoon, rather than into the ocean.

11. Merced of 1573 (AGN Tierras, vol. 3343, exp. 4, Mapoteca 2378); merced of 1580 (AGN Tierras, vol. 2729, exp. 4, Mapoteca 1903); merced of 1583 (AGN Tierras, vol. 2737, exp. 25, Mapoteca 1964); merced of 1585 (AGN Tierras, vol. 3343, exp. 20, Mapoteca 2391); merced of 1586 (AGN Tierras, vol. 2737, exp. 25, Mapoteca 1965); merced of 1598 (AGN Tierras, vol. 2764, exp. 26, Mapoteca 2084)

12. This same logographic place-name also marks the Mapa de Huilotepec (Barlow 1943). The published version of this indigenous manuscript is a photograph of poor quality. Nonetheless, the artist does not seem to be the same as the artist of the Tehuantepec Relación Geográfica and mercedes maps.

13. In 1573, he painted a map for a land grant case overseen by the alcalde mayor Gaspar Maldonado; in 1580, the year of the Relación Geográfica map, he also painted a merced map for the alcalde mayor Juan de Torres de Lagunas; in 1583 he worked for the alcalde mayor Cristóbal Holgado; in 1585 and 1586 for the alcalde mayor Hernando de Vargas; and in 1598 for the alcalde mayor Gaspar de Vargas.

14. The map of 1573 is written upon by the loose hand of the alcalde mayor Gaspar Maldonado, the map of 1580 and the 1580 Relación map by their scribes, that of 1583 by Cristóbal Holgado, those of 1585 and 1586 by Hernando de Vargas, and that of 1598 by Gaspar de Vargas.

15. In this respect, the map is like the Relación Geográfica map of Cempoala discussed in chapters 5 and 6.

16. Italicized words are written by a second hand, that of the scribe of the grant's diligencias.

17. The text of the map lists the name of the hill as "Huixachtla"; Bernardino de Sahagún gives us a slightly different version, "Uixachtecatl" (1950–1963: book 7, ch. 9–12).

18. Gibson says that *tlalmilli* (land plots of the calpolli) "were often elongated strips" (1964: 269). See also Cline (1986).

19. The unusual trilobed and wasp-waisted shape of these hills seems to have been a regional variation of the tepetl symbol. Similar shapes grace the Codex Misantla (Mena 1911) as well as a merced map of 1573 from nearby Colipa (AGN Tierras, vol. 2672, exp. 18, Mapoteca 1535).

Bibliography

ABBREVIATIONS

AGN Archivo General de la Nación, Mexico City

CDI *Colección de documents inéditos, relativos al descubrimiento, conquista y organización de las antiguas posesiones españolas de América y Oceanía, sacados de los archivos del reino, y muy especialmente del de Indias.*

HMAI *Handbook of Middle American Indians*

INAH Instituto Nacional de Antropología é Historia

PNE *Papeles de Nueva España*

RGS *Relaciones geográficas del siglo XVI*

UNAM Universidad Nacional Autónoma de México

NATIVE PICTORIAL SOURCES

Codex Aubin. 1963. *Historia de la nación mexicana: reproducción a todo color del códice de 1576 (Códice Aubin)* Charles Dibble, ed. and trans. Madrid: Ediciones J. Porrúa Turanzas.

Codex Azcatitlan. 1949. "El Códice Azcatitlan." Robert Barlow, ed. *Journal de la Société des Américanistes,* (n.s.) 38: 101–35, and album.

Codex Becker 1. 1961. *Códices Becker 1/2.* Karl Anton Nowotny, ed. Graz, Austria: Akademische Druck-u. Verlagsanstalt.

Codex Becker 2. 1961. *Códices Becker 1/2.* Karl Anton Nowotny, ed. Graz, Austria: Akademische Druck-u. Verlagsanstalt.

Codex Bodley. 1960. *Interpretación del Códice Bodley 2858.* Alfonso Caso, ed. Mexico City: Sociedad Mexicana de Antropología.

Codex Borbonicus. 1974. *Codex Borbonicus, Bibliothèque de l'Assemblèe nationale, Paris (Y 120): vollstandige Faksimile-Ausg. des Codex im Originalformat.* Jacqueline de Durand-Forest and Karl Anton Nowotny, eds. Codices selecti, vol. 44. Graz, Austria: Akademische Druck-u. Verlagsanstalt.

Codex Borgia. 1976. *Codex Borgia: Biblioteca apostolica vaticana (Messicano Riserva 28).* Codices selecti, vol. 62. Graz, Austria: Akademische Druck-u. Verlagsanstalt.

Codex Boturini. 1964. *Antigüedades de México,* vol. 2: 7–29. José Corona Núñez, ed. Mexico City: Secretaría de Hacienda y Crédito Público.

Codex of Cholula. 1967–1968. "The Codex of Cholula: a preliminary study." Bente Bittmann Simons. *Tlalocan* 5, nos. 3–4: 267–339.

Codex Colombino. 1966. *Interpretación del Códice Colombino/ Interpretation of the Codex Colombino* [by Alfonso Caso]. *Las glosas del Códice Colombino/ The Glosses of the Codex Colombino* [by Mary Elizabeth Smith]. Mexico City: Sociedad Mexicana de Antropología.

Codex Fejérváry-Mayer. 1971. *Codex Fejérváry-Mayer: M 12014 City of Liverpool Museums.* C. A. Burland, ed. Codices selecti, vol. 26. Graz, Austria: Akademische Druck-u. Verlagsanstalt.

Codex of Huamantla. 1984. *Códice de Huamantla.* Carmen Aguilera, ed. Tlaxcala, Mexico: Instituto Tlaxcalteca de la Cultura.

Codex Ixtlilxóchitl. 1976. *Codex Ixtlilxóchitl: Bibliothèque national, Paris (Ms. Mex. 55-710).* Jacqueline de Durand-Forest, ed. Fontes rerum Mexicanarum, vol. 8. Graz, Austria: Akademische Druck-u. Verlagsanstalt.

Codex Kingsborough. 1912. *Códice Kingsborough: Memorial de los Indios de Tepetlaoztoc.* Francisco Paso y Troncoso, ed. Madrid: Fototipia de Hauser y Menet.

Codex Mendoza. 1992. *The Codex Mendoza.* Frances F. Berdan and Patricia R. Anawalt, eds. 4 vols. Berkeley: University of California Press.

Codex Mexicanus. 1952. "Commentaire du Codex Mexicanus, nos. 23–24 de la Bibliothèque Nationale de Paris." Ernest Mengin, ed. *Journal de la Société des Américanistes* (n.s.) 41: 387–498, atlas.

Codex Misantla. 1911. "Códice 'Misantla' publicado é interpretado." Ramón Mena, ed. *Memorias de la sociedad científica "Antonio Alzate"* 30, nos. 10–12: 389–95.

Codex Ríos. 1964. *Antigüedades de México,* vol. 3: 7–313. José Corona Núñez, ed. Mexico City: Secretaría de Hacienda y Crédito Público.

Codex Sánchez Solís [Codex Egerton]. 1965. *Codex Egerton 2895, British Museum, London.* C. A. Burland, ed. Graz, Austria: Akademische Druck-u. Verlagsanstalt.

Codex Selden. 1964. *Interpretación del Códice Selden 3135 (A.2)/Interpretation of the Codex Selden.* Alfonso Caso, ed. Mexico City: Sociedad Mexicana de Antropología.

Codex Vienna [Codex Vindobonensis]. 1963. *Codex Vindobonensis Mexicanus 1: Österreichische Nationalbibliothek, Wien.* Otto Adelhofer, ed. Codices selecti, vol. 5. Graz, Austria: Akademische Druck-u. Verlagsanstalt.

Codex Xolotl. 1980. *Códice Xolotl.* Charles E. Dibble, ed. Mexico City: Instituto de Investigaciones Históricas and UNAM.

Codex of Yanhuitlan. 1940. *Códice de Yanhuitlán.* Wigberto Jiménez Moreno and Salvador Mateos Higuera, eds. Mexico City: Museo Nacional.

Codex Zouche-Nuttall. 1975. *The Codex Nuttall: A Picture Manuscript from Ancient Mexico.* Zelia Nuttall, ed. New York: Dover.

Florentine Codex. 1979. *Códice Florentino.* El Manuscrito 218–220 de la colección Palatino de la Biblioteca Medicea Laurenziana. Bernardino de Sahagún. Facsimile edition, 3 vols. Florence: Gunti Barbéra and AGN.

Historia Tolteca-Chichimeca. 1976. *Historia tolteca-chichimeca.* Paul Kirchhoff, Lina Odena Güemes, and Luis Reyes García, eds. Mexico City: INAH.

Lienzo de Tlaxcala. 1892. *Antigüedades mexicanas publicadas por la Junta Colombina de México en el cuarto centenario del descubrimiento de América.* 2 vols. Alfredo Chavero, ed. Mexico City: Oficina Tipográfica de la Secretaría de Fomento.

Lienzo de Yolotepec. 1890. *Monumentos del arte mexicano,* vol. 3, plate 317. Antonio Peñafiel. Berlin: A. Asher

Lienzo de Zacatepec 1. 1973. *Picture Writing from Ancient Southern Mexico: Mixtec Place Signs and Maps,* pp. 264, 266–90. Mary Elizabeth Smith. Norman: University of Oklahoma Press.

Lienzo de Zacatepec 2. 1973. *Picture Writing from Ancient Southern Mexico: Mixtec Place Signs and Maps,* pp. 298, 300–6. Mary Elizabeth Smith. Norman: University of Oklahoma Press.

Lienzos de Tuxpan. 1970. *Los lienzos de Tuxpan.* José Luis Melgarejo Vivanco. Mexico City: Editorial la Estampa Mexicana.

Mapa de Cuauhtinchan no. 2. 1981. *Los mapas de Cuauhtinchan y la historia cartográfica prehispánica.* Keiko Yoneda. Mexico City: AGN.

Mapa de Huilotepec. 1943. "The Mapa de Huilotepec." Robert H. Barlow. *Tlalocan* 1, no. 2: 155–7.

Mapa de Quinantzin. 1983. *Aztec Art,* plates 41–42, pp. 202–5. Esther Pasztory. New York: Harry Abrams.

Mapa de Sigüenza. 1964. *Catálogo de la colección de códices,* pp. 54–5. John B. Glass. Mexico City: Museo Nacional de Antropología and INAH.

Matrícula de Tributos. 1980. *Matrícula de Tributos (Códice de Mocteuzuma).* Frances Berdan and Jacqueline de Durand-Forest, eds. Codices selecti, vol. 67. Graz, Austria: Akademische Druck-u. Verlagsanstalt.

Tira de Peregrinación. *See* Codex Boturini.

OTHER SOURCES

Academia española. 1726–1739. *Diccionario de la lengua castellana, en que se explica el verdadero sentido de las voces. . . .* 6 vols. Madrid: F. del Hierro.

Acosta, José de. 1940. *Historia natural y moral de las Indias.* Edmundo O'Gorman, ed. Mexico City: Fondo de Cultura Económica.

Actas de cabildo de la ciudad de México. 1889. Vol. 6. Mexico City: Municipio Libre.

Acuña, René. 1981–. see *Relaciones geográficas del siglo XVI* [RGS].

Aguilera, Carmen, ed. 1984. *Códice de Huamantla.* Tlaxcala, Mexico: Instituto Tlaxcalteca de la Cultura.

———. 1990. "Glifos toponímicos en el Mapa de México-Tenochtitlan hacia 1550 (área de Chiconauhtla)." *Estudios de cultura náhuatl* 20: 163–72.

Akerman, James R., and David Buisseret. 1985. *Monarchs, Ministers & Maps.* Chicago: The Newberry Library.

Alcina Franch, José. 1973. "Juan de Torquemada, 1564–1624." HMAI, vol. 13: 256–75. Austin: University of Texas Press.

Alpers, Svetlana. 1983. *The Art of Describing: Dutch Art in the Seventeenth Century.* Chicago: University of Chicago Press.

Alvarado, Francisco de. 1962. *Vocabulario en lengua mixteca.* Wigberto Jiménez Moreno, ed. Mexico City: Instituto Nacional Indigenista and INAH, Secretaría de Educación Pública.

Anales de Cuauhtitlan. n.d. Unpublished translation from the Nahuatl. Thelma Sullivan.

Anawalt, Patricia Rieff. 1981. *Indian Clothing before Cortés.* Norman: University of Oklahoma Press.

Anders, Ferdinand, and Maarten Jansen. 1988. *Schrift und Buch im alten Mexiko.* Graz: Akademische Druck- und Verlagsanstalt.

Anders, Ferdinand, Maarten Jansen, and Gabina Aurora Pérez Jiménez. 1992. *Crónica Mixteca: El rey 8 Venado, Garra de Jaguar, y la dinastía de Teozacualco-Zaachila.* Mexico City: Fondo de Cultura Económica.

Anderson, Arthur J. O. 1963. "Materiales colorantes prehispanicos." *Estudios de cultura náhuatl* 6: 73–83.

Anderson, Arthur J. O., Frances F. Berdan, and James Lockhart. 1976. *Beyond the Codices: The Nahua View of Colonial Mexico.* Berkeley: University of California Press.

Andrews, J. Richard. 1975. *Introduction to Classic Nahuatl.* Austin: University of Texas Press.

Anghiera, Pietro Martire d'. 1964–1965 [1530]. *Décadas del Nuevo Mundo [De Novo Orbe Decades].* 2 vols. Mexico City: J. Porrúa.

Antigüedades de México. 1964. José Corona Núñez, ed. 7 vols. Mexico City: Secretaría de Hacienda y Crédito Público.

Apenas, Ola. 1947. *Mapas antiguos del Valle de México.* Mexico City: UNAM, Instituto de Historia.

Arana Osnaya, Evangelina, and Mauricio Swadesh. 1965. *Los elementos del mixteco antiguo.* Mexico City: Instituto Nacional Indigenista and INAH.

Arnheim, Rudolf. 1983. *The Power of the Center: A Study of Composition in the Visual Arts.* Berkeley and Los Angeles: University of California Press.

Artigas Hernández, Juan B. 1979. *La piel de la arquitectura: Murales de Santa María Xoxoteco.* Mexico City: Escuela Nacional de Arquitectura, UNAM.

Aveni, Anthony. 1980. *Skywatchers of Ancient Mexico.* Austin: University of Texas Press.

Bagrow, Leo. 1985. *History of Cartography.* Revised and enlarged by R. A. Skelton. Chicago: Precedent Publishing.

Bailey, Joyce Waddell. 1972. "Map of Texúpa (Oaxaca, 1579): A Study of Form and Meaning." *Art Bulletin* 54, no. 4: 452–79.

Bakewell, Peter J. 1971. *Silver Mining and Society in Colonial Mexico: Zacatecas, 1546–1700.* Cambridge: Cambridge University Press.

Ballesteros Gaibrois, Manuel. 1973. "Antonio de Herrera, 1549–1625." HMAI, vol. 13: 240–55. Austin: University of Texas Press.

Barlow, Robert H. 1943. "The Mapa de Huilotepec." *Tlalocan* 1, no. 2: 155–7.

———. 1947. "Glifos toponímicos de los códices mixtecos." *Tlalocan* 2, no. 3: 285–6.

———, ed. 1949a. "El Códice Azcatitlan." *Journal de la Société des Américanistes* (n.s.) 38: 101–35, and album.

———. 1949b. "The Extent of the Empire of the Culhua Mexica." *Ibero-Americana,* no. 28. Berkeley and Los Angeles: University of California Press.

Barlow, Robert, and Byron McAfee. 1949. *Diccionario de elementos fonéticos en escritura jeroglífica: Códice Mendocino.* Publicaciones del Instituto de Historia, Primera Serie 9. Mexico City: UNAM.

Barrell, John. 1987. *The Dark Side of the Landscape: The Rural Poor in English Painting 1730–1840.* Cambridge: Cambridge University Press.

Baudot, Georges. 1983. *Utopía e historia en México: Los primeros cronistas de la civilización mexicana (1520–1569).* Vincente González Loscertales, trans. Madrid: Espasa-Calpe.

Becker, Jerónimo. 1917. *Los estudios geográficos en España.* Madrid: J. Ratés.

Berdan, Frances F. 1992. "Glyphic Conventions of the Codex Mendoza." Codex Mendoza (1992), vol. 1: 93–102.

———. n.d. "A Comparitive Analysis of Aztec Tribute Documents: Submitted to the Forty-first International Congress of Americanists, Mexico City, 1974." Unpublished.

Berlin, Heinrich. 1947. *Fragmentos desconocidos del Códice de Yanhuitlan y otras investigaciones mixtecas.* Mexico City: Antigua Librería Robredo.

Berlin, Heinrich, Gonzalo de Balsalobre, and Diego de Hevia y Valdés. 1988. *Idolatría y superstición entre los indios de Oaxaca.* Mexico City: Ediciones Toledo.

Boone, Elizabeth Hill. 1991. "Migration Histories as Ritual Performance." *To Change Place: Aztec Ceremonial Landscapes,* pp. 121–51. Davíd Carrasco, ed. Boulder: University of Colorado Press.

———. 1992a. "The Aztec Pictorial History of the Codex Mendoza." Codex Mendoza (1992), vol. 1: 35–54.

———. 1992b. "Glorious Imperium: Understanding Land and Community in Moctezuma's Mexico." *Moctezuma's Mexico: Visions of the Aztec World,* pp. 159–73. D. Carrasco and E. Matos Montezuma, eds. Nowot: University Press of Colorado.

Boone, Elizabeth Hill, and Walter D. Mignolo, eds. 1994. *Writing without Words: Alternative Literacies in Mesoamerica and the Andes.* Durham, NC: Duke University Press.

Bönisch, Fritz. 1967. "The geometrical accuracy of 16th and 17th century topographical surveys." *Imago Mundi* 21: 62–9.

Borah, Woodrow. 1943. "Silk Raising in Colonial Mexico." *Ibero-Americana,* no. 20. Berkeley and Los Angeles: University of California Press.

———. 1983. *Justice by Insurance: The General Indian Court of Colonial Mexico and the Legal Aides of the Half-Real.* Berkeley: University of California Press.

Borah, Woodrow, and Sherburne F. Cook. 1979. "A Case History of the Transition from Precolonial to the Colonial Period in Mexico: Santiago Tejúpan." *Social Fabric and Spatial Structure in Colonial Latin America,* pp. 409–32. David Robinson, ed. Syracuse: Department of Geography, Syracuse University.

Braun, Georg, and Frans Hogenberg. 1965 [1572]. *Civitates Orbis Terrarum.* 3 vols. Introduction by R. A. Skelton. Amsterdam: Theatrum Orbis Terrarum.

Brotherston, Gordon, and Ana Gallegos. 1990. "El Lienzo de Tlaxcala y el manuscrito de Glasgow (Hunter 242)." *Estudios de cultura náhuatl* 20: 117–40.

Brown, Lloyd Arnold. 1979. *The Story of Maps.* New York: Dover.

Bryson, Norman. 1985. *Vision and Painting: The Logic of the Gaze.* New Haven, CT: Yale University Press.

Buisseret, David, ed. 1992. *Monarchs, Ministers and Maps: The Emergence of Cartography as a Tool of Government in Early Modern Europe.* Chicago: The University of Chicago Press.

Bunbury, Edward H., and Charles R. Beazley. 1911. "Ptolemy." *Encyclopedia Britannica,* 11th edition, vol. 23: 621–4.

Burgoa, Francisco de. 1934. *Geográfica descripción.* 2 vols. Publicaciones del AGN, vols. 25–26. Mexico City: Talleres Gráficos de la Nación.

Burkhart, Louise M. 1989. *The Slippery Earth: Nahua-Christian Moral Dialogue in Sixteenth-Century Mexico.* Tucson: University of Arizona Press.

Burland, C. A. 1960. "The Map as a Vehicle of Mexican History." *Imago Mundi* 15: 11–18.

Burrus, Ernest J., S.J. 1973. "Religious Chroniclers and Historians: A Summary with Annotated Bibliography." HMAI, vol. 13: 138–85. Austin: University of Texas Press.

Byland, Bruce E. and John M. D. Pohl. 1990. "Mixtec Landscape Perception and Archeological Settlement Patterns." *Ancient Mesoamerica* 1: 113–31.

———. 1994. *In the Realm of 8 Deer: The Archaeology of the Mixtec Codices.* Norman: University of Oklahoma Press.

Calnek, Edward. 1972. "Settlement Pattern and Chinampa Agriculture at Tenochtitlan." *American Antiquity* 37, no. 1: 104–15.

———. 1973. "The Localization of the 16th-century Map called the Maguey Plan." *American Antiquity* 38, no. 1: 190–95.

Campbell, R. Joe. 1985. *A Morphological Dictionary of Classic Nahuatl.* Madison, WI: Seminary of Hispanic Medieval Studies.

Campbell, Tony. 1987. "Portolan Charts from the Late Thirteenth Century to 1500." *The History of Cartography: Cartography in Prehistoric, Ancient, and Medieval Europe and the Mediterranean,* vol. 1: 371–463. J. B. Harley and David Woodward, eds. Chicago: University of Chicago Press.

Carrasco, Pedro. 1971. "Social Organization of Ancient Mexico." HMAI, vol. 10: 349–75. Austin: University of Texas Press.

Caso, Alfonso. 1949. "El Mapa de Teozacoalco." *Cuadernos americanos* 47, no. 5: 145–81.

———. 1955. "Vida y aventuras de 4 Viento 'Serpiente de Fuego.'" *Miscelánea de estudios dedicados a Fernando Ortiz,* vol. 1: 289–98. Havana: Ucar, García S.A.

———. 1956. "Los barrios antiguos de Tenochtitlan y Tlatelolco." *Memorias de la Academia Mexicana de la Historia* 15, no. 1: 7–63.

———. 1958. "Lienzo de Yolotepec." *Sobretiro de la Memoria de el Colegio Nacional* 3, no. 4: 41–55.

———. 1977–1979. *Reyes y reinos de la Mixteca.* 2 vols. Mexico City: Fondo de Cultura Económica.

Caso, Alfonso, and Mary Elizabeth Smith. 1966. *Interpretación del Códice Colombino/Interpretation of the Codex Colombino* [by Alfonso Caso]. *Las glosas del Códice Colombino/The Glosses of the Codex Colombino* [by Mary Elizabeth Smith]. Mexico.

Castillo F., Victor M. 1972. "Unidades nahuas de medida." *Estudios de cultura náhuatl* 10: 195–223.

Catálogo de Ilustraciones. 1979–. Centro de Información Gráfica del Archivo General de la Nación. Mexico City: AGN.

Cervantes de Salazar, Francisco. 1984. *México en 1554.* Joaquín García Icazbalceta, trans. Mexico City: UNAM.

Chance, John K. 1978. *Race and Class in Colonial Oaxaca.* Stanford, CA: Stanford University Press.

Chavero, Alfredo, ed. 1892. *Antigüedades mexicanas publicadas por la Junta Colombina de Mexico en el cuarto centenario del descubrimiento de América.* 2 vols. Mexico City: Oficina Tipográfica de la Secretaría de [illegible] [[illegible]].

Chevalier, François. 1963. *Land and Society in Colonial Mexico: The Great Hacienda.* Lesley Byrd Simpson, ed. and trans. Berkeley and Los Angeles: University of California Press.

Clanchy, Michael. 1979. *From Memory to Written Record in England 1066–1307.* Cambridge, MA: Harvard University Press.

Clark, James Cooper, ed. 1938. *Codex Mendoza.* 3 vols. London: Waterlow & Sons.

Clendinnen, Inga. 1991. *Aztecs.* Cambridge: Cambridge University Press.

Cline, Howard F. 1949. "Civil Congregations of the Indians in New Spain, 1598-1606." *Hispanic American Historical Review* 29: 349 69.

———. 1959. "The Patiño Maps of 1580 and Related Documents." *El México antiguo* 9: 633–92.

———. 1962. "The Ortelius Maps of New Spain, 1579, and related contemporary materials, 1560-1610." *Imago Mundi* 16: 98–115.

———. 1964. "The Relaciones Geográficas of the Spanish Indies, 1577-1586." *Hispanic American Historical Review* 44: 341-374.

———. 1966a. "Colonial Mazatec Lienzos and Communities." *Ancient Oaxaca,* pp. 270–97. John Paddock, ed. Stanford, CA: Stanford University Press.

———. 1966b. "The Oztoticpac Lands Map of Texcoco, 1540." *Quarterly Journal of the Library of Congress* 23, no. 2: 75–115.

———. 1972a. "A Census of the Relaciones Geográficas 1579-1612." HMAI, vol. 12: 324–69. Austin: University of Texas Press.

———. 1972b. "Ethnohistorical Regions of Middle America." HMAI, vol. 12: 166–82. Austin: University of Texas Press.

———. 1972c. "The Oztoticpac Lands Map of Texcoco, 1540." *A La Carte: Selected Papers on Maps and Atlases.* Walter W. Ristow, ed. Washington: Library of Congress.

———. 1972d. "The Relaciones Geográficas of the Spanish Indies, 1577–1648." HMAI, vol. 12: 183–242. Austin: University of Texas Press.

———. 1972e. "The Relaciones Geográficas of Spain, New Spain, and the Spanish Indies: An Annotated Bibliography." HMAI, vol. 12: 370–95. Austin: University of Texas Press.

Cline, S. L. 1986. *Colonial Culhuacan 1580–1600: A Social History of an Aztec Town.* Albuquerque: University of New Mexico Press.

Códice Franciscano: *Siglo XVI.* 1941. Nueva colección de documentos para la historia de México, vol. 2. Mexico City: Editorial Salvador Chavez Hayhoe.

Códice Mendieta: Documentos franciscanos, siglos XVI y XVII. 1971. 2 vols. Guadalajara: Edmundo Aviña Levy.

Coe, Michael D. 1962. *Mexico.* New York and Washington: Frederick A. Praeger.

———. 1993. *Breaking the Maya Code.* New York and London: Thames and Hudson.

Coe, Michael D., and Richard Diehl. 1980. *In the Land of the Olmec: The People of the River.* 2 vols. Austin: University of Texas Press.

Coe, Michael D., and Gordon Whittaker. 1982. *Aztec Sorcerers in Seventeenth Century Mexico.* Albany: State University of New York, Institute for Mesoamerican Studies.

Colección de documents inéditos, relativos al descubrimiento, conquista y organización de las antiguas posesiones españolas de América y Oceanía, sacados de los archivos del reino, y muy especialmente del de Indias [CDI]. 1864–1886. J. F. Pacheco, F. de Cardenas, and L. Torres de Mendoza, eds. 42 vols. Madrid.

Collier, George A., Renato I. Rosaldo, and John D. Wirth, eds. 1982. *The Inca and Aztec States, 1400–1800: Anthropology and History.* New York: Academic Press.

Conway, Jill Ker. 1990. *The Road from Coorain.* New York: Vintage Books.

Cook, Sherburne F. 1949. "The Historical Demography and Ecology of the Teotlalpan." *Ibero-Americana,* no. 33. Berkeley and Los Angeles: University of California Press.

Cook, Sherburne F., and Lesley Byrd Simpson. 1948. "The Population of Central Mexico in the Sixteenth Century." *Ibero-Americana,* no. 31. Berkeley and Los Angeles: University of California Press.

Cook de Leonard, Carmen. 1974. "Reconstrucción geográfico-politica del reino de Cozcatlan." *Proceedings of the Forty-first International Congress of Americanists,* vol. 2. 117–30.

Cordero, Karen. n.d. "Mural Painting in the Open Chapel at Actopan and at Santa María Xoxoteco." Paper written for Professor Mary Miller, Yale University, June 1982.

Cortés, Hernán. 1986. *Letters from Mexico.* Anthony Pagden, ed. and trans. New Haven: Yale University Press.

Cortesão, Armando and Avelino Teixeira da Mota, eds. 1960–1962. *Portugaliae monumenta cartographica.* 6 vols. Lisbon.

Cuesta Domingo, Mariano. 1983. *Alonso de Santa Cruz y su obra cosmográfica.* Madrid: Consejo Superior de Investigaciones Científicas, Instituto "Gonzalo Fernández de Oviedo."

Cuevas, Mariano, ed. 1914. *Documentos inéditos del siglo XVI para la historia de México.* Mexico City: Museo Nacional.

Dahlgren, E. W. 1892. *Map of the World by Alonzo de Santa Cruz.* Stockholm: Royal Printing Office.

Dahlgren de Jordan, Barbro. 1954. *La mixteca: su cultura é historia prehispánicas.* Mexico City: Imprenta Universitaria.

Davies, Nigel. 1980. *The Aztecs.* Norman: University of Oklahoma Press.

Derrida, Jacques, 1976. "The Battle of Proper Names." *Of Grammatology,* pp. 107–18. G. Spivak, trans. Baltimore: Johns Hopkins University Press.

Díaz del Castillo, Bernal. 1956. *The Discovery and Conquest of Mexico.* A. P. Maudslay, trans. New York: Farrar, Straus and Cudahy.

Díaz de Salas, Marcelo, and Luis Reyes García. 1970. "Testimonio de la fundación de Santo Tomás Ajusco." *Tlalocan* 6, no. 3: 193–212.

Dibble, Charles E. 1971. "Writing in Central Mexico." HMAI, vol. 10: 322–31. Austin: University of Texas Press.

———. 1973. "The Syllabic-Alphabetic Trend in Mexican Codices." *Acts of the Fortieth International Congress of Americanists* (1972), vol. 1: 373–78.

———, ed. 1980. *Códice Xolotl.* Mexico City: Instituto de Investigaciones Históricas and UNAM.

Dilke, O. A. W. 1971. *The Roman Land Surveyors: An Introduction to the Agrimensores.* Newton Abbot: David and Charles.

———. 1987. "Itineraries and Geographical Maps in the Early and Late Roman Empires." *The History of Cartography: Cartography in Prehistoric, Ancient, and Medieval Europe and the Mediterranean,* vol. 1: 234–57. J. B. Harley and David Woodward, eds. Chicago: University of Chicago Press.

Domínguez, Miguel. 1943. *Coscomatepec de Bravo: Apuntes para la historia veracruzana.* Mexico.

Durán, fray Diego. 1977. *Book of Gods and Rites and the Ancient Calendar.* Fernando Horcasitas and Doris Heyden, eds. and trans. Norman: University of Oklahoma Press.

———. 1984. *Historia de las indias de Nueva España é islas de la tierra firme.* 2 vols. Angel Ma. Garibay K., ed. 2nd ed. Mexico City: Editorial Porrúa.

Dyckerhoff, Ursula. 1985. "Umerziehung in Neu-Spanien." *Mexicon* 7: 10–13.

Edmonson, Munro S., ed. 1974. *Sixteenth-Century Mexico: The Work of Sahagún.* Albuquerque: University of New Mexico Press.

Edwards, Clinton R. 1969. "Mapping by Questionnaire: An Early Spanish Attempt to Determine New World Geographical Positions." *Imago Mundi* 23: 17–28.

Edwards, Emily. 1966. *Painted Walls of Mexico.* Austin: University of Texas Press.

Elliott, J. H. 1970. *The Old World and the New, 1492–1650.* Cambridge: Cambridge University Press.

Elsasser, A. B. 1974? *The Alonso de Santa Cruz Map of Mexico City and Environs. Dating from 1550.* Berkeley: Lowie Museum of Anthropology.

Emmart, Emily Walcott. 1940. *The Badianus Manuscript (Codex Barberini, Latin 241), Vatican Library: An Aztec Herbal of 1552.* Baltimore: Johns Hopkins University Press.

Ettlinger, L. D. 1952. "A Fifteenth-Century View of Florence." *Burlington Magazine* 94: 160–7.

Faber, Harold. 1993. "On Patrol Against Enemies Foreign or Domestic: Where 3 States Meet, a Connecticut Town's Perambulators Remain Ever Vigilant." *The New York Times,* 6 January, B1, B4.

Farriss, Nancy M. 1984. *Maya Society under Colonial Rule: The Collective Enterprise of Survival.* Princeton, NJ: Princeton University Press.

Furst, Jill Leslie. 1977. "The Tree Birth Tradition in the Mixteca, Mexico." *Journal of Latin American Lore* 4, no. 1: 183–226.

———. 1978. *Codex Vindobonensis Mexicanus I: A Commentary.* Albany: State University of New York, Institute for Mesoamerican Studies.

Galarza, Joaquín. 1967. "Prénoms et noms de lieux exprimés par des glyphes et des attributs chrétiens dans les manuscrits pictographiques mexicains." *Journal de la Société des Américanistes,* n.s., 56: 533–85.

———. 1972. *Lienzos de Chiepetlan.* Mexico City: Mission Archéologique et Ethnologique Française au Mexique.

———. 1979. *Estudios de escritura indígena tradicional azteca-náhuatl.* Mexico: AGN.

Galarza, Joaquín, and B. Torres. 1986. "Acatl: carrizo, signo de la escritura azteca: el glifo y la planta." *Journal de la Société des Américanistes,* n.s., 72: 33–54.

García Icazbalceta, Joaquín. 1886. *Bibliografía mexicana del siglo XVI: Primera parte.* Mexico: Librería de Andrade y Morales.

García Zambrano, Angel J. 1992. "El poblamiento de México en la época de contacto, 1520-1540." *Mesoamérica* 24 (Dec.): 239–96.

Garibay K., Angel María. 1953–1954. *Historia de la literatura náhuatl.* 2 vols. Mexico City: Editorial Porrúa.

———. 1979. *Teogonía é historia de los mexicanos: Tres opúsculos del siglo XVI.* Mexico City: Editorial Porrúa.

Gerhard, Peter. 1968. "Descripciones geográficas (pistas para investigadores)." *Historia mexicana* 68: 618–27.

———. 1972a. "Colonial New Spain, 1519–1786: Historical Notes on the Evolution of Minor Political Jurisdictions." HMAI, vol. 12: 63–137. Austin: University of Texas Press.

———. 1972b. *A Guide to the Historical Geography of New Spain 1519–1821.* Cambridge: Cambridge University Press.

———. 1977. "Congregaciones de indios en la Nueva España antes de 1570." *Historia mexicana* 103: 347–95.

———. 1986. *Geografía histórica de la Nueva España 1519–1821.* Stella Mastrangelo, trans. Mexico City: UNAM. (Revised, Spanish edition of Gerhard 1972b.)

Gibson, Charles. 1952. *Tlaxcala in the Sixteenth Century.* New Haven, CT: Yale Historical Publications, Miscellany, 56.

———. 1955. "The Transformation of the Indian Community of New Spain 1500–1800." *Journal of World History* 2: 581–607.

———. 1964. *The Aztecs under Spanish Rule.* Stanford, CA: Stanford University Press.

Gillespie, Susan D. 1989. *The Aztec Kings: The Construction of Rulership in Mexica History.* Tucson: University of Arizona Press.

———. n.d. "Completing the Past: Triples in Aztec Tradition." Paper given at Dumbarton Oaks Symposium: Native Traditions in the Postconquest World, October 3, 1992, Washington, D.C.

Glass, John B. 1964. *Catálogo de la colección de códices.* Mexico City: Museo Nacional de Antropología and INAH.

———. 1975a. "Annotated References." HMAI, vol. 15: 537–724. Austin: University of Texas Press.

———. 1975b. "A Survey of Native Middle American Pictorial Manuscripts." HMAI, vol. 14: 3–80. Austin: University of Texas Press.

Glass, John B., with Donald Robertson. 1975. "A Census of Native Middle American Pictorial Manuscripts." HMAI, vol. 14: 81–252. Austin: University of Texas Press.

Gómez Canedo, Lino. 1977. *Evangelización y conquista: Experiencia franciscana en hispanoamérica.* Mexico City: Editorial Porrúa.

———. 1982. *La educación de los marginados durante la época colonial.* Mexico City: Editorial Porrúa.

———. 1983. "Franciscans in the Americas: A Comprehensive View." *Franciscan Presence in the Americas,* pp. 5–45. Francisco Morales, ed. Washington, D.C.: Academy of American Franciscan History.

González de Cossío, Francisco, ed. 1952. *El libro de las tasaciones de pueblos de la Nueva España, siglo XVI.* Mexico City: AGN.

González Muñoz, María del Carmen. 1971. "Estudio Preliminar." *Geografía y descripción universal de las Indias,* pp. v–xlviii. Juan López de Velasco. Biblioteca de Autores Españoles, vol. 248. Marcos Jiménez de la Espada, ed. Madrid: Atlas.

Goodman, David C. 1988. *Power and Penury: Government, Technology and Science in Philip II's Spain.* Cambridge: Cambridge University Press.

Goody, Jack. 1987. *The Interface between the Written and the Oral.* Cambridge: Cambridge University Press.

Gould, Peter, and Rodney White. 1974. *Mental Maps.* Harmondsworth: Penguin Books.

Gruzinski, Serge. 1987. "Colonial Indian maps in sixteenth-century Mexico." *Res: Anthropology and Aesthetics* 13: 46–61.

———. 1993. *The Conquest of Mexico.* E. Corrigan, trans. Cambridge: Polity Press.

Guevara, Felipe de. 1788. *Comentarios de la pintura. . .* .Madrid: Don G. Ortega, hijos de Ibarra y compañia.

Guzman, Eulalia. 1939. "The Art of Map-making among the Ancient Mexicans." *Imago Mundi* 3: 1–6.

Haggard, J. Villasana. 1941. *Handbook for Translators of Spanish Historical Documents.* Austin: Archive Collection, The University of Texas.

Hanke, Lewis. 1965. *The Spanish Struggle for Justice in the Conquest of America.* Boston: Little, Brown.

Haring, C. H. 1975. *The Spanish Empire in America.* New York: Harcourt Brace Jovanovich.

Harley, J. B. 1988. "Silences and Secrecy: the Hidden Agenda of Cartography in Early Modern Europe." *Imago Mundi* 40: 57–76.

Harley, J. B. and David Woodward, eds. 1987. *The History of Cartography: Cartography in Prehistoric, Ancient, and Medieval Europe and the Mediterranean,* vol. 1. Chicago: University of Chicago Press.

Harvey, H. R. 1972. "The Relaciones Geográficas, 1579–1586: Native Languages." HMAI, vol. 12: 279–323. Austin: University of Texas Press.

———. 1991. "The Oztoticpac Lands Map: A Reexamination." *Land and Politics in the Valley of Mexico: A Two Thousand Year Perspective,* pp. 163–85. H. R. Harvey, ed. Albuquerque: University of New Mexico Press.

Harvey, P. D. A. 1980. *The History of Topographical Maps: Symbols, Pictures and Surveys.* London and New York: Thames and Hudson.

———. 1987. "Local and Regional Cartography in Medieval Europe." *The History of Cartography: Cartography in Prehistoric, Ancient, and Medieval Europe and the Mediterranean,* vol. 1: 464–501. J. B. Harley and David Woodward, eds. Chicago: University of Chicago Press.

———. 1994. *Maps in Tudor England.* Chicago: University of Chicago Press.

Haskett, Robert. 1987. "Indian Town Government in Colonial Cuernavaca: Persistence, Adaptation, and Change." *Hispanic American Historical Review* 67, no. 2: 203–31.

———. 1988. "Living in Two Worlds: Cultural Continuity and Change among Cuernavaca's Colonial Indigenous Ruling Elite." *Ethnohistory* 35, no. 1: 34–59.

———. 1991. *Indigenous Rulers: An Ethnohistory of Town Government in Colonial Cuernavaca.* Albuquerque: University of New Mexico Press.

Hassig, Ross. 1988. *Aztec Warfare.* Norman: University of Oklahoma Press.

Haverkamp-Begemann, Egbert. 1969. "The Spanish Views of Anton van den Wyngaerde." *Master Drawings,* vol. 7: 375–99.

Helgerson, Richard. 1986. "The Land Speaks: Cartography, Chorography, and Subversion in Renaissance England." *Representations,* vol. 16: 50–85.

Herrera y Tordesillas, Antonio de. 1944? [1601–1615]. *Historia general de los hechos de los Castellanos en las islas i tierra-firme del Mar Océano.* 15 vols. Asunción del Paraguay: Editorial Guarania.

Hoberman, Louisa S. 1972. "City Planning in Spanish Colonial Government: The Response of Mexico City to the Problem of Floods, 1607–1637." Ph.D. diss., Columbia University.

Hunt, Eva. 1972. "Irrigation and the Socio-Political Organization of Cuicatec Cacicazgos." *The Prehistory of the Tehuacan Valley,* vol. 4: 162–259. Austin: University of Texas Press.

Instrucciones que los virreyes de Nueva España dejaron a sus sucesores. 1867. Mexico City: Imprenta Imperial.

Ixtlilxóchitl, Fernando de Alva. 1985. *Obras históricas.* Edmundo O'Gorman, ed. 2 vols. Mexico City: UNAM, Imprenta Universitaria.

Jansen, Maarten E. R. G. N. 1979. "Apoala y su importancia para la interpretación de los códices Vindobonensis y Nuttall," *Actes du XLII Congrès International de Américanistes,* vol. 7: 161–72.

———. 1982. *Huisi Tacu: Estudio interpretativo de un libro mixteco antiguo: Codex Vindobonensis Mexicanus 1.* Amsterdam: CEDLA Incidentele Publicaties 24.

Jervis, W. W. 1937. *The World in Maps: A Study in Map Evolution.* New York: Oxford University Press.

Jiménez de la Espada, Marcos. 1885–1897. *Relaciones geográficas de Indias: Perú.* 4 vols. Madrid: Ministerio de Fomento.

———. 1965. "Antecedentes." *Relaciones geográficas de Indias—Perú.* Biblioteca de Autores Españoles, vol. 183: 5–117.

Jiménez Moreno, Wigberto, ed. 1962. *Vocabulario en lengua mixteca.* Mexico City: Instituto Nacional Indigenista and INAH, Secretaría de Educación Pública.

Jiménez Moreno, Wigberto, and Salvador Mateos Higuera, eds. 1940. *Códice de Yanhuitlán.* Mexico City: Museo Nacional.

Kagan, Richard L. 1986. "Philip II and the Art of the Cityscape." *Journal of Interdisciplinary History* 17, no. 1: 115–35.

———, ed. 1989. *Spanish Cities of the Golden Age: The Views of Anton van den Wyngaerde.* Berkeley: University of California Press.

———. n.d. "*Urbs* and *Civitas* in Sixteenth- and Seventeenth-Century Spain." Nebenzahl Lecture, Newberry Library, Chicago, 1991.

Kamen, Henry. 1983. *Spain 1469–1714: A Society of Conflict.* London and New York: Longman.

Karttunen, Frances. 1983. *An Analytical Dictionary of Nahuatl.* Austin: University of Texas Press.

Kelemen, Pál. 1951. *Baroque and Rococo in Latin America.* New York: Macmillan.

Kirchhoff, Paul. 1943. "Mesoamerica." *Acta Americana* 1: 92–107.

———. 1956. "Land Tenure in Ancient Mexico." *Revista mexicana de estudios antropológicos* 14, no. 1: 351–61.

Kirchhoff, Paul, Lina Odena Güemes, and Luis Reyes García. 1976. *Historia tolteca-chichimeca.* Mexico City: INAH.

Klor de Alva, Jorge. 1989. "Language, Politics, and Translation: Colonial Discourse and Classic Nahuatl in New Spain." *The Art of Translation: Voices from the Field,* pp. 143–62. R. Warren, ed. Boston: Northeastern University Press.

Koeman, C. 1964. *The History of Abraham Ortelius and his Theatrum Orbis Terrarum.* Lausanne: Sequoia S.A.

Kubler, George. 1942. "Mexican Urbanism in the Sixteenth Century." *The Art Bulletin* 24, no. 2: 160–71.

———. 1948. *Mexican Architecture of the Sixteenth Century.* 2 vols. New Haven, CT: Yale University Press.

———. 1961. "On the Colonial Extinction of the Motifs of Pre-Columbian Art." *Essays in Pre-Columbian Art and Archaeology,* pp. 14–34, 485–86. Samuel K. Lothrop, ed. Cambridge: Harvard University Press.

———. 1984. *The Art and Architecture of Ancient America: The Mexican, Maya, and Andean Peoples.* 3rd edition. New York: Penguin Books.

———. 1985. "The Colonial Plan of Cholula." *Studies in Ancient American and European Art: The Collected Essays of George Kubler,* pp. 92–101. Thomas Reese, ed. New Haven, CT: Yale University Press.

Kubler, George, and Charles Gibson. 1951. *The Tovar Calendar.* Memoirs of the Connecticut Academy of Arts and Sciences, vol. 11. New Haven, CT: The Connecticut Academy of Arts and Sciences.

Kubler, George, and Martín Soria. 1959. *Art and Architecture in Spain and Portugal and Their American Dominions, 1500 to 1800.* Harmondsworth: Penguin Books.

Lafuente, Antonio. 1984. *La geometrización de la tierra: observaciones y resultados de la expedición geodesica hispanofrancesa al virreinato del Peru (1735–1744).* Madrid: Consejo Superior de Investigaciones Científicas, Instituto "Arnau de Villanova."

Lamb, Ursula. 1974. "The Spanish Cosmographic Juntas of the Sixteenth Century." *Terrae Incognitae* 6: 51–62.

Las Casas, fray Bartolomé de. 1967. *Apologética historia sumaria.* 2 vols. Edmundo O'Gorman, ed. Mexico City: UNAM, Instituto de Investigaciones Históricas.

———. 1974. *In Defense of the Indians.* Stafford Poole, trans. DeKalb: Northern Illinois University Press.

———. 1986. *Historia de las Indias.* 3 vols. Agustín Millares Carlo, ed. 2nd ed. Mexico City: Fondo de Cultura Económica.

Latorre y Setien, Germán. 1913. "Los geógrafos españoles del siglo XVI." *Boletín del Instituto de Estudios Americanistas* 1, no. 2: 29–51.

———. 1916. *La cartografía colonial americana.* Seville: Establecimiento Tipográfico de la Guía Oficial.

Leibsohn, Dana. n.d. "Between Church and Hill: The Territory of Christian Signs on Native Maps." Paper presented at the Forty-seventh International Congress of Americanists, July 11, 1991. New Orleans.

———. 1993. "The *Historia Tolteca-Chichimeca:* Recollecting Identity in a Nahua Manuscript." Ph.D. diss., University of California, Los Angeles.

Lenz, Hans. 1949. "Las fibras y las plantas del papel indígena mexicano." *Cuadernos americanos* 45, no. 3: 157–69.

León Portilla, Miguel. 1982. *Aztec Thought and Culture.* Jack Emory Davis, trans. Norman: University of Oklahoma Press.

León-Portilla, Miguel, and Carmen Aguilera. 1986. *Mapa de México-Tenochtitlan y sus contornos hacia 1550.* Mexico City: Celanese Mexicana, S.A.

Leonard, Irving. 1964. *Books of the Brave.* New York: Gordian Press.

Lind, Michael. 1994. "The Obverse of the Codex of Cholula: Defining the Borders of the Kingdom of Cholula." *Caciques and Their People: A Volume in Honor of Ronald Spores,* pp. 87–100. J. Marcus and J. F. Zeitlin, eds. Anthropological Papers, Museum of Anthropology, University of Michigan, no. 89.

Linné, S. 1948. *El valle y la ciudad de México en 1550.* Stockholm, Sweden: Statens Etnografiska Museum.

Liss, Peggy K. 1984. *Mexico Under Spain, 1521–1556: Society and the Origins of Nationality.* Chicago: University of Chicago Press.

Lockhart, James. 1991. *Nahuas and Spaniards: Postconquest Central Mexican History and Philology.* UCLA Latin American Studies, vol. 76. Stanford, CA: Stanford University Press.

———. 1992. *The Nahuas After the Conquest.* Stanford, CA: Stanford University Press.

Lockhart, James, Frances Berdan, and Arthur J. O. Anderson. 1986. *The Tlaxcalan Actas.* Salt Lake City: University of Utah Press.

Lombardo de Guzmán, don Guillén. 1934. "Reparto y medida de tierras en el siglo XVII." *Boletín del Archivo General de la Nación* 5, no. 3 (May–June): 321–31.

López de Gómara, Francisco. 1966. *Cortés: The Life of the Conquerer by His Secretary.* Lesley Bird Simpson, trans. and ed. Berkeley: University of California Press.

López de Velasco, Juan. 1894. *Geografía y descripción universal de las Indias, recopilada por el Cosmógrafo-cronista . . . desde el año 1571 al de 1574, publicada por primera vez en el Boletín de la Sociedad Geográfica de Madrid.* Justo Zaragoza, ed. Madrid: Fortanet.

———. 1971. *Geografía y descripción universal de las Indias.* Biblioteca de Autores Españoles, vol. 248. Marcos Jiménez de la Espada, ed. Madrid: Atlas.

López Piñero, J. M. 1979. *Ciencia y técnica en la sociedad española de los siglos XVI y XVII.* Barcelona: Labor Universitaria.

McAlister, Lyle N. 1984. *Spain and Portugal in the New World: 1492–1700.* Minneapolis: University of Minnesota Press.

McAndrew, J. 1965. *The Open-Air Churches of Sixteenth Century Mexico.* Cambridge, MA: Harvard University Press.

Manrique, Jorge Alberto. 1982. "La estampa como fuente del arte en la Nueva España." *Anales del Instituto de Investigaciones Estéticas,* UNAM, vol. 50, no. 1: 55–60.

Marcus, Joyce. 1992. *Mesoamerican Writing Systems: Propaganda, Myth and History in Four Ancient Civilizations.* Princeton, NJ: Princeton University Press.

Martin, Cheryl English. 1985. *Rural Society in Colonial Morelos.* Albuquerque: University of New Mexico Press.

Martínez, Andrea. 1990. "Las pinturas del manuscrito de Glasgow y el Lienzo de Tlaxcala." *Estudios de cultura náhuatl* 20: 141–62.

Mathes, W. Michael. 1985. *The America's First Academic Library: Santa Cruz de Tlatelolco.* Sacramento: California State Library Foundation.

Maudslay, Alfred P. 1913. "A Note on the Position and Extent of the Great Temple Enclosure of Tenochtitlan, and the Position, Structure and Orientation of the Teocalli of Huitzilopochtli." *Acts of the International Congress of Americanists,* vol. 18, part 2: 173–75. London.

Medina, José Toribio. 1912. *La imprenta en México (1539–1821).* 8 vols. Santiago de Chile: Casa de Medina.

Melgarejo Vivanco, José Luis. 1970. *Los lienzos de Tuxpan.* México: Editorial la Estampa Mexicana.

Melville, Elinor G. K. 1994. *A Plague of Sheep: Environmental Consequences of the Conquest of Mexico.* New York and Cambridge: Cambridge University Press.

Mena, Ramón. 1911. "Códice 'Misantla' publicado é interpretado." *Memorias de la sociedad científica "Antonio Alzate,"* vol. 30, nos. 10–12: 389–95.

Mendieta, fray Gerónimo de. 1971. *Historia eclesiástica indiana.* Joaquin García Icazbalceta, ed. Mexico City: Editorial Porrúa.

Menéndez-Pidal, Gonzalo. 1944. *Imagen del mundo hacia 1570.* Madrid: Consejo de la Hispanidad.

Mignolo, Walter D. 1987. "El mandato y la ofrenda: La *Descripción de la ciudad y provincia de Tlaxcala,* de Diego Muñoz Camargo, y las Relaciones de Indias." *Nueva revista de filología hispánica,* 35, no. 2: 451–84.

Mignolo, Walter D. 1991. "Colonial Situations, Geographical Discourses and Territorial Representations: Toward a Diatopical Understanding of Colonial Semiosis." *Dispositio* 14, no. 36–38: 93–140.

Miller, Arthur. 1991. "Transformations of Time and Space: Oaxaca, Mexico, circa 1500–1700." *Images of Memory: On Remembering and Representation,* pp. 141–257. S. Kuchler and W. Melion, eds. Washington, D.C.: Smithsonian Institution Press.

Molina, Alonso de. 1977 [1571]. *Vocabulario en lengua castellana y mexicana y mexicana y castellana.* Mexico City: Editorial Porrúa.

Montellano, Marcela. 1988. "Molino de papel de Culhuacán." *Boletín del Consejo de Arqueología.* Mexico City: INAH.

Montêquin, François-Auguste de. 1974. "Maps and Plans of Cities and Towns in Colonial New Spain, The Floridas, and Louisiana: Selected Documents from the Archivo General de Indias of Sevilla." Ph.D. diss., University of New Mexico, Albuquerque.

———. 1990. "The Planning of Spanish Cities in America: Characteristics, Classification, and Main Urban Features." *The Working Paper Series of the Society for American City and Regional Planning History,* vol. 212. Denver: The Society for American City and Regional Planning History.

Motolinía, fray Toribio de Benavente. 1971. *Memoriales o libro de las cosas de la Nueva España y de los naturales de ella.* Edmundo O'Gorman, ed. Mexico City: UNAM, Instituto de Investigaciones Históricas.

Moyssén, Xavier. 1965. "Pinturas murales en Epazoyucan." *INAH Boletín* 22: 20–7.

Multhauf, Robert P. 1958. "Early Instruments in the History of Surveying: Their Use and Their Invention." *Surveying and Mapping* 18, no. 4: 399–415.

Mundy, Barbara E. 1993. "The Maps of the Relaciones Geográficas of New Spain, 1579– c. 1584: Native Mapping in the Conquered Land." Ph.D. diss., Yale University.

———. n.d. "Mesoamerican Cartography." *The History of Cartography,* vol. 2, book 3. David Woodward and G. Malcolm Lewis, eds. Chicago: University of Chicago Press, forthcoming.

Muñoz Camargo, Diego. 1981. *Descripción de la ciudad y provincia de Tlaxcala de las Indias y del mar océano para el buen gobierno y ennoblecimiento dellas* [Relación Geográfica of Tlaxcala]. René Acuña, ed. Mexico City: UNAM and Instituto de Investigaciones Filológicas.

———. 1986. *Historia de Tlaxcala.* Germán Vázquez, ed. Crónicas de América, 26. Madrid: Historia 16.

Musset, Alain. 1989. "Etudes de deux cartes de Relations geographiques de 1580: Cuzcatlan." Descifre de las escrituras mesoamericanas: Codices, pinturas, estatuas, ceramica. *BAR International Series,* vol. 518, pp. 99–121.

Nesmith, Robert I. 1955. "The Coinage of the First Mint of the Americas at Mexico City, 1536–1572." *Numismatic Notes and Monographs,* vol. 131. New York: American Numismatic Society

Nicholson, H. B. 1966. "The Problem of the Provenience of the Members of the 'Codex Borgia Group': A Summary." *Summa antropológica en homenaje a Roberto J. Weitlaner,* pp. 145–158. Mexico City: INAH.

. 1971a. "Pre-Hispanic Central Mexican Historiography." *Investigaciones contemporáneas sobre Historia de México,* pp. 38–95. Austin: University of Texas Press.

———. 1971b. "Religion in Pre-Hispanic Central Mexico." HMAI, vol. 10: 395–446. Austin: University of Texas Press.

———. 1973. "Phoneticism in the Late Pre-Hispanic Central Mexican Writing System." *Mesoamerican Writing Systems,* pp. 1–46. Elizabeth P. Benson, ed. Washington, D.C.: Dumbarton Oaks.

———. 1982. "The Mixteca-Puebla Concept Revised." *The Art and Iconography of Late Post-Classic Central Mexico,* pp. 227–54. Elizabeth H. Boone, ed. Washington, D.C.: Dumbarton Oaks.

Normann, Anne Whited. 1985. "Testerian Codices: Hieroglyphic Catechisms for Religious Conversion of the Natives of New Spain." Ph.D. diss., Tulane University.

Nowotny, Karl Anton. 1959. "Die Hieroglyphen des Codex Mendoza: Der Bau einer mittelamerikanischen Wortschrift." *Mitteilungen aus dem Museum für Völkerkunde in Hamburg,* vol. 25: 97–113.

———, ed. 1961a. *Códices Becker 1/2.* Graz, Austria: Akademische Druck-u. Verlagsanstalt.

———. 1961b. *Tlacuilolli. Die mexikanischen Bilderhandschriften. Stil und Inhalt mit einem Katalog der Codex-Borgia-Gruppe.* Ibero-Amerikanische Bibliothet, Monumenta Americana, 3. Berlin.

Nuttall, Zelia. 1921. "Royal Ordinances Concerning the Laying Out of New Towns." *Hispanic American Historical Review* 4: 743–53.

———. 1922. "Royal Ordinances Concerning the Laying Out of New Towns." *Hispanic American Historical Review* 5: 249–54.

———. 1928. "La observación del paso del sol por el zenit por los antiguos habitantes de la América tropical; su restablecimiento como un factor patriótico y educativo en las escuelas modernas situadas en la misma zona." *Publicaciones de la Secretaría de Educación Pública* 17, no. 20.

———, ed. 1975. *The Codex Nuttall: A Picture Manuscript from Ancient Mexico.* New York: Dover.

O'Gorman, Edmundo. 1940. "Estudio Preliminar." *Historia natural y moral de las Indias.* pp. ix–lxxxv. José de Acosta. Mexico City: Fondo de Cultura Económica.

———. 1961. *The Invention of America.* Bloomington: Indiana University Press.

Oettinger, Marion, Jr. 1983. *Lienzos Coloniales: una exposición de pinturas de terrenos communales de México, siglos XVI–XIX.* Mexico City: UNAM, Instituto de Investigaciones Antropológicas.

Oettinger, Marion, Jr., and Fernando Horcasitas. 1982. "The Lienzo of Petlacala: A Pictorial Document from Guerrero, Mexico." *Transactions of the American Philosophical Society,* (n.s.) vol. 72, part 7.

Offner, Jerome A. 1983. *Law and Politics in Aztec Texcoco.* Cambridge and New York: Cambridge University Press.

Orozco y Berra, Manuel. 1864. *Geografía de las lenguas y carta etnográfica de México.* Mexico City: Imprenta de J.M. Andrade y F. Escalante.

———. 1871. *Materiales para una cartografía mexicana.* Mexico City: Imprenta del Gobierno.

———. 1881. *Apuntes para la historia de la geografía en México.* Anales del Ministerio de Fomento de la República Mexicana, vol. 4. Mexico City: Imprenta de Francisco Díaz de Léon.

Ortelius, Abraham. 1991 [1579]. Theatrum Orbis Terrarum. Facsimile of 1595 edition. Florence: Giunti.

Ortiz, Alfonso. 1972. "Ritual Drama and the Pueblo World View." *New Perspectives on the Pueblos,* pp. 135–61. Alfonso Ortiz, ed. Albuquerque, University of New Mexico Press.

Ortroy, Fernand Gratien van. 1963. *Bibliographie de l'oeuvre de Pierre Apian, par F. van Ortroy.* Amsterdam: Meridian Publishing.

Oviedo y Valdés, Gonzalo Fernández. 1944–1945 [1535, 1557]. *Historia general y natural de las Indias.* 14 vols. Asunción de Paraguay: Editorial Guarania.

Owens, Craig. [illegible] no. 10 (October). [illegible]

Paddock, John. 1982. "Confluence in Zapotec and Mixtec Ethnohistories: The 1580 Mapa de Macuilxochitl." Native American Ethnohistory. Joseph W. and Judith Bradley Whitecotton, eds. *Papers in Anthropology* 23, no. 2: 345–57. Norman: Department of Anthropology, University of Oklahoma.

———. 1983. *Lord 5 Flower's Family: Rulers of Zaachila and Cuilapan.* Vanderbilt University Publications in Anthropology, no. 29. Nashville: Vanderbilt University.

Palm, Erwin Walter. 1975. "Estilo cartográfico y tradición humanista en las Relaciones Geográficas de 1579–1581." *Acts of the Fortieth International Congress of Americanists* (1972), vol. 3: 195–203.

Papeles de Nueva España [PNE]. 1905–. Segunda Serie: Geografía y Estadística, vols. 1, 3–7. Francisco del Paso y Troncoso, ed. Madrid: Impresores de la Real Casa.

Parker, Geoffrey. 1987. *The Army of Flanders and the Spanish Road, 1567–1659.* Cambridge: Cambridge University Press.

———. 1992. "Maps and Ministers: The Spanish Hapsburgs." *Monarchs, Ministers and Maps: The Emergence of Cartography as a Tool of Government in Early Modern Europe,* pp. 124–52. David Buisseret, ed. Chicago: The University of Chicago Press.

Parmenter, Ross. 1982. *Four Lienzos of the Coixtlahuaca Valley.* Studies in Pre-Columbian Art & Archaeology, no. 26. Washington D.C.: Dumbarton Oaks.

Parsons, Jeffrey R. 1970. "An Archaelogical Evaluation of the Códice Xolotl." *American Antiquity* 35, no. 4: 431–40.

———. 1976. "The Role of Chinampa Agriculture in the Food Supply of Aztec Tenochtitlan." *Cultural Change and Continuity: Essays in Honor of James Bennett Griffin,* pp. 233–57. Charles E. Cleland, ed. New York: Academic Press.

Pasztory, Esther. 1983. *Aztec Art.* New York: Harry Abrams.

Peñafiel, Antonio. 1885. *Nombres geográficos de México.* Mexico City: Oficina tipográfica de la Secretaría de Fomento.

———. 1900. *Códice Mixteco: Lienzo de Zacatepec.* Mexico City: Secretaría de Fomento.

———. 1890. *Monumentos del arte mexicano.* 3 vols. Berlin: A. Asher.

———. 1897. *Nomenclatura geográfica de México: Etimologías de los nombres de lugar.* Mexico City: Oficina tipográfica de la Secretaría de Fomento.

Peterson, David A. 1987. "The Real Cholula." *Notas mesoamericanas* 10: 71–117.

Peterson, Jeanette Favrot. 1985. "The Garden Frescos of Malinalco: Utopia, Imperial Policy, and Acculturation in Sixteenth-Century Mexico." Ph.D. diss., University of California, Los Angeles.

———. 1993. *The Paradise Garden Murals of Malinalco.* Austin: University of Texas Press.

Pezzat Arzave, Delia. 1990. *Elementos de paleografía novohispana.* Mexico City: Facultad de Filosofía y Letras, UNAM.

Phelan, John Leddy. 1970. *The Millennial Kingdom of the Franciscans in the New World.* 2nd rev. ed. Berkeley and Los Angeles: University of California Press.

Pierce, Donna L. 1981. "Identification of the Warriors in the Frescos of Ixmiquilpan." *Research Center for the Arts Review* 4: 1–8.

Piña Chán, Román. 1972. *Primer informe de exploraciones arqueológicas.* Toluca: Gobierno de Estado.

———. 1973. *Segundo informe de exploraciones arqueológicas.* Toluca: Gobierno de Estado.

———, ed. 1975. *Teotenango, el antiguo lugar de la muralla.* 2 vols. Mexico City: Dirección de Turismo.

Pinto, John. 1976. "Origins and Development of the Ichnographic City Plan." *Journal of the Society of Architectural Historians* 35: 35–50.

Pita Moreda, María Teresa. 1988. "La expansión de la Orden por Nueva España." *Actas del primero Congreso Internacional sobre los Dominicos y el Nuevo Mundo,* pp. [illegible]. Madrid: Editorial Deimos.

Pohl, John M. D. 1994. *The Politics of Symbolism in the Mixtec Codices.* Vanderbilt University Publications in Anthropology, no. 46. Nashville: Vanderbilt University.

Prem, Hanns J. 1970. "Aztec Hieroglyphic Writing System—Possibilities and Limits." *Acts of the Thirty-eighth International Congress of Americanists* (1968), vol. 2: 159–65.

———. 1992. "Aztec Writing," HMAI supplement, vol. 5: 53–69. Austin: University of Texas Press.

Price, Derek J. 1955. "Medieval Land Surveying and Topographical Maps." *Geographical Journal* 121: 1–10.

———. 1964. "Precision Instruments: To 1500." *A History of Technology,* vol. 3: 582–619. Charles Singer et al., eds. Oxford: The Clarendon Press.

Prown, Jules David. 1980. "Style as Evidence." *Winterthur Portfolio* 15: 197–210.

———. 1982. "Mind in Matter," *Winterthur Portfolio* 17: 1–19.

Pulido Rubio, José. 1923. *El piloto mayor de la casa de la contratación de Sevilla.* Seville: Tipógrafo Zarzuela.

Rauh, James. 1972. "Some Problems Concerning Lienzos de Zacatepec, #1 and #2." *Acts of the Thirty-eighth International Congress of Americanists* (1968), vol. 2: 51–65.

Relaciones geográficas del siglo XVI [RGS]. 1981–1988. Vol. 1: Guatemala; vols. 2 and 3: Antequera; vols. 4 and 5: Tlaxcala; vols. 6–8: México; vol. 9: Michoacan; vol. 10: Nueva Galicia. René Acuña, ed. Mexico City: UNAM, Instituto de Investigaciones Antropológicas.

Relaciones histórico-geográficas de la gobernación de Yúcatan: Mérida, Valladolid y Tabasco. 1983. Mercedes de la Garza, ed. 2 vols. Mexico City: UNAM, Instituto de Investigaciones Filologicas, Centro de Estudios Mayas.

Reparaz Ruiz, Gonzalo de. 1950. "The Topographical Maps of Portugal and Spain in the 16th Century." *Imago Mundi* 7: 75–82.

Reyes, Antonio de los. 1976. *Arte en lengua mixteca.* Vanderbilt University Publications in Anthropology, no. 14. Nashville: Vanderbilt University.

Reyes García, Luis. 1977. *Cuauhtinchan del siglo XII al XVI: Formación y desarrollo histórico de un señorío prehispánico.* Wiesbaden: Franz Steiner Verlag GMBH.

———. 1978. *Documentos sobre tierras y señorio en Cuauhtinchan.* Colección Cientifica Fuentes 57. Mexico City: INAH, Centro de Investigaciones Superiores.

Reyes-Valerio, Constantino. 1989. *El pintor de conventos: Los murales del siglo XVI en la Nueva España.* Mexico City: INAH.

Ricard, Robert. 1982. *The Spiritual Conquest of Mexico.* Lesley Byrd Simpson, trans. Berkeley and Los Angeles: University of California Press.

Riese, Frauke Johanna. 1981. "Indianische Landrechte in Yukatan um die Mitte des 16. Jahrhunderts: Dokumentenanalyse und Konstruktion von Wirklichkeitsmodellen am Fall des Landvertrages von Mani." Ph.D. diss., Hamburgisches Museum für Völkerkunde.

Robertson, Donald. 1959a. *Mexican Manuscript Painting of the Early Colonial Period: The Metropolitan Schools.* New Haven, CT: Yale University Press.

———. 1959b. "The Relaciones Geográficas of Mexico." *Acts of the Thirty-third International Congress of Americanists* (1958), vol. 2: 540–47.

———. 1972a. "The Pinturas (Maps) of the Relaciones Geográficas, With a Catalog." HMAI, vol. 12: 243–78. Austin: University of Texas Press.

———. 1972b. "Provincial Town Plans from Late Sixteenth Century Mexico." *Acts of the Thirty-eighth International Congress of Americanists* (1968), vol. 4: 123–29.

———. 1974. "The Treatment of Architecture in the Florentine Codex of Sahagún." *Sixteenth-Century Mexico: The Work of Sahagún,* pp. 151–64. Munro S. Edmonson, ed. Albuquerque: University of New Mexico Press.

Robertson, Janice Lynn. 1982. "Glyphs in the Codex Mendoza: The Language of the Glyphs." Master's essay, Columbia University.

Robinson, Arthur H., and Barbara Bartz Petchenik. 1976. *The Nature of Maps: Essays toward Understanding Maps and Mapping.* Chicago: University of Chicago Press.

Roys, Ralph L. 1933. "The Book of Chilam Balam of Chumayel." *Carnegie Institution of Washington, Publication no. 438.* Washington, D.C.: Carnegie Institution.

———. 1939. "The Titles of Ebtun." *Carnegie Institution of Washington, Publication no. 505.* Washington, D.C.: Carnegie Institution.

———. 1943. "The Indian Background of Colonial Yucatan." *Carnegie Institution of Washington, Publication no. 548.* Washington, D.C.: Carnegie Institution.

Ruiz Naufal, Víctor M., Ernesto Lemoine, and Arturo Gálvez Medrano. 1982. *El territorio mexicano.* 2 vols. Mexico City: Instituto Mexicano del Seguro Social.

Sáenz de Santa María, Carmelo. 1972. "La 'reducción a poblados' en el siglo XVI en Guatemala." *Anuario de estudios americanos* 29: 187–228.

Sahagún, Bernardino de. 1950–1963. *Florentine Codex: General History of the Things of New Spain.* 13 vols. Arthur Anderson and Charles Dibble, trans. Santa Fe: School of American Research and the University of Utah.

———. 1979. *Códice Florentino.* El Manuscrito 218–220 de la colección Palatino de la Biblioteca Medicea Laurenziana. Facsimile edition, 3 vols. Florence: Gunti Barbéra and AGN.

Sanders, William T. 1971. "Settlement Patterns in Central Mexico." HMAI, vol. 10: 3–44. Austin: University of Texas Press.

Sanders, William T., Jeffrey R. Parsons, and Robert S. Santley. 1979. *The Basin of Mexico: Ecological Processes in the Evolution of a Civilization.* New York: Academic Press.

Sandys, John E. 1911. "Pliny." *Encyclopedia Britannica,* 11th edition, vol. 21: 841–44.

Santa Cruz, Alonso de. 1918. *Islario general de todas las islas del mundo.* Madrid: Imprenta del Patronato de Huérfanos de Intendencia é Intervención Militares.

———. 1921. *Libro de las longitudines y manera que hasta agora se ha tenido en el arte de navegar.* Seville: Tipógrafo Zarzuela.

———. 1951. *Crónica de los Reyes Católicos.* Seville: Escuela de Estudios Hispano-Americanos, Serie 7a, no. 49.

Saralegui y Medina, Manuel de. 1914. *Alonso de Santa Cruz: Inventor de las cartas esféricas de navegación.* Madrid: Los Hijos de M.G. Hernández.

Schäfer, Ernesto. 1935–1947. *El consejo real y supremo de las Indias.* 2 vols. Seville: Escuela de Estudios Hispano-Americanos.

Scholes, France V., and Eleanor B. Adams. 1957. *Información sobre los tributos que los indios pagaban a Moctezuma, año de 1554.* Documentos para la historia del México colonial, vol. 4. Mexico City: José Porrúa é hijos.

———.1958. *Sobre el modo de tributar los indios de Nueva España a su majestad, 1561–1564.* Documentos para la historia del México colonial, vol. 5. Mexico City: José Porrúa é hijos.

Scholes, France V., and Ralph L. Roys. 1968. *The Maya Chontal Indians of Acalan-Tixchel: A Contribution to the History and Ethnography of the Yucatan Peninsula.* 2nd ed. Norman: University of Oklahoma Press.

Schroeder, Susan. 1991. *Chimalpahin and the Kingdoms of Chalco.* Tucson: University of Arizona Press.

Schuller, Rodolfo R. 1912. "Acerca del 'Yslario General' de Alonso de Santa Cruz." International Congress of Americanists, *Proceedings of the Eighteenth Session,* part 1: 115–[illegible]. London.

Schulz, Juergen. 1978. "Jacopo de' Barbari's View of Venice: Map Making, City Views, and Moralized Geography Before the Year 1500." *The Art Bulletin* 60, no. 3: 425–74.

———. 1987. "Maps as Metaphors: Mural Map Cycles of the Italian Renaissance." *Art and Cartography,* pp. 97–122. David Woodward, ed. Chicago: University of Chicago Press.

Schwaller, John Frederick. 1987. *The Church and Clergy in Sixteenth-Century Mexico.* Albuquerque: University of New Mexico Press.

Seler, Eduard, ed. 1892. *Historische Hieroglyphen der Azteken im Jahr 1803 im Königreiche Neu-Spanien gesamlet von Alexander von Humboldt.* Berlin.

———. 1904. "The Mexican Picture Writings of Alexander von Humboldt in the Royal Library at Berlin." *Smithsonian Institution, Bureau of American Ethnology Bulletin* 28: 123–229.

———. 1960–1961. *Gesammelte Abhandlungen zur Amerikanischen Sprach- und Altertumskunde.* 5 vols. Graz, Austria: Akademische Druck-u. Verlagsanstalt.

———. 1986. *Plano Jeroglífico de Santiago Guevea.* Carlos Enrique Delgado, trans. Mexico City: Ediciones Guchachi' Reza. Translation of "Das Dorfbuch von Santiago Guevea" in Seler 1960, vol. 3: 157–93.

———. 1988. *Comentarios al Códice Borgia.* 3 vols. Mariana Frenk, trans. Mexico City: Fondo de Cultura Económica.

Siméon, Rémi. 1986. *Diccionario de la lengua náhuatl o mexicana.* Josefina Olivia de Coll, trans. Mexico City: Siglo Veintiuno.

Simons, Bente Bittmann. 1967–1968. "The Codex of Cholula: a preliminary study." *Tlalocan* 5, nos. 3–4: 267–339.

———. 1968. *Los mapas de Cuauhtinchan y la Historia tolteca-chichimeca.* Serie Investigaciones 15. Mexico City: INAH.

Simpson, Lesley Byrd. 1934. "The Civil Congregation." *Studies in the Administration of the Indians in New Spain,* vol. 2. Berkeley: University of California Press.

———. 1952. *Exploitation of Land in Central Mexico in the 16th Century.* Berkeley: University of California Press.

———. 1982. *The Encomienda in New Spain.* Berkeley and Los Angeles: University of California Press.

Singer, Charles Joseph, et al., eds. 1957–1958. *A History of Technology.* Oxford: The Clarendon Press.

Sisson, Edward B. 1983. "Recent Work on the Borgia Group Codices." *Current Anthropology* 24, no. 5 (December): 653–56.

Smith, Mary Elizabeth. 1973a. *Picture Writing from Ancient Southern Mexico: Mixtec Place Signs and Maps.* Norman: University of Oklahoma Press.

———. 1973b. "The Relationship between Mixtec Manuscript Painting and the Mixtec Language: A Study of Some Personal Names in Codices Muro and Sánchez Solís." *Mesoamerican Writing Systems,* pp. 47–98. Elizabeth P. Benson, ed. Washington D.C.: Dumbarton Oaks.

———. 1983. "Codex Selden: A Manuscript from the Valley of Nochixtlan?" *The Cloud People,* pp. 238–45. K. Flannery and J. Marcus, eds. New York: Academic Press.

Smith, Michael. 1984. "The Aztlan Migration of the Nahuatl Chronicles: Myth or History?" *Ethnohistory* 31, no. 3: 153–86.

Snyder, John P. 1993. *Flattening the Earth: Two Thousand Years of Map Projections.* Chicago: University of Chicago Press.

Somolinos D'Ardois, Germán. 1960. "Vida y Obra de Francisco Hernández." In Francisco Hernández, *Obras Completas,* vol. 1, pp. 97–373. Mexico City: UNAM.

Soria, Martín S. 1956. *La pintura del siglo XVI en Sudamérica.* Buenos Aires: Instituto de Arte Americano é Investigaciones Estéticas.

Soustelle, Jacques. 1970. *Daily Life of the Aztecs.* Patrick O'Brian, trans. Stanford, CA: Stanford University Press.

Spores, Ronald. 1965. "The Zapotec and Mixtec at Spanish Contact," HMAI, vol. 3: 962–87. Austin: University of Texas Press.

———. 1967. *The Mixtec Kings and Their People.* Norman: University of Oklahoma Press.

———. 1984. *The Mixtecs in Ancient and Colonial Times.* Norman: University of Oklahoma Press.

Stevenson, Edward Luther. 1909. "Early Spanish Cartography in the New World." *American Antiquarian Society, Proceedings,* (n.s.) vol. 19: 369–419.

Stevenson, Edward Luther. 1927. "The Geographical Activities of the Casa de Contratación." *Annals of the Association of American Geographers,* vol. 17: 39–59.

———, trans. 1932. *Geography of Claudius Ptolemy.* New York.

Street, Brian V. 1984. *Literacy in Theory and Practice.* Cambridge: Cambridge University Press.

Sullivan, Thelma D. 1987. *Documentos tlaxcaltecas del siglo XVI en lengua náhuatl.* Mexico City: UNAM.

Taylor, E. G. R. 1956. *The Haven-Finding Art: A History of Navigation from Odysseus to Captain Cook.* London: Hollis and Carter.

———. 1964. "Cartography, Survey, and Navigation 1400–1750." *A History of Technology,* vol. 3: 530–57. Charles Singer, et al. eds. Oxford: The Clarendon Press.

———. 1968. *Tudor Geography, 1485–1583.* New York: Octagon Books.

Taylor, William B. 1972. *Landlord and Peasant in Colonial Oaxaca.* Stanford, CA: Stanford University Press.

Thompson, J. Eric S. 1950. "Maya Hieroglyphic Writing: Introduction." *Carnegie Institution of Washington, Publication 589.* Washington, D.C.: Carnegie Institution.

———. 1972. *A Commentary on the Dresden Codex: A Maya Hieroglyphic Book.* Philadelphia: The American Philosophical Society.

Todorov, Tzvetan. 1984. *The Conquest of America.* New York: Harper and Row.

Tomassi de Magrelli, Wanda. 1978. *La cerámica funeraria de Teotenango.* Mexico City: Biblioteca Enciclopédica del Estado de México.

Torquemada, Juan de. 1986. *Monarquía indiana.* 3 vols. Mexico City: Editorial Porrúa.

Torres Lanzas, Pedro. 1900. *Relación descriptiva de los mapas y planos de México y Floridas existentes en el Archivo General de Indias.* 2 vols. Seville.

Toulmin, Stephen. 1960. *The Philosophy of Science.* New York: Harper and Row.

Toussaint, Manuel. 1965. *Pintura colonial en México.* Mexico City: Imprenta Universitaria.

Toussaint, Manuel, Federico Gómez de Orozco, and Justino Fernández. 1938. *Planos de la ciudad de México.* XVI Congreso Internacional de Planificación y de la Habitación. Mexico City: Instituto de Investigaciones Estéticas de la Universidad Nacional Autónoma.

Townsend, Richard. 1979. *State and Cosmos in the Art of Tenochtitlan.* Studies in Pre-Columbian Art & Archaeology, no. 20. Washington, D.C.: Dumbarton Oaks.

Troike, Nancy P. 1974. "The Codex Colombino Becker." Ph.D. diss., University of London.

———. 1978. "Fundamental Changes in the Interpretations of the Mixtec Codices." *American Antiquity* 43, no. 4: 553–68.

———. 1982. "The Interpretation of Postures and Gestures in the Mixtec Codices." *The Art and Iconography of Late Post-Classic Central Mexico,* pp. 175–206. Elizabeth H. Boone, ed. Washington, D.C.: Dumbarton Oaks.

Urcid Serrano, Javier. 1992. "Zapotec Hieroglyphic Writing." Ph.D. diss., Yale University.

van Zantwijk, Rudolf. 1985. *The Aztec Arrangement: The Social History of Pre-Spanish Mexico.* Norman: University of Oklahoma Press.

Viñas y Mey, Carmelo y Ramon Paz. 1949–1971. *Relaciones histórico geográfico-estadísticas de los pueblos de España hecha por iniciativa de Felipe II.* 5 vols. Madrid: Consejo Superior de Investigaciones Científicas.

Von Winning, Hasso. 1987. *La iconografía de Teotihuacan: Los dioses y los signos.* Mexico City: UNAM.

Warren, J. Benedict. 1973. "An Introductory Survey of Secular Writings in the European Tradition on Colonial Middle America, 1503–1818." HMAI, vol. 13: 42–137. Austin: University of Texas Press.

Weismann, Elizabeth Wilder. 1985. *Art and Time in Mexico.* New York: Harper and Row.

Whitecotton, Joseph W. 1977. *The Zapotecs: Princes, Priests, and Peasants.* Norman: University of Oklahoma Press.

———. 1982. "Zapotec Pictorials and Zapotec Naming: Towards an Ethnohistory of Ancient Oaxaca." Native American Ethnohistory. Joseph W. and Judith Bradley Whitecotton, eds. *Papers in Anthropology* 23, no. 2: 285–336. Norman: Department of Anthropology, University of Oklahoma.

Wieser, Franz R. v. 1908. *Die Karten von Amerika in dem Islario General des Alonso de Santa Cruz.* Innsbruck: Verlag der Wagner'schen Universitäts-Buchhandlung.

Williams, Barbara J. 1984. "Mexican Pictoral Cadastral Registers: An Analysis of the Códice de Santa María Asunción and the Codex Vergara." *Explorations in Ethnohistory: Indians of Central Mexico in the Sixteenth Century,* pp. 103–25. H. R. Harvey and Hanns J. Prem, eds. Albuquerque: University of New Mexico Press.

Woodward, David. 1987. "Medieval *Mappaemundi.*" *The History of Cartography: Cartography in Prehistoric,*

Ancient, and Medieval Europe and the Mediterranean, vol. 1: 286–370. J. B. Harley and David Woodward, eds. Chicago: University of Chicago Press.

Yoneda, Keiko. 1981. *Los mapas de Cuauhtinchan y la historia cartográfica prehispánica.* Mexico City: AGN.

Zavala, Silvio. 1935. *La encomienda indiana.* Madrid: Centro de Estudios Históricos, Sección hispanoamericana, 2.

Zeitlin, Judith Francis. 1989. "Ranchers and Indians on the Southern Isthmus of Tehuantepec: Economic Change and Indigenous Survival in Colonial Mexico." *Hispanic American Historical Review* 69: 23–60.

Zorita, Alonso de. 1963. *Life and Labor in Ancient Mexico: The Brief and Summary Relation of the Lords of New Spain.* Benjamin Keen, ed. and trans. New Brunswick, NJ: Rutgers University Press.

Index

In entries for Relaciones Geográficas maps, modern names are cited after sixteenth-century versions.

Acambaro (Acambaro, Guanajuato), 74, 75 fig. 36, 217
Acapa, 132
Acapistla (Yecapixtla, Morelos)
  place-names of, 142 table 7, 149–52, 158
  Relación Geográfica map, 69 fig. 31, 69–71, 71, 83, 127, 176, 217
  San Juan Bautista Yecapixtla (monastery), 70
Acapulco, 26
Achichipico, 151
Acolhua, 94
Acolman, merced map, 185
acordados, 183, 184, 186, 188, 210, 233–34
Actopan (Veracruz), merced map of, 189–95
agriculture, 183, 188, 195, 205
Alba, Fernando Alvarez de Toledo, third duke of, 2
Albrecht V, duke of Bavaria, 4
alcaldes, 62, 63, 65, 66
alcaldes mayores
  as mapmakers, 34, 39–4, 57
  office, 32–4, 66, 93
  relation to indigenes, 33–4, 40, 64–7
  role in mercedes (land grants), 183, 185
  See also corregidores; *and names of individuals*
alcaldías mayores, 32
Alcázar, 1, 8, 144
alphabetic writing
  literacy in, 30, 32, 77, 164–67, 169
  on mercedes (land grant) maps, 187, 192; determining meaning of, 188, 194–95, 202, 207
  on Relaciones Geográficas maps, 167–70; determining meaning of, 136–37, 166, 169–70, 176, 178; used for Nahuatl, 132, 165, 170–71, 172–75, 216, 231–32, 245 n. 28; used for Spanish, 171–72, 175–79
  replacing logographic writing, xix, xx, 135–38, 164–67, 170–75, 179, 186, 197–99, 210–11, 215
altepetl, xxiii
  organization of, 74–5, 105–6, 118, 129–30, 176
  symbols for, 118–19, 141–44, 149
Amaquemecan, 121–26
Ameca (Ameca, Jalisco), 176, 177 fig. 84, 178, 217
Amiztlan, 154–55
Amoltepec (Santiago Amoltepec, Oaxaca)
  Relación Geográfica map, 112–13, 113 fig. 51 [and plate 6], 115, 116, 177, 187, 214, 218
  place-names of, 149, 159–61, 244 n. 20
Analytical Dictionary of Nahuatl, An, xxiii
Andrés, don (ruler of Oztotlatlauhca), 132, 171
Apian, Philipp, 4
Apoala Valley, 102–4, 107, 114, 139
Aquino, don Pablo de, 63, 131, 132, 171
architecture on maps, 98, 100 table 5
  to define spaces, xiv, 4, 22, 43, 88, 120, 175, 203

architecture on maps *(cont.)*
as symbols of communities, 49, 118–19, 122, 125, 126, 127–28, 145
See also churches on maps
Archivo General de la Nación, Mexico (AGN), 181, 202, 211
armadillos, 47
Arnheim, Rudolf, 117
artists, indigneous. *See* native painters of Relaciones Geográficas maps
astrolabes, 14, 19, 52
astronomy, 55–6
Atezca, merced map of, 189–95
atlas-chronicles, 15–18
atlas, Escorial, 2–8, 6 fig. 6
Atlatlahuca. *See* Atlatlauca
atlatls, 44
Atlatlauca and Malinaltepec (San Juan Bautista Atatlahuca and Maninaltepec, Oaxaca), 36–7, 38 fig. 13, 218
Atlatlauca and Suchiaca (Atlatlahuca, Mexico), 172, 218
Atzaqualco, xvi
audiences of indigenous maps
indigenous, xvi, 71–2, 110–11, 121
literacy defining, 163–67
Spanish, 71–2, 119, 179, 204, 210, 215–16
Audiencia of New Spain, 183
Augustinians, 64 fig. 29, 70, 73, 82–3, 85–6
Austria, don Juan de, 132, 171
Ayapanco, 151–52
Aztec empire, 92, 105, 132, 137, 140–41, 144, 158, 214
See also Culhua-Mexica

Barcelona, map of, 1 fig. 5, 2–3
Bavaria, 4
Benavente, fray Toribio de, 78
bilingualism, 165
Bonifacio, don Agustín, 209
Bonifacio, Domingo, 62
books in New Spain, 38, 47, 49, 78
botanical surveys, 19
boundaries on maps
on community maps, 107, 108, 111, 117–18
on Relaciones Geográficas maps, 70 fig. 32, 112–16, 135, 161, 166, 169, 178
to separate communities, 57, 106
Braun, Georg, 5
Bravo de Lagunas, Constantino, 66

caballerías, 186
cabeceras, 76, 127–28, 176
cabildos, 62–5, 92, 97, 111, 114, 121, 138
Cabot, Sebastian, 13
Cabrera, Juan de, 178
cacicazgos, 105
cadastral maps, 207
Cahuitan. *See* Cuahuitlan
Çaldivar, Vincente de, 178
calendar, Mesoamerican, 40 fig. 15, 41, 107, 162
calpolli, xvi, xxiii, 74–6, 105, 118–21, 127–30
Camino Real, 188, 189, 191
Cangas y Quiñónes, Suero de, 51
Cano, Martín, 62, 63 fig. 28, 86–9, 202–9, 210
Cape Verde islands, 14
Casa de Contratación, 12–13
casas de la comunidad, 88, 203
cartography, European, xi, xiii, xvi, xix, 4–5
in the New World, xi, xx, 9
cartography, indigenous colonial
of boundaries, 107, 108, 111–18
cadastral maps, 207
cartographic histories, 106–17
community maps, xiv, 91–2, 106–7, 144, 176–78, 187–88, 214
form of, 112, 116–17
as legal documents, 111, 184
mercedes (land grants), 181–83, 188, 189–95, 197–202, 203–7, 210–11
projections, xiv–xvi, 110–111, 115–117
social settlement maps, 107, 118–33
See also cartography, pre-Hispanic; Relaciones Geográficas, maps of; *and names of individual maps*
cartography, Mexican (after 1600), 58, 178, 216
cartography, Netherlandish, 2
cartography, pre-Hispanic, 91
norms of, xi–xii, xiv, 101–4, 106, 107–8
relation to colonial maps, xix–xx, 92, 97–101, 104, 112, 132–33
See also cartography, indigenous colonial; Relaciones Geográficas, maps of; *and names of individual maps*
cartography, Spanish
in Europe, xi, 1–8

in New Spain, xiii–xiv, 19–20, 22–7, 55–8, 213–15 (*see also* Spanish colonials and Relaciones Geográficas maps)
in the New World, xx, 11–15, 17–20, 21–2
norms, xiv, 4–5
world maps, 50
Caso, Alfonso, 114, 161
Casteñeda, Francisco de, 33, 34
Castile, 49
catechisms, 85, 240 n. 12
Cempoala (Zempoala, Hidalgo)
organization of, 63, 129–30, 176, 242 n. 33
in pre-Hispanic period, 92, 94, 132, 175
Relación Geográfica map, 82 table 3, 94, 95 fig. 42 [and plate 5], 129–33, 177, 214, 218; place-name, 142 table 7, 145; written inscr-iptions of, 165, 170 table 9, 175, 243 n. 35
Cempoala (Veracruz), 192
cenecehuilli, 112, 116
Cerro de la Estrella, 207
Cervantes, Hernando de, 31
Chalchiuhmomoztli, 126
Chalchiuhtlicue, 101
Chalco, 122, 125
Charles V (king of Spain), 1, 7, 14
Chávez, Gabriel de, 33, 39–44, 50, 65, 67, 92
Chichimecs, 39 fig. 14, 43 fig. 18, 44, 238 n. 11
Chicoalapa (Chicoloapan de Juarez, Mexico), 87, 155–56, 157 fig. 75, 218, 244 n. 13
Chicoloapan. *See* Chicoalapa
Chila. *See* Matlatlan and Chila
Chimalhuacan Atengo (Santa María Chimalhuacan, Mexico), 69, 73 fig. 35, 169, 218
Chimalpahin Quauhtlehuanitzin, don Domingo de San Antón Muñón, 122–23, 126
Chocho people, 158
Cholula (Cholula de Rivadabia, Puebla), xvi
maps of, 119–21, 126, 127–28
monastery of, 56, 76, 81
organization of, 119–21, 127–29
in pre-Hispanic period, 94–96, 108, 119–20, 242 nn. 22–3
Relación Geográfica map, 69, 72 fig. 34 [and plate 3], 76, 82, 126–29, 218; place-name, 149; written inscriptions of, 165, 170 table 9
Chontal language, 165
chorography. *See* maps, chorographic
chronicles of the New World, 11–12, 15–18
models, 16–17
questionnaires as sources, 11, 16–17, 20
church cloths, 82
church, Roman Catholic, 68–9, 89, 97
See also mendicants; secular clergy
churches on maps, 82, 88, 132, 169, 170, 197
influence of European prints, 78–9
as symbols of communities, 22, 49, 69–75, 118–19, 127, 128, 187, 203–5
Cíbola, Seven Cities of, 9
Cihuayocan (Tetlistaca), 157, 244 n. 14
circular maps, 117
city plans. *See* maps, chorographic
civilization, notions of, 44, 67, 238 n. 11
Civitates Orbis Terrarum, 5
clergy. *See* mendicants; secular clergy
Cline, Howard F., xxiii
coastal charts. *See* maps, nautical
Coatepec Chalco (Coatepec, Mexico), 69, 87, 218
Coatzacoalcos. *See* Coatzocoalco
Coatzocoalco (Coatzacoalcos River; near Tuzande-petl, Veracruz), 24 fig. 8, 25, 51–2, 53 fig. 26, 218
Codex Borgia, 96
Codex en Cruz, 94
Codex Kingsborough, 79–80, 81 fig. 39
Codex Mendoza
creation of, xiv, xvii, 184
map of Tenochtitlan (fol. 2r), xiv–xvii, xv fig. 2, 118, 119, 130
place-names in, xiv, 141–45, 158, 169
tribute list (fols. 17–55), 140–45, 141 fig. 65
Codex of Cholula, 127–28
Codex Sierra, 96
Codex Xolotl, 94
Codex Zouche-Nuttall, 101–4, 102 fig. 46, 107–8, 114, 138, 139, 174, 241 n. 9
Códice Franciscano, 85
coinage, 49
Columbus, Christopher, 9, 16
Comaltepec, 158
commerce, 188, 195
commoners, 66, 77, 165
community, understandings of, 91, 105–6, 112, 130, 214, 241 n. 12
community kingdoms, 105–6, 141–44
See also altepetl

community maps, xiv, 91–2, 106–7, 144, 173, 176–78, 187–88, 214
compasses, 52, 238 n. 16
Comuneros movement, 7
conquistadores, xi, 33, 183
Conway, Jill Ker, xii
Coqui Pilla, 174, 231–32
Coqui Piziatuo, 174, 231–32
corregidores
  as mapmakers, 34, 47–9, 57
  office, 31, 32–4, 66, 93
  relation to indigenes, 33–4, 64–7
  role in mercedes (land grants), 183, 185, 209
  See also alcaldes mayores; *and names of individuals*
corregimientos, 33
Cortés, Hernán, xii, 16, 140, 173, 245 n. 1
Cortés map of Tenochtitlan, xii fig. 1, xiii–xiv, xvi–xvii, 4, 22, 35
cosmógrafo-cronista mayor, 17–18
Cosmographiae universalis, 41 fig. 16, 41–3
cosmography, 11, 13, 31, 57, 59
Council of the Indies, 17, 18, 27
Coxcatlan. *See* Cuzcatlan
Creoles, xx, 25, 29, 32–3, 61, 62, 66, 68, 166
Cristóbal, don (of Suchitepec), 171
cross staffs, 14
Cuahuitlan (Cahuitan, Oaxaca), 84, 219
Cuananá, 112
Cuatototla, 154
Cuauhtinchan, 96, 108, 116
  See also Historia Tolteca-Chichimeca
Cuazcotzin, 132
Cuenca, Juan de la, 178
Cuepopan, xvi
Cuicatecs, 23, 74, 77
Cuicatlan, 77
Cuitlahuac, 205, 209
Culhuacan (Culhuacan, Distrito Federal, Mexico), xvi, 76
  murals, 82–3
  Relación Geográfica map, 64 fig. 29, 65, 82, 87–9, 97, 207, 219; place-name of, 87, 142 table 7, 166–67; written inscriptions, 62, 175
Culhua-Mexica, xiv, 105, 106, 140, 144, 149, 175, 214
  See also Aztec empire
Cuzcatlan (Coxcatlan, Puebla), 69, 70 fig. 32, 71 fig. 33, 84, 127, 169, 219

Dante, Egnazio, 4
da Vinci, Leonardo, 4
demography, 57, 96–7, 106, 127, 129, 188
De Novo Orbe Decades, 16
desagüe, 59, 238 n. 24
determinatives, 149, 155
Deventer, Jacob van, 2, 4
diligencias, 183
doctrinas, 23
Domínguez, Francisco, 19–20, 22, 25, 27, 55, 56, 237 n. 14
Dominicans, 65, 73, 76, 79, 82 table 3, 145
double-consciousness of artists, 72, 215–16

eclipses, observations of, 15, 17, 18–19, 22
  in New Spain, 27, 32, 55–6
economy of New Spain, 183, 195
  See also trade
education, indigenous, 65, 73–7, 76–7, 79–84
8 Deer Jaguar Claw, 139 fig. 64, 140
elites, indigenous
  and church, 66–7, 68–9, 73–6, 84–5
  and colonial government, 66–7, 74, 88
  consciousness of status, 88–9, 97
  education of, 65, 76–7
  literacy of, 164–65
  painters among, 65, 84–5, 89, 107
  portrayals on Relaciones Geográficas maps, 114–15, 130–32, 135, 170–71, 187, 208 fig. 101
  as rulers, 62–4, 105, 107–8 (*see also* government of New Spain, indigenous)
Elizabeth I (queen of England), 4
Elotepec, San Juan, 26, 36, 113–14
  See also Peñoles, Los
encomenderos, 33, 66, 67, 81 fig. 39
encyclopedias, 16–17, 77
England, 4
Epazoyuca (Epazoyucan, Hidalgo), 82 table 3, 129, 130, 131 fig. 61, 132, 219, 242 n. 34
Epazoyucan. *See* Epazoyuca
equinox, 56
Escorial, 3, 9
Escorial atlas, 2–8, 6 fig. 6
Esquivel, Pedro de, 2–8, 6 fig. 6, 23, 213

Estetla, Santa Catarina, Oaxaca, 25–6, 36
 See also Peñoles, Los
ethnicity, 62, 87–9, 131–32, 239 n. 1
evangelization, 23, 74, 77, 82, 83–4, 85
exploration in the New World, xi, 12–13

Ferdinand (king of Castile and Aragon), 8, 14
Fernandez de Córdova, Francisco, 136, 178
fiscales, 66
5 Flower, 162–63
5 Wind, 107
Flanders, 5, 78
floods, 59
Florence, maps of, 41 fig. 16, 44, 45
Florentine Codex, 77
foundings of towns, xiv, 107–8, 116, 119, 126, 242 n. 30
fortresses, 43–4, 45 fig. 19, 158
Franciscans, 38, 73, 77, 82 table 3, 86, 127
frescos. *See* monasteries, mural programs of
friars. *See* mendicants
frieze symbol, 159

Gallegos, Gonzalo, 62
Gante, fray Pedro de, 76
García Ruiz, Hernán, 171–72
gender of Relaciones Geográficas artists, 239 n. 5
genealogies on maps, 108, 112, 114–15
Germany, 5
Geografía y descripción universal de las Indias, 18
geographic maps. *See* maps, geographic
Geography, 3–4, 13, 57
geography in Relación Geográfica questionnaire, 20–2, 35
global circumference, 15
gobernadores
 individuals, 77, 131
 office, 62–4, 65
gobierno of New Spain, xviii, 29
goniometers, 20
González, Juan, 66
government of New Spain
 indigenous, 62–4, 176; responses to mercedes (land grants), 185, 202, 209, 245 n. 2
 local colonial, 29, 32–4, 93, 175–76; relationship to indigenes, 33, 66–7, 74, 112, 208–9; role in mercedes, 183, 185, 209
 viceregal, 17, 86–7, 129–130, 183–84
 See also cabildos; *and individual names of offices*
Gregory XIII (pope), 4
Guatulco, 26
Guaxtepec (Oaxtepec, Morelos)
 monastery, 76, 79 fig. 37, 82
 Relación Geográfica map, 68 fig. 30 [and plate 2], 69, 76, 78–9, 80, 82, 169–70, 171–72, 175, 219; place-name, 68 fig. 30, 142 table 7, 145
Gueguetlan (Santo Domingo Huehuetlan, Puebla), 55–6, 165, 170 table 9, 171, 172 fig. 82, 219
Guevara, Felipe de, 3
Gueytlalpa (Hueytlalpan, Puebla), 150 fig. 68, 153, 219
Gueytlalpa group, 152–55, 164
 See also individual names
Gulf coast, maps of, 12, 24 fig. 8, 50–3, 55
Guzmán, don Francisco de, 63, 131, 132, 171

Hernández, don Juan (of Gueguetlan), 171
Hernández family (of Suchitepec), 171
Hernández, Francisco, 19–20, 237 n. 13
hill symbol (tepetl), 135, 145, 146–49, 155, 189, 206, 210, 211
Historia general y natural de las Indias, 16
Historia Tolteca-Chichimeca, 96
 Fols. 26v–27r, 119–21, 120 fig. 54, 126, 127–28
 Fols. 32v–33r, 108–11, 110 fig. 50, 115–16, 130
histories on maps, 106–8, 112, 132–33, 173–75
Hogenberg, Frans, 5
Holy Roman Emperors, 5
Horcasitas, Fernando, 216
Huajoloticpac, Santiago, Oaxaca, 26
 See also Peñoles, Los

Huehuetlan. *See* Gueguetlan
Hueitzacualco. *See* Teozacoalco
Hueytlalpan. *See* Gueytlalpa
Huitepec (Huiztepec), San Antonio, 26, 36
 See also Peñoles, Los
Huixachtla, 206–7
Humboldt Fragment II, 207, 208 fig. 101

iconography, indigenous
 forms of, 97–101
 as indicator of native artist, 61, 87–9, 105, 189, 195

iconography *(cont.)*
meaning of, 62, 87–9, 97
on Relaciones Geográficas maps, 100 table 5, 189
See also place-names, indigenous; symbols on maps, indigenous
illusionism, 77–80, 82–3
images' role in New Spain, 30, 40, 58–9, 61–2, 84–7, 97
Indies, xx, 13
instrucción for observing eclipses, 18–9, 22
colonial responses, 27, 32, 55–6
instruments for mapmaking, 14, 19–20
Ircio Tapayoltzin, Martín de, 63, 131–32
Isabella (queen of Castile), 8, 14
Islario general de todas las islas del mundo, 16
isolarii, 16, 22
Italy, 4
itinerary maps, 35–7
Itzcoatl, 132, 175
Itzcuintepec, 36
Itztacalco, 205
Ixcatlan (Santa María Ixcatlan, Oaxaca), 47–9, 47 fig. 21, 48 fig. 22, 73, 220
Ixhuacán, 189
Ixtapalapa (Ixtapalapa, Distrito Federal, Mexico)
mercedes maps of, 202–9
Relación Geográfica map of, 62, 63 fig. 28, 86, 88–9, 181, 202–5, 220
Ixtepeji. *See* Ixtepexic
Ixtepexic (Santa Catarina Ixtepeji, Oaxaca), 64–5, 84, 220

Jalapa. *See* Xalapa de la Vera Cruz
Jojupango. *See* Jujupango
Jonotla. *See* Xonotla
Juan, Jaime, 27, 56
Jujupango (Jojupango, Puebla), 151 fig. 69, 153, 154–55, 220

Kagan, Richard, 5, 31
Kartunnen, Frances, xxiii
Kubler, George, 127

labor, 76
land grants. *See* mercedes (land grants)
landholding
in indigenous communities, 126, 187–88
on mercedes maps, 186–87, 194, 207
on Relaciones Geográficas maps, 130, 177–78, 211
land use, xx, 195
See also agriculture, ranching
landscape painting, 44, 46 fig. 20, 79, 82–3
languages. *See names of individual languages*
Las Casas, fray Bartolomé de, 65, 166
latitude, 4, 12, 13–14, 20, 22, 52, 55–6
Ledesma, Bartolomé de, 49 fig. 23, 78
libraries, 38, 92
Lienzo of Petlacala, 216
Lienzo of Yolotepec, 161
Lienzo of Zacatepec, 107–8, 109 fig. 49, 115, 216, 241 nn. 13–14
lienzos, 108
line of demarcation, 14–15
literacy, alphabetic, 30, 32, 77, 164–67, 169
Lockhart, James, 118
logographic writing
characteristics of, 101–4, 138–39, 243 n. 1
expressing indigenous languages, 84, 139–40, 164, 166–67, 243 n. 5
literacy in, 30, 165–66, 184
and oral traditions, 139–40
relationship to alphabetic writing, 135, 169
Spanish attitudes towards, 58, 166, 181
See also alphabetic writing; place-names, indigenous
longitude
measurement, 4, 12, 13, 15, 19, 23
political ramifications, 14–15
questionnaires for determining, 15, 17, 18–19, 22; colonial responses, 27, 32, 55–6
López de Velasco, Juan
as chronicler, 11–12, 17–18
as cosmographer, 3, 6 fig. 6, 11–12, 17–19, 31
and eclipse observations, 18–19, 32
and Relación Geográfica questionnaire, 20–3, 29, 67
and Relaciones Geográficas responses, 23–7, 50, 56, 213–15
López de Zárate, Juan, 34
Los Peñoles. *See* Peñoles, Los
Low Countries. *See* Netherlands, Spanish
Loya, Francisco de, 62
Luna, don Antonio de, 66
lunar eclipses. *See* eclipses, observations of

macanas, 44
macehualtin, 66, 77, 165
Macuilsuchil (San Mateo Macuilxochitl, Oaxaca)
  Relación Geográfica map of, 162 fig. 79 [and plate 8], 216, 220; place-name, 142 table 7, 161–63; inscriptions of, 165, 170 table 9, 173–75, 216, 231–32
  town of, 216
Macuilxochitl. *See* Macuilsuchil
Macupilco, San Miguel (vicinity of Santa María Xadan, Oaxaca), 170–71, 220
Madrid, 7, 15
Malinalco, Mexico, 85–6, 240 n. 10
manuscripts, pre-Hispanic, xix, 65, 84, 164, 170, 240 n. 14
  relation to Relaciones Geográficas maps, 87–8, 93–7, 203
  as sources for Relaciones Geográficas texts, 39–41, 44
  survival of, 85, 92–3, 104
  See also cartography, pre-Hispanic
Maninaltepec. *See* Atlatlauca and Malinaltepec
Map of Chichimec History, 121–26, 122 fig. 56, 128
map printing, xi, 3, 5, 34, 38–9
map projections, 5, 43
  Albertian, xiii–xiv, 2, 4, 31
  Euclidean, xiii–xiv, xvi, xix, 3, 4, 12
  ichnographic, 4
  indigenous, xvi, 110–11, 115–17
  panoramic, xiii–xiv, 4
  using latitude and longitude, 12, 13–15, 19, 22, 31, 50
Mapa Quinantzin, 94
Mapa Tlotzin, 94
mapping. *See* cartography
maps and political ideologies
  indigenous, 106–7, 176
  Spanish, 1–3, 5, 7–8, 176, 213–14
maps, chorographic
  defined, xiii–xiv, 3–5, 45
  as models for Relaciones Geográficas maps, 12, 22–3, 31, 34–5, 38–44, 91
  political ideology expressed in 5, 7–8
maps, definition, xii–xiii, 235 n.1
maps, geographic
  defined, 3–4
  political ideology expressed in, 5, 7–8
  Ptolemaic, 12, 22
  See also maps, survey
maps, itinerary, 35–7
maps, nautical, 13, 22
  instruments used in making, 14, 50, 51–3
  and Relación Geográfica corpus, 25, 35, 50–5
maps, survey
  methods of, 3, 19–20, 57–8
  survey projects, 2–3, 4; in New Spain 12, 18, 19–20, 25, 27, 56
  See also Escorial atlas; maps, geographic
Marina, doña (interpreter), 140
mariners, 13, 14, 50–5
Martire d'Anghiera, Pietro, 16
Matlactonatico. *See* Tenanpulco and Matlactonatico
Matlatlan and Chila (Chila, Puebla), 152 fig. 70, 153, 220
Matrícula de Tributos, 140
Maya, 117, 140
media of maps, 65, 108, 110, 136 fig. 62
  pigments, 77, 189, 239 n.6
Medina, Juan de, 50
mendicants, 23
  Augustinians, 64 fig. 29, 70, 73, 82–3, 85–6
  Dominicans, 65, 73, 76, 79, 82 table 3, 145
  Franciscans, 38, 73, 77, 82 table 3, 86, 127
  and images, 84–7
  influence upon Relaciones Geográficas artists, 68–76
  and native education, 65, 73, 76–7, 79–84
  See also names of individuals
Mendieta, fray Jerónimo de, 85, 86
Mendoza, Antonio de (viceroy of New Spain), xvii, 183, 184
Mendoza, Codex. *See* Codex Mendoza
Mendoza, don Diego de, 63–4, 131, 132, 171
Mendoza, Luis de, 209
mensuration, 57–8, 112, 116, 186, 238 n. 23
mercedes (land grants), 62, 181–85, 214, 245 n. 1
  indigenous responses to, 185, 202, 209, 245 n. 2
  procedures for, 183, 186, 209, 233–34, 245 n. 5
  role of colonial officials, 183, 185, 209
  See also mercedes (land grant) maps
mercedes (land grant) maps
  commissions of, 181, 189, 193–94, 201
  influence on Relaciones Geográficas maps, 181–83, 185–91, 195, 200–3, 210–11
  of Ixtapalapa, 202–9

mercedes (land grant) maps *(cont.)*
purpose of, 184–185, 195
of Tarímbaro, 186–87
of Tehuantepec region, 197–202, 210, 246 nn. 12–13
of Xalapa region, 189–95, 210, 246 nn. 7–8
Mesoamerica, region of, xvi, xvii fig. 3
Metztitlan. *See* Meztitlan
Mexica. *See* Culhua-Mexica
Mexicatzingo (Mexicaltzingo, D. F., Mexico), 62, 205
Mexico City, 15, 23, 26–7, 56, 76, 184
Meztitlan (Metztitlan, Hidalgo)
Relación Geográfica map, 39–44, 39 fig. 14, 43 fig. 18, 50, 221
Relación Geográfica text, 39–40, 40 fig. 15, 44, 65, 67, 92
Miahualtepec, 158
Minas de Temazcaltepec. *See* Temazcaltepec, Minas de
Minas de Zumpango. *See* Zumpango, Minas de
mines, 20, 177, 184
Miral, Juan del, 178
Miranda, Francisco de, 33, 34, 94, 215
Misantla. *See* Mizantla
miscegenation, 86
Misquiahuala (Mixquihuala, Hidalgo), 65–6, 82 table 3, 135–37, 136 fig. 62 [and plate 7], 211, 221
inscriptions on map, 178
place-name, 143 table 7, 149
Mixquictla (Mixquitla/Mixquiteca), Santiago, 127, 128
Mixquihuala. *See* Misquiahuala
Mixtec
place-names, 102–4, 112–13, 144, 158–61, 165–66, 243 n. 5
polities, 76, 105–6, 245 n. 32
Mizantla (Misantla, Veracruz), 83 fig. 40, 84, 178, 210–11, 221
Mochitlan. *See* Muchitlan
Molina, Alonso de, 46 fig. 20, 47
monasteries
mural programs of, 82–3, 85–6, 239 n. 8, 240 nn. 9–11
schools in, 73–4, 76–7, 80–1, 82
See also churches on maps; mendicants
Mondéjar, second marqués of (Luis Hurtado de Mendoza), 17
Montjuich, 1 fig. 5, 3
Montúfar, Alonso de (archbishop of Mexico), 49
Moras, Pedro de, 176
Moteuczoma, xxiii, 144, 208 fig. 101
Motolinía, fray Toribio de Benavente, 78
Moyotlan, xvi
Muchitlan (Mochitlan, Guerrero), 145–49, 146 fig. 66, 166–67, 221
Muñoz Camargo, Diego, 76, 112, 173
Münster, Sebastian, 41 fig. 16, 41–3
murals. *See* monasteries, mural programs of

Nahuatl
displacement by Spanish, 137–38, 166–67
as lingua franca, 65, 140, 144, 158, 165, 167
polities, 74–6, 108, 118
spelling of, xxiii
written alphabetically on maps, 132, 165, 170–71, 172–75, 216, 245 n. 28
See also place-names, indigenous
native painters of Relaciones Geográficas maps
artistic influences: European, 77–81, 84, 87; indigenous, 78–81, 87, 94–7, 112–16, 126–33, 214
and Catholic church, 68–76, 89; mendicants, 73–4, 76–7, 80–3, 84–5; secular clergy, 74 table 2, 83–4
education, 76–7, 80–1, 84, 145
identified as native painters, 30 table 1, 61, 189, 195
images, attitude toward, 62, 84–7
named, 62, 202
other works, 82–3, 181, 189–95, 196–202, 203–9, 216
relation to local governments, 65, 209
status, 65, 77
use of styles, 61, 77–81, 87–9, 209, 210
Naturalis historia, 16
navigation, 20
nautical charts. *See* maps, nautical
Netherlands, Spanish, 2, 9
New Fire ceremonies, 207
New World
attempts to map, 11–15, 18–23
exploration of, 12–13
Nezahualcoyotl, 94
9 Lizard, 107
Núñez de la Cerda, Melchior, 45–7

Núñez, Juan, 62
Núñez, Sancho, 178
Nuremberg map of Tenochtitlan. *See* Cortés map of Tenochtitlan
ñuu, 105, 149, 158

Oaxaca de Antequera, 23, 27, 36
Oaxtepec. *See* Guaxtepec
Obregón, Luis, 63, 132
Ocoñaña, 115 fig 52, 173–75, 231–32
Oettinger, Marion, Jr., 216
oficial de pintor, 86, 202, 208
1 Flower, 107
oral traditions, 139–40, 172–73
orientation, 58, 176–77
Ortelius, Abraham, 5
Otomi language, 131–32, 161, 164
Ovando y Godoy, Juan de, 18, 20
Oviedo y Valdés, Gonzalo Fernández de, 16
Oztotlatlauhca, 132

Pacheco, don Bartolomé, 171
Pachuca, 132, 177
Padilla, Juan de, 65
padrón de leguas, 17
[illegible]
Papantla (Papantla de Olarte, Veracruz), 143 table 7, 153, 153 fig. 71, 221
Pazulco, 152
Palm, Erwin W., 57
paper, 65
Peñoles, Los (Santa Catarina Estetla, San Antonio Huitepec, Santa María Peñoles, Santiago Huajolotipac, San Juan Elotepec, San Pedro Totomachapan, Oaxaca), 34, 113
    Relación Geográfica map of, 25 fig. 9, 25–6, 35–6, 215, 221
personal names, xvi, 107, 139–40
    in Cempoala map, 131–32
    in Suchitepec maps, 170–71
Philip II (king of Spain)
    and images, 85
    as patron of maps of Spain, 1–8, 144, 213
    as patron of New World maps, 11, 23
    portraits of, 9
pictographs. *See* logographic writing
pictorial works, indigenous. *See* manuscripts, pre-Hispanic
picture-writing. *See* logographic writing
pigments, 77, 189, 239 n.6
pintura, definition, 59, 239 n. 25
    See also manuscripts, pre-Hispanic
place-names, indigenous
    in Codex Mendoza, xiv, 101 fig. 45, 141–45, 150–51, 158, 162, 169
    in Codex Zouche-Nuttall, 101–4, 138, 139
    elements of, 99 table 4, 101
    in Historia Tolteca-Chichimeca, 108, 115–16, 119–20
    logographic writing for, 89, 101, 104, 105, 138–40
    in Map of Chichimec History, 121–22, 125
    pre-Hispanic use of, 101, 138–39, 159
    on Relaciones Geográficas maps, 100 table 5, 138, 145; Mixtec names, 112–13, 158–61, 165–66; Nahuatl names, 54 fig. 27, 63 fig. 28, 70 fig. 32, 71 fig. 33, 73 fig. 35, 87, 145–58, 162, 167–70, 178, 195, 203; Otomi names, 161, 164; Totonac names, 161, 164; Zapotec names, 161–64
    replacement with Spanish names, 166–67
    spelling of, xxiii
planimetry, 116
[illegible]
plants on maps, 46–7, 96 fig. 43, 145–46, 158, 173, 207
Pliny the Elder, 16
pohua, 140
Pole Star, 14, 19, 22, 52, 55
political ideology expressed in maps
    indigenous, 106–7, 176
    Spanish, 1–3, 5, 7–8, 176, 213–14
Pomar, don Juan Bautista de, 65, 135
population. *See* demography
Portugal, 3, 14
Prem, Hanns J., 149
principales
    individuals, 63, 64, 77
    office, 63–4, 66, 89
prints, European
    influence on Relaciones Geográficas maps, 44–9, 78, 145
    techniques used in, 44, 47, 48
    used to teach native painters, 77, 87, 145
Ptolemy, Claudius, 18
    and chorography, 3–4, 5, 43

Ptolemy, Claudius *(cont.)*
- and geography, 3–4, 5, 22
- influence on sixteenth-century cartography, 3–5, 12, 38, 57
- limitations of, 13–14
- and longitude and latitude, 13–14, 15

Puebla de los Angeles, 23, 27
Pueblo Indians, 117
pueblos de españoles, 23, 32

quadrants, on astrolabes, 14, 19
quarries, 20
Quatototla, 154
Quauxoloticpac, 36
questionnaires
- on eclipses, 15, 17, 18–19, 22; colonial responses, 27, 32, 55–6
- use by crown, 11, 17
- *See also* Relaciones Geográficas, questionnaire of

Ramírez de Castro, Diego, 65
ranching, 178, 184, 188, 209, 210
Real de Arriba. *See* Temazcaltepec, Minas de
regidores, 63, 66
Relaciones Geográficas, maps of
- audience (*see* audiences of indigenous maps)
- authorship, xviii, 30 table 1 (*see also* native painters of Relaciones Geográficas maps; Spanish colonials and Relaciones Geográficas maps)
- corpus: catalogued, 217–26; defined, xviii–xx, 235 n. 9
- models for, 31, 34–5; cartographic histories, 106–7, 112–17, 174–75; chorographic maps, 38–44; community maps, 91–2, 106–7, 173, 176–78, 187–88, 214; itineraries, 35–7; land grant maps, 181–83, 185–91, 195, 200–3; nautical maps, 35, 50–5, 195–96; prints, 31, 34, 44–9, 145; social settlement maps, 107, 126–33
- pre-Hispanic iconography in, 97–101
- pre-Hispanic traditions in, 92–7
- relation to texts, xix, 65, 67, 68–9, 166
- *See also names of individual Relaciones*

Relaciones Geográficas, questionnaire, xviii
- creation, xx, 1, 2, 11–12, 18, 20
- items of, 20–3, 29, 32, 35, 50, 55–6, 61, 67, 91, 126
- text of, 21 fig. 7, 227–30

Relaciones Geográficas, texts of, xix
- authorship, 2, 23, 29–32, 61, 62–5, 66, 132, 138
- content, xix, 67, 92
- relation to maps, xix
- *See also names of individual Relaciones*

Relaciones Geográficas of Spain, 2, 236 n. 8, 237 n. 1
república de indios, 61, 86–7, 88
Reverendi patris fratris . . . , 49 fig. 23, 78
Río Grande, Oaxaca, 37, 38
Robertson, Donald, xxiii, 87–9, 94
Rome, 16, 45
Ruiz Zuazo, Juan, 74
ruling elite. *See* elites, indigenous

Sacromonte, 126
Sahagún, fray Bernardino de, 77, 86
saints, 89
Salinas, don Francisco, 77
San Agustín, Pedro de, 62, 64 fig. 29, 65, 82–3, 87–9, 97
San Francisco, Tlaxcala, 173
San Gabriel, Cholula, 56
San José de los Naturales, 76–7, 86
San Tomás, Hidalgo. *See* Tetlistaca
Sanchez, Juan, 178
Santa Cruz, Alonso de
- attempts to determine longitude, 13–15
- as chronicler, 11–12, 15–17, 18
- as cosmographer, 11–15, 18, 23
- maps of, 12, 13–14, 16, 18
- questionnaires of, 15, 16–17, 20

Saxton, Christopher, 4
scale on maps, 20
schools, 73, 74, 76–7, 80–1, 82
screenfolds. *See* manuscripts, pre-Hispanic
scribes
- individuals, 33, 45–7, 65
- office, 67, 164, 245 n. 6
- work on Relaciones Geográficas maps, 135, 169, 176

secular clergy, 34, 35–6, 74 table 2, 83–4
Seron, Martín, 178
settlement patterns, 74–6, 91, 118, 129–30
Seven Cities of Cíbola, 9
Seville, 13

'Sgrooten, Christopher, 2
shadow squares, on astrolabes, 14
slaves, 61
social settlement maps, 107, 118–26
    and Relaciones Geográficas maps, 126–33
Solís, Gaspar de, 172
solstices, 56
Spanish colonials and Relaciones Geográficas maps
    attributions to, 31, 41, 53–4
    images, attitudes toward, 30–1
    mapmakers: Chávez, Gabriel de, 33, 39–44, 50; Núñez de la Cerda, Melchior, 45–7; Stroza Gali, Francisco de, 50–5; Treviño, Diosdado, 34, 35–6; Velázquez de Lara, Gonzalo, 47 fig. 21, 48–9
Spanish language, use in Relaciones Geográficas texts, 32
    displacement of Nahuatl, 137, 166–67
    literacy in, 164–66
    written on Relaciones Geográficas maps, 136–38, 171–72, 175–79
Spanish Netherlands. *See* Netherlands, Spanish
spatial perceptions
    changes, xix–xx
    cultural differences, xii–xiii, xvii
Stroza Gali, Francisco de, 21 fig. 8, 50–5
style, meaning of, xix, 87–9, 209, 210, 215
Suchitepec (Santa María Xadan, Oaxaca), 163 fig. 80, 164–65, 170–71, 215, 222
Suchitepec group, 165, 170–71
Suchitlan (Xochitlan), 151
survey maps. *See* maps, survey
symbols on maps, European, 22, 73, 78, 87
    See also churches on maps
symbols on maps, indigenous, 97, 100 table 5, 105, 111
    for altepetl, 118–19, 141–44, 149
    for tepetl (hill), 135, 145, 146–49, 155, 189, 206, 210, 211
    of topography, 38 fig.13, 79–80, 97–100, 159, 171–72
    See also logographic writing; place-names, indigenous

Tamagazcatepec, San Bartolomé (vicinity of Santa María Xadan, Oaxaca), 170–71, 222
Tapayoltzin, Martín de Ircio, 63, 131–32
Tarímbaro, merced map of, 185 fig. 86, 186–87
Tecama (Tecameca), San Pablo, 127, 128
technology of maps, xix
    of latitude measurement, 14, 19–20, 55–6
    of longitude measurement, 15, 19
    of nautical maps, 14, 50, 51–3
    of survey maps, 3, 19–20, 57–8
Tecolutla (Tecolutla, Veracruz), 153, 154, 154 fig. 72, 222
Tecomahuaca, Oaxaca, 49
tecpan (palaces) on maps, 118, 125, 128
Tecpilpan (Cempoala), 63, 129, 130 fig. 60, 131, 132
Tecuanipan (Amaquemecan), 123–25
Tecuicuilco (Teococuilco, Oaxaca), 36–7, 37 fig. 12, 222
Tehuantepec (Santo Domingo Tehuantepec, Oaxaca), 53–5
    land use in, 184, 195
    mercedes maps of, 197–202, 210, 246 nn. 12–13
    Relación Geográfica map A, 54 fig. 27, 54–5, 181, 195–97, 200–1, 202, 214, 222; place-name, 143 table 7, 164, 196 fig. 92a, 197
    Relación Geográfica map B, 52 fig. 25, 53 fig. 26, 53–4, 222
Tejúpan. *See* Texupa
Tejupilco de Hidalgo. *See* Texupilco
Temascaltepec. *See* Temazcaltepec
Temazcaltepec, Minas de (Real de Arriba, Mexico), 45–6, 184, 222
Temazcaltepec (Temascaltepec, Mexico), 46, 222
Templo Mayor of Tenochtitlan, xiii, 56, 141
Tenampulco. *See* Tenanpulco and Matlactonatico
Tenango de Arista. *See* Teutenango
Tenanpulco and Matlactonatico (Tenampulco, Puebla), 153–55, 155 fig. 73, 223
Tenayuca, xiv
Tenoch, xvi
Tenochtitlan, 106, 140, 205
    maps of, xiii–xvii, 22, 35, 118, 130
    social struture, xiv–xvi
Teococuilco. *See* Tecuicuilco
Teopan, xvi
Teotihuacan, 215
Teotitlan del Camino (Teotitlan del Camino, Oaxaca), 33, 223, 241 n. 4
Teozacoalco (San Pedro Teozacoalco, Oaxaca)
    region, 106
    Relación Geográfica map, 26, 26 fig. 10 [and

Teozacoalco *(cont.)*
    plate 1], 112–17, 214, 216, 223; audience of, 165–66; genealogy of, 114–15, 174; place-names on, 159, 161
  Relación Geográfica text, 31, 74
Teozapotlan, 173
Tepeapulco, 56
Tepechichilco, 156
Tepechocotlan, San Lucas, 146, 147 fig. 67
Tepetlaoztoc, 81 fig. 39
Tepeyacac, 151
Tequisistlan. *See* Tequizistlan
Tequizistlan (Tequisistlan, Mexico), 33, 223, 241 n. 4
Tescaltitlan (Santiago Texcaltitlan, Mexico), 45 fig. 19, 46–7, 223
Testera, fray Jacobo de, 85
Tetetla (Tetetla de Ocampo, Puebla), 66, 223
Tetlistaca (San Tomás, Hidalgo), 94, 96 fig. 43, 158, 223
  place-names of, 156–57, 164
Tetzcoco, 94, 135, 206, 215
Tetzcoco, Lake, 73 fig. 35, 203, 205–6
Tetzcoco School, 94, 240 n. 3
Teutenango (Tenango de Arista, Mexico), 143 table 7, 167–69, 168 fig. 81, 223, 244 n. 26
Texcaltitlan. *See* Tescaltitlan
Texupa (Santiago Tejúpan, Oaxaca)
  population, 106
  Relación Geográfica map, 80 fig. 38 [and plate 4], 224; and pre-Hispanic painting, 94–6; landscape on, 79–80, 98, 101; place-names, 143 table 7, 144, 158–59, 161, 166–67
Texupilco (Tejupilco de Hidalgo, Mexico), 46, 224
Tezontepec, 82 table 3
Theatrum Orbis Terrarum, 5
13 Flower, 107
Tianguisnahuetl (Tianquiznauaca), San Miguel, 127, 128
Tianquismanalco, 157
Tilantongo, 101, 114, 174
Tlacaxipehualiztli, 56
Tlacolula (Oaxaca), 161
Tlacolula (Veracruz), 189
Tlacotalpa (Tlacotalpan, Veracruz), 50–3, 51 fig. 24, 53 fig. 26, 224
Tlacotalpan. *See* Tlacotalpa
Tlacotepec (vicinity of Santa María Xadan, Oaxaca), 170–71, 224
Tlapanaltepec, 202
Tlaquilpa (Cempoala), 63, 129, 130 fig. 60, 131–32
Tlatelolco, 38, 57, 77
tlatoani, 140
Tlaxcala (Tlaxcala, Tlaxcala), 76, 112, 173, 174 fig. 83, 224
tlaxilacalli, 123–26
tlayacatl, 123–26
topography on maps, 4–5, 22, 106
  indigenous symbols for, 38 fig.13, 79–80, 97–100, 159, 171–72
toponyms. *See* place-names, indigenous
Tordesillas, treaty of, 14
Torre de Lagunas, Juan de, 195
Totomachapan, San Pedro, 26, 36
  See also Peñoles, Los
Totonac language, 161, 164
Toulmin, Stephen, xvii
town councils. *See* cabildos; government of New Spain
trade, 13, 20, 38, 106, 188, 195
translation, 64–5, 67, 74, 140
Treviño, Diosdado, 25 fig. 9, 34, 35–6
tribute, 33, 66, 92, 106, 175, 176, 214
  on *Codex Mendoza*, 140–41
Tropics, latitude in 14
Tuscany, 45
Tuzandepetl. *See* Coatzocoalco
Tuzantla (Tuzantla, Michoacan), 46, 224
Tzaquala (Cempoala), 63, 129, 130 fig. 60, 131, 132
Tzaqualtitlan Tenanco (Amaquemecan), 123–25

units of land measurement, 57–8
urbanization, New World, xx
urban plans, 127, 128, 169, 210
utopias, 86

Valley of Mexico
  land grants in, 184
  pre-Hispanic manuscripts from, 93–4
Vatican, maps, 4
Vázquez de Coronado, Francisco de, 9
Vázquez, Juan, 178
Velasco, Luis de (viceroy of New Spain), 86, 184
Velázquez de Lara, Gonzalo, 47 fig. 21, 48–9
Venice, xiii
Vera Cruz Vieja, 191

Veracruz, 192
Villa de Espiritu Santo. *See* Coatzocoalco
vista de ojos, 57
Vocabulario en lengua mexicana . . . , 46 fig. 20, 47

warfare, 56, 158
  depictions of warriors, 44, 45 fig. 19
  in Relación Geográfica questionnaire, 20, 67
witnesses, role in mercedes (land grants), 183, 194, 202, 209
women, 77, 86, 207, 239 n. 5
world maps, xi, 13–14, 15, 50
Wyngaerde, Anton van den
  maps of Spain, 1 fig. 5, 2–6, 8, 12, 23, 34–5, 144, 213
  working methods, 2–3

Xadan, Santa María. *See* Suchitepec
Xalpa, fort of, 43–4
Xalapa de la Vera Cruz (Jalapa Enríquez, Veracruz)
  land use in, 188
  mercedes maps, 189–95, 210, 246 nn. 7–8
  Relación Geográfica map, 181, 182 fig. 85, 188–94, 214, 224
  Relación Geográfica text, 66
Xelitla, 11
Xicotencatl, 173
Xilotepec, 101, 139
Ximénez de Bohorquez, Juana, 209
Xixictla, Santa María, 127, 128 table 6
Xochimilco, 206
Xomimitl, xvi
Xonotla (Jonotla, Puebla), 66, 225
Xoxopango. *See* Jujupango

Yecapixtla. *See* Acapistla
Yoca Xonaxi Palala, 174, 231–32
Yolotepec, 161
Yopico, xvi

Zaachila, 173
Zacatepec, Lienzo of, 107–8, 109 fig. 49, 115, 241 nn. 13–14
Zacatepec, Oaxaca, 108
Zacatlan (Zacatlan, Puebla), 153, 155, 156 fig. 74, 225
Zempoala. *See* Cempoala
Zapotec
  place-names, 161–64
  polities, 106
  speakers, 65, 216
Zárate, Juan de, 65
zenith passage, solar, 14, 22, 55–6
Zozopastepec (vicinity of Sta. María Xadan, Oaxaca), 170–71, 225
Zultepec, 181
Zumárraga, fray Juan de (archbishop of Mexico), 92
Zumpango, Minas de (Zumpango del Río, Guerrero), 97, 98 fig. 44, 100, 111, 175, 225